STUDENT'S SOLUTION GUIDE

Third Edition

BIOCHEMISTRY

by Geoffrey Zubay

Prepared by

Eugene M. Gregory
Virginia Polytechnic Institute and State University

Thomas Sitz
Virginia Polytechnic Institute and State University

Wm. C. Brown Publishers
Dubuque, Iowa•Melbourne, Australia•Oxford, England

A Times Mirror Company

ISBN 0-697-14789-4

Printed in the United States of America by Wm. C. Brown Communications, Inc., 2460 Kerper Boulevard, Dubuque, Iowa, 52001

10 9 8 7 6 5 4 3 2

Contents

The authors would like to thank Drs. R. E. Ebel, W. G. Niehaus, J. L. Hess, and Malcolm Potts for their contribution in reviewing this text.

2 Thermodynamics in Biochemistry

Problems

1. How do conditions in the cell limit the manipulation of reaction conditions.

   Can't ↑T, concentration, volume, or extremes in pH in cell

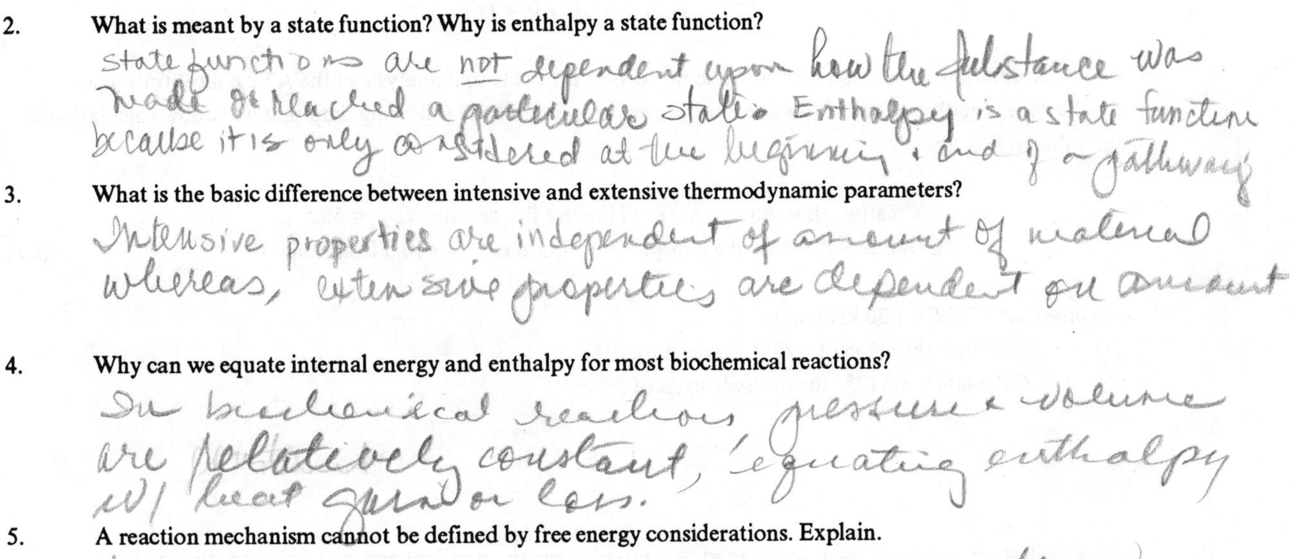

2. What is meant by a state function? Why is enthalpy a state function?

   state functions are not dependent upon how the substance was made or reached a particular state. Enthalpy is a state function because it is only considered at the beginning and end of a pathway

3. What is the basic difference between intensive and extensive thermodynamic parameters?

   Intensive properties are independent of amount of material whereas, extensive properties are dependent on amount

4. Why can we equate internal energy and enthalpy for most biochemical reactions?

   In biochemical reactions pressure & volume are relatively constant, equating enthalpy w/ heat gain or loss.

5. A reaction mechanism cannot be defined by free energy considerations. Explain.

   Free energy is a state function & thus independent of the reaction

6. Transfer of a hydrophobic molecule (e.g., a hydrophobic amino acid side chain) from an aqueous to a nonaqueous environment is entropically favorable. Explain.

7. As we will see in chapter 14, oxaloacetate is formed by oxidation of malate. The reaction

$$\text{L-malate} + NAD^+ \rightarrow \text{Oxaloacetate} + NADH + H^+$$

1) Coupling occurs or 2) the principle of mass action is in operation

has a $\Delta G^{\circ\prime}$ of +7.0 kcal/mole. Suggest reasons that the reaction proceeds in the direction of oxaloacetate production in the cell.

8. Cite three factors that make ATP ideally suited to transfer energy in biochemical systems.

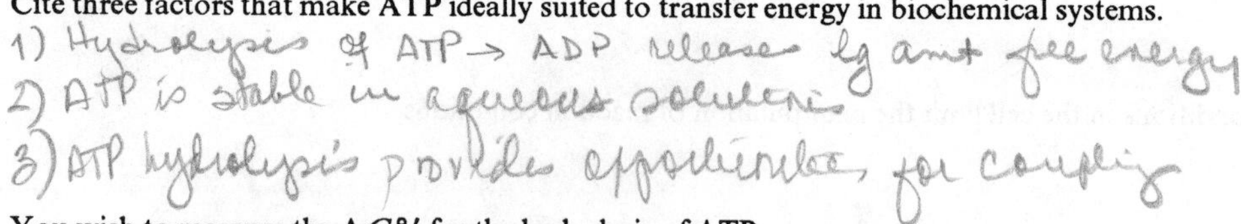

9. You wish to measure the $\Delta G^{\circ\prime}$ for the hydrolysis of ATP,

$$ATP + H_2O \rightarrow ADP + P_i + H^+$$

but the equilibrium for the hydrolysis lies so far toward products that analysis of the ATP concentration at equilibrium is neither practical nor accurate. However, you have the following data that will allow calculation of the value indirectly.

$$\text{Creatine phosphate} + ADP + H^+ \rightarrow ATP + \text{creatine} \quad K'_{eq} = 59.5 \qquad \textbf{(P1)}$$
$$\text{Creatine} + P_i \rightarrow \text{creatine phosphate} + H_2O \quad \Delta G^{\circ\prime} = +10.5 \text{ kcal/mole} \qquad \textbf{(P2)}$$

Assume that $2.3RT = 1.36$ kcal/mole.

(a) Calculate the value of $\Delta G^{\circ\prime}$ for reaction (P1).

(b) Calculate the $\Delta G^{\circ\prime}$ for the hydrolysis of ATP.

P1 $\Delta G^{\circ\prime} = -2.3\,RT \log K'_{eq}$
$= -1.36$ kcal/mole $(\log 59.5)$
$= -1.36\,(1.77)$
$= -2.41$

P2 $\Delta G^{\circ\prime} =$

10. As we will see in a later chapter, mitochondria establish a proton gradient across the inner mitochondrial membrane. Translocation of protons from the matrix (or inner surface of the inner membrane) to the outer surface of the inner membrane establishes the gradient. Translocation of the protons is an endergonic process. Explain. (Information in chapter 15 is necessary to answer this problem.)

11. What is the minimum number of moles of protons that must be translocated across a membrane to drive phosphorylation of ADP → ATP if $\Delta\Psi$ is –150 mV and ΔpH is 0.5? Assume that phosphorylation of ADP requires 11 kcal/mole.

12. The hydrolysis of lactose (D-galactosyl-β-1, 4-D-glucose) to D-galactose and D-glucose occurs with $\Delta G°'$ of -4 kcal/mole.

(a) Calculate K'_{eq} for the hydrolytic reaction.

-4 kcal/mole = - 2.3 RT (log Keq)
-4/-1.36 = log Keq
2.94 = log Keq
Keq =

(b) What are the $\Delta G°'$ and K'_{eq} for the synthesis of lactose from D-galactose and D-glucose?

Synthesis opposite hydrolysis: ΔG°' = +4 kcal/mole
log Keq = 4/-1.36 = Keq = antilog 2.94

?

(c) Lactose is synthesized in the cell from UDP-galactose plus D-glucose, and is catalyzed by lactose synthase. Given that $\Delta G°'$ of hydrolysis of UDP-galactose is -7.3 kcal/mole, calculate $\Delta G°'$ and K'_{eq} for the reaction

UDP ΔG°' = -7.3 kcal/mole
lactose ΔG°' = +4 kcal/mole

UDP-galactose + D-glucose ⇌ lactose + UDP

ΔG°' = -3.3 kcal/mole
Keq = ?

13. Dihydrolipoamide dehydrogenase catalyzes the reaction

Dihydrolipoamide + NAD^+ ⇌ lipoamide + NADH + H^+

?

$\Delta G°'$ is + 1.38 kcal/mole. Calculate the steady-state ratio of lipoamide/dihydrolipoamide if the ratio of $NADH/NAD^+$ is (a) 1:10 and (b) 10:1.

A) 1.38 kcal/mole = - 1.36 kcal/mole log [

14. For each of the following reactions, calculate $\Delta G°'$ and indicate whether the reaction is thermodynamically favorable as written.

a. Glycerate-1, 3-bisphosphate + creatine ⇌ phosphocreatine + 3-phosphoglycerate   -11.8 kcal/mole

b. Glucose-6-phosphate ⇌ glucose-1-phosphate   + 3.3 kcal/mole

c. Phosphoenolpyruvate + ADP ⇌ pyruvate + ATP   - 14.8 kcal/mole

d. Glycerol phosphate + ADP → glycerol + ATP

15. Although ATP is an important phosphate donor, in biosynthetic reactions the AMP portion of the molecule is often transferred to an acceptor with the release of pyrophosphate. Such a transfer occurs as an intermediate step in the reaction:

$$R—COO^- + ATP + CoASH \rightarrow R—CO—SCoA + AMP + pyrophosphate \quad (P7)$$

Inorganic pyrophosphatases hydrolyze the pyrophosphate, yielding two molecules of inorganic phosphate. Assume that the hydrolysis of R—CO—SCoA to R—COO$^-$ + CoASH proceeds with $\Delta G°'$ of –10 kcal/mole and that hydrolysis of a pyrophosphate anhydride bond yields –7.5 kcal/mole. Calculate the K'_{eq} for reaction (P7) in the presence and in the absence of inorganic pyrophosphatase. What role do pyrophosphatases play in biosynthetic reactions dependent on adenylate transfer?

Solutions

1. Elevated temperature, increased pressure, large concentrations of reactants, and extremes of pH may be applied to chemical reactions carried out *in vitro.* For example, the synthesis of urea from ammonia and CO_2 requires application of several hundred atmospheres of pressure at elevated temperature. Reactions within the cell occur under a much more restricted range of temperature, pressure, pH, and reactant concentration than is possible in *in vitro* reactions. Temperatures of 25+/–15°C, near neutral pH, and micro- to millimolar concentrations of reactant are conditions typically found in the cell. Urea is formed in the liver using ammonium ions provided by amino acids, and CO_2 (HCO_3^-) provided by oxidative metabolism. The process utilizes energy from metabolism carried out at normobaric pressure and comparatively mild temperature. Urea synthesis is possible *in vivo* because of the presence of enzymes that efficiently catalyze the reactions.
2. State function describes the thermodynamic parameters of the system under consideration at a particular moment. Changes in those parameters change the thermodynamic state of the system being considered. Only the difference between initial and final states, not the path taken to achieve these states, is important in most thermodynamic considerations. Enthalpic contributions defining the thermodynamic state are considered only at the initial and final states. Enthalpy is independent of pathway and is therefore a state function.
3. Intensive thermodynamic properties are independent of the amount of material in the state (e.g., temperature, density) whereas extensive properties depend on the amount of material (energy, mass).
4. Consider that $dH = d(E + PV)$

$$dH = dq - dw + PdV + VdP$$

At constant pressure and minuscule volume changes associated with most biochemical reactions, the terms dw, PdV, and VdP may be considered zero. Thus, $dH = dq$ and enthalpy (H) is equal to heat gain or loss at constant pressure (q_p).

5. A reaction mechanism defines the specific chemical reactions that occur during formation of product(s) from reactant(s). Free energy changes are state functions and as such are independent of pathway. One can predict free energy changes based on a reaction mechanism but not vice versa. The reaction

$$CH_3—CHOH—COO^- + NAD^+ \rightarrow CH_3—CO—COO^- + NADH + H^+$$

could proceed either by direct transfer of the H:$^-$(hydride) to NAD^+ or by transfer of the H:$^-$ to an acceptor on the enzyme with subsequent hydride transfer to the NAD^+ cofactor. The mechanisms are entirely different, but the two reactions are thermodynamically equivalent (final state and initial state in each case are the same). Data support direct transfer of the hydride in this case.

6. The water molecules surrounding a hydrophobic molecule form a clathrate structure and are more highly ordered than water molecules in bulk solution. Removal of the hydrophobic molecule to a nonaqueous environment may increase the order of the hydrophobic groups, but that unfavorable entropic contribution is more than compensated by the increased disorder of the water molecules that were previously in the more highly ordered structure (favorable entropic contribution).

7. The reaction shown describes the formation of oxaloacetate from the oxidation of malate in the TCA or Krebs cycle. Given the large (+) $\Delta G°'$ value for the reaction, the reaction lies strongly in favor of the reactants at standard state equilibrium. However, conditions in the cell impose a steady state whose conditions are far removed from the thermodynamic standard state. The reaction will proceed toward oxaloacetate formation in the cell if the concentration of products is kept small. This is accomplished in the mitochondria in two ways.
   (a) NADH is oxidized by the mitochondrial electron transport system, thereby diminishing the product (NADH) concentration and replenishing the reactant (NAD^+) concentration.
   (b) Oxaloacetate is condensed with acetyl-CoA in the citrate synthase-catalyzed formation of citrate. This condensation is thermodynamically favorable (approximately -8 kcal/mole). Coupling citrate synthase and malate dehydrogenase through the common intermediate oxaloacetate is an example of an exergonic (thermodynamically favorable) reaction driving an endergonic (thermodynamically unfavorable) reaction.

8. ATP has two phosphoanhydride bonds, each of which releases 7.5 — 12 kcal/mole upon hydrolysis. The high free energy of hydrolysis may be attributed in part to:
   (a) decreased negative charge density in the products compared to the triphosphate portion of ATP;
   (b) increased resonance stabilization of the products compared to ATP. (See discussion in text.)

   Cleavage of the phosphoanhydride bonds of ATP may be coupled to biochemical reactions:
   (a) $ATP \rightarrow ADP + P_i$
   (b) $ATP \rightarrow AMP$ + pyrophosphate

   The standard state free energy of hydrolysis of the phosphoanhydride bonds occupies a position in the middle of free energies of hydrolysis of organophosphate compounds that are encountered in metabolism. Mixed phosphoric anhydrides (e.g., glycerate-1, 3-bisphosphate), phosphoenolates (phosphoenolpyruvate), and phosphoramides (phosphocreatine) may be used to phosphorylate ADP. ATP is consumed in phosphoryl or adenyl transfer reactions. ATP thus may be considered a "high energy phosphate" transfer agent central to metabolism.

   From thermodynamic considerations, the hydrolysis of ATP,

$$(1)\ ATP + H_2O \rightarrow ADP + P_i + H^+ \quad \Delta G°' -7.5 \text{ kcal/mole}$$

may be coupled to the phosphorylation of a compound, for example, glucose.

$$(2)\ \text{glucose} + P_i \rightarrow \text{glucose-6-phosphate } H_2O \quad \Delta G°' +3.3 \text{ kcal/mole}$$

Each of the reactions (1 and 2) may be considered a separate partial reaction of

$$\text{glucose} + ATP \rightarrow \text{glucose-6-phosphate} + ADP \quad G°' -4.5 \text{ kcal/mole.}$$

However, neither ATP nor other phosphorylated biological molecules undergo hydrolysis to free phosphate, with the free phosphate being in turn added to an acceptor. Specific enzymes are required to decrease the activation energy of phosphoryl transfer. From a kinetic consideration, the phosphate from ATP is transferred directly to the substrate in the enzyme active site or through an enzyme-bound intermediate during catalysis. In either of the processes described, the final thermodynamic states are identical but the kinetic considerations differ markedly.

9a. The standard state free energy change can be calculated from the equilibrium constant,

$$\Delta G^{\circ\prime} = -2.3RT\log K'_{eq}$$
$$\Delta G^{\circ\prime} = -1.36 \text{ kcal/mole } (\log 59.5)$$
$$\Delta G^{\circ\prime} = -2.4 \text{ kcal/mole}$$

9b. The hydrolysis of ATP can be determined by the combination of reactions (P3) and (P4) as follows:

(1) Creatine phosphate + ADP + H^+ → ATP + creatine + $\Delta G^{\circ\prime} = -2.4$ kcal/mole (P3)
(2) Creatine + P_i → creatine phosphate + H_2O $\Delta G^{\circ\prime} = +10.5$ kcal/mole (P4)
(3) H^+ + ADP + P_i → ATP + H_2O $\Delta G^{\circ\prime} = +7.9$ kcal/mole (P5)

Therefore, hydrolysis is the reverse of reaction (P5).

$$\text{ATP} + H_2O \rightarrow \text{ADP} + P_i + H^+ \; \Delta G^{\circ\prime} = -7.9 \text{ kcal/mole}$$

10. Translocation of protons (or other ions) from a region of lower concentration to a region of higher concentration requires the input of metabolic energy. The movement of ions down a concentration gradient is a thermodynamically favorable situation. Protons are translocated from the inside of the inner membrane to the outer surface of the inner membrane. Translocation in this direction is against a concentration and electrical gradient. In the mitochondria, the controlled oxidation of substrates by molecular oxygen releases energy (exergonic process), some of which is coupled to the translocation of protons against the concentration and electrical gradient (an endergonic process). The energy stored in the proton gradient has two contributing components: the H^+ concentration differential and the membrane charge, which is negative on the matrix (inner face) side of the inner membrane.

11. The difference in free energy of protons across a diffusion barrier is defined by the relationship

$$\Delta G_j = z_j F \Delta \Psi - 2.3RT\text{pH}$$

where

z_j is ion valency with a sign indicating charge (+1 for proton)
$\Delta\Psi$ is membrane potential in volts
F is Faraday constant (23.06 kcal V^{-1} eq^{-1})

$$2.3RT = 1.36 \text{ kcal/mole}$$
$$\Delta G_j = (+1 \text{ eq/mole})(23.06 \text{ kcal } V^{-1} eq^{-1})(-0.15 \text{ V}) - (1.36 \text{ kcal/mole})(0.5)$$
$$\Delta G_j = (-4.1 \text{ kcal/mole})$$

To release 11 kcal of energy, a minimum of 2.7 moles of protons per mole ADP phosphorylated must be translocated across the membrane under the conditions described.

12a.

$$\text{Lactose} + \text{HOH} \rightarrow \text{D-galactose} + \text{D-glucose} \quad \Delta G^{\circ\prime} = -4.0 \text{ kcal/mole}$$

The equilibrium constant is related to the free energy change by the expression

$$\Delta G^{\circ\prime} = -2.3RT \log K'_{eq}$$

$$\text{Log } K'_{eq} = \frac{(-4 \text{ kcal/mole})}{(-1.36 \text{ kcal/mole})} = 2.94$$

$$K'_{eq} = 8.7 \times 10^{2}$$

12b. The free energy for the synthesis of lactose from D-galactose plus D-glucose is +4 kcal/mole because the reaction considered is the reverse of that examined in (12a).

$$\text{D-galactose} + \text{D-glucose} \rightarrow \text{lactose} + \text{HOH}$$

K'_{eq} for the synthesis of lactose from the component parts is

$$\text{Log } K'_{eq} = \frac{(4 \text{ kcal/mole})}{(-1.36 \text{ kcal/mole})}$$

$$K'_{eq} = 1.1 \times 10^{-3} \text{ (inverse of 870)}$$

12c. Consider the reactions

UDP-galactose + HOH → UDP + galactose $\Delta G^{\circ\prime}$	= –7.3 kcal/mole
D-galactose + D-glucose → lactose + HOH $\Delta G^{\circ\prime}$	= +4.0 kcal/mole
UDP-galactose + D-glucose → lactose + UDP $\Delta G^{\circ\prime}$	= –3.3 kcal/mole

The equilibrium constant of the final reaction shown is calculated from the overall free energy change of the coupled reactions:

$$K'_{eq} = 2.7 \times 10^{2}$$

In this example, a thermodynamically unfavorable reaction (formation of a glycosidic bond between D-galactose and D-glucose) is coupled to a thermodynamically favorable reaction (hydrolysis of UDP-galactose).

13. The expression

$$\Delta G = \Delta G^{\circ\prime} + 2.3RT \log [(\text{products})/(\text{reactants})]$$

defines the free energy change for a reaction that occurs at other than standard conditions. (Products)/(reactants) represents the concentration ratio of reaction products and reactants. At steady state, ΔG is 0 and

$$\Delta G^{\circ\prime} = -2.3RT \log [(\text{P})/(\text{R})]$$

Let D_{ox} and D_{red} represent the concentration of oxidized and reduced lipoamide, respectively.

$$\Delta G^{\circ\prime} = -2.3RT \log \frac{[(D_{ox}) (\text{NADH})]}{[(D_{red}) (\text{NAD}^{+})]}$$

If the ratio of NADH/NAD^+ is 1:10:

$$+1.38 \text{ kcal/mole} = -1.36 \text{ kcal/mole} \log \frac{[(0.1)(D_{ox})]}{(D_{red})}$$

$$-1.01 = \log(0.1)(D_{ox})/(D_{red})$$

$$(D_{ox})/(D_{red}) = 0.98$$

If ratio of NADH/NAD^+ is 1:10, ratio of lipoamide/dihydrolipoamide is 0.98. Use the same expression as shown above, but substitute the ratio of NADH/NAD^+ equal to 10. If NADH/NAD^+ ratio is 10:1, the ratio of lipoamide/dihydrolipoamide is 9.8×10^{-3}. Alternatively, the antilog of the expression

$$-1.01 = \log\frac{[(NADH)(D_{ox})]}{[(NAD^+)(D_{red})]}$$

may be solved. Now,

$$9.8 \times 10^{-2} = \frac{[(NADH)(D_{ox})]}{[(NAD^+)(D_{red})]}$$

$$(9.8 \times 10^{-2})\frac{(NAD^+)}{(NADH)} = \frac{(D_{ox})}{(D_{red})}$$

for any ratio of NAD^+/NADH.

14a. Glycerate-1, 3-bisphosphate H_2O → glycerate-3-phosphate + P_i ($\Delta G°'$ = –11.8 kcal/mole)

Creatine + P_i → phosphocreatine + H_2O $\Delta G°'$ = +10.3 kcal/mole

Glycerate-1, 3bisphosphate + creatine → phosphocreatine + glycerate-3-phosphate $\Delta G°'$ –1.5 kcal/mole

Thermodynamically favorable as written.

14b. D-glucose-6-P_i + H_2O → D-glucose + P_i $\Delta G°'$ = –3.3 kcal/mole

D-glucose + P_i → D-glucose-1-P_i + H_2O $\Delta G°'$ = +5.0 kcal/mole

D-glucose-6- P_i → D-glucose-1-P_i $\Delta G°'$ = +1.7 kcal/mole

Thermodynamically unfavorable as written.

(The phosphate linkage to the 6-OH group of glucose is a phosphate ester (a phosphorylated alcohol). The phosphate linkage to the C-1 position is a phosphorylated hemiacetal OH, thus the difference in free energy of hydrolysis of the two compounds.)

14c. Phosphoenolpyruvate (PEP) + $H_2O \rightarrow$ pyruvate + P_i $\Delta G^{\circ\prime} = -14.8$ kcal/mole

$$H^+ + ADP + P_i \rightarrow ATP + H_2O \quad \Delta G^{\circ\prime} = +7.5 \text{ kcal/mole}$$

$$H^+ + PEP + ADP \rightarrow ATP + \text{pyruvate} \quad \Delta G^{\circ\prime} = -7.3 \text{ kcal/mole}$$

Thermodynamically favorable as written.

14d. Glycerol-3-phosphate + $H_2O \rightarrow$ glycerol + P_i $\Delta G^{\circ\prime} = -2.2$ kcal/mole

$$H^+ + ADP + P_i \rightarrow ATP + H_2O \quad \Delta G^{\circ\prime} = +7.5 \text{ kcal/mole}$$

$$H^+ + \text{Glycerol-3-phosphate} + ADP \rightarrow ATP + \text{glycerol} \quad \Delta G^{\circ\prime} = +5.3 \text{ kcal/mole}$$

Thermodynamically unfavorable as written.

15. In the absence of pyrophosphatase, the reactions are

$$H^+ + R\text{—}COO^- + CoASH \rightarrow R\text{—}CO\text{—}SCoA + H_2O \quad \Delta G^{\circ\prime} = +10 \text{ kcal/mole}$$

$$H_2O + ATP \rightarrow AMP + PP_i + H_2O \quad \Delta G^{\circ\prime} = -7.5 \text{ kcal/mole}$$

$$R\text{—}COO^- + ATP + CoASH \rightarrow R\text{—}CO\text{—}SCoA + AMP + PP_i$$

$$\Delta G^{\circ\prime} = +2.5 \text{ kcal/mole}$$

$$\Delta G^{\circ\prime} = -2.3RT\log K'_{eq}$$

K'_{eq} is 1.5×10^{-2} $\qquad +2.5/-1.36 = \log K'_{eq}$

In the presence of pyrophosphatase, the reactions are

$$H^+ + R\text{—}COO^- + CoASH \rightarrow R\text{—}CO\text{—}SCoA + H_2O \quad \Delta G^{\circ\prime} = +10 \text{ kcal/mole}$$

$$H_2O + ATP \rightarrow AMP + PP_i + H^+ \quad \Delta G^{\circ\prime} = -7.5 \text{ kcal/mole}$$

$$H_2O + PP_i \rightarrow 2\ P_i + H^+ \quad \Delta G^{\circ\prime} = -7.5 \text{ kcal/mole}$$

$$R\text{—}COO^- + CoASH + ATP + H_2O \rightarrow R\text{—}CO\text{—}SCoA + AMP + 2\ P_i + H^+ \quad \Delta G^{\circ\prime} = -5.0 \text{ kcal/mole}$$

The $K'_{eq} = 4.8 \times 10^3$

Pyrophosphatase removes the pyrophosphate from the reaction and shifts the equilibrium toward R—CO—SCoA formation. The shift is energetically equivalent to the hydrolysis of a second mole of ATP.

Some of the values for $\Delta G^{\circ\prime}$ were based on data in *Handbook of Biochemistry*, 2d ed., CRC Press, Boca Raton, Fla., 1970.

3 The Building Blocks of Proteins: Amino Acids, Peptides, and Polypeptides

Problems

1. (a) A 10 mM solution of a weak monocarboxylic acid has a pH of 3.00. Calculate the values for K_a and pK_a for this carboxylic acid.

   (b) You add 0.06 g NaOH (M_r = 40) to 1,000 ml of the acid solution in part (a). Calculate the final pH, assuming no volume change.

2. Given the pK_a values in the text, predict how the titration curves for glutamic acid and glutamine would differ.

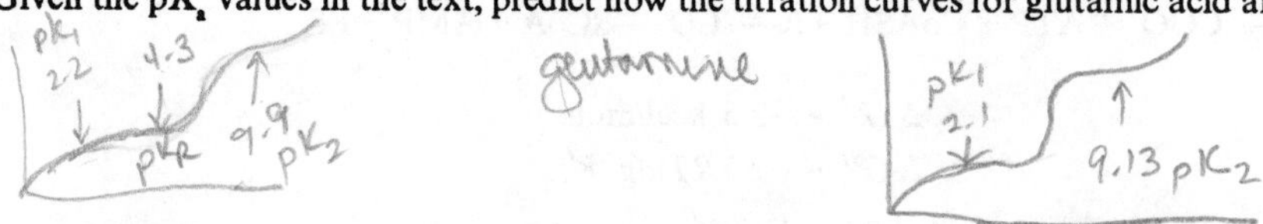

3. You have 50 ml of 10 mM fully protonated histidine. How many millimoles of base must be added to bring the histidine solution to a pH equivalent to the pI?

4. Calculate the isoelectric point for histidine, aspartic acid, and arginine. Calculate the fractional charge for each ionizable group on aspartate at pH equal to pI. Do the results verify the isoelectric point of aspartic acid?

5. Which of the naturally occurring amino acid side chains are charged at pH 2? pH 7? pH 12? (Consider only those amino acids whose side chains have > 10% charge at the pH examined.)

6. Amino acids are sometimes used as buffers. Indicate the appropriate pH values of a buffer containing aspartic acid, histidine, or serine.

7. Ten ml of a 10 mM solution of lysine was adjusted to pH 11.20. Draw the structures of the principal ionized forms present in solution. Use the pK_a values shown in table 3.3 in the text and calculate the concentration of each principal form.

8. For the tripeptide shown below, the numbers in parentheses are the pK_a values of the ionizable groups.

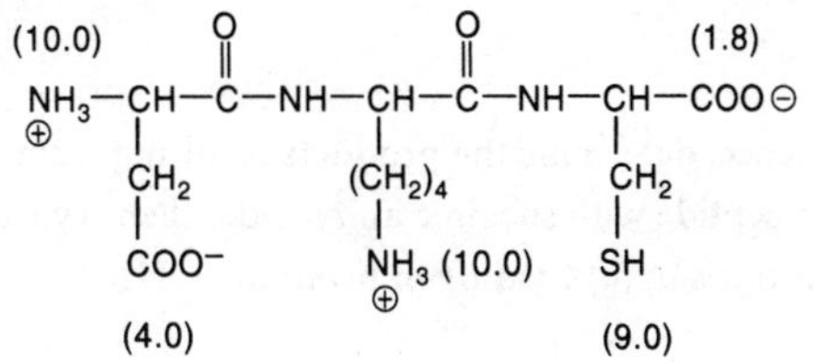

```
(10.0)        O               O             (1.8)
              ‖               ‖
NH3—CH—C—NH—CH—C—NH—CH—COO⊖
⊕        |              |              |
        CH2         (CH2)4          CH2
          |              |              |
       COO−       NH3 (10.0)      SH
                       ⊕
       (4.0)                          (9.0)
```

(a) Estimate the net charge at pH 1 and pH 14.

(b) Estimate the isoelectric pH.

9. Polyhistidine is insoluble in water at pH 7.8 but is soluble at pH 5.5. Explain the observation. Would you expect the polymer to be soluble at pH 10?

10. Protamines are basic proteins whose sulfate salts are often used as a precipitating agent added to crude cellular extracts. What types of biomolecules would you expect to be precipitated by protamine sulfate at pH 7? What is the physical basis for the precipitation?

11. A mixture of alanine, glutamic acid, and arginine was chromatographed on a weakly basic ion-exchange column (positively charged) at pH 6.1. Predict the order of elution of the amino acids from the ion-exchange column. Are the amino acids separated from each other? Explain.

Suppose you have a weakly acidic ion-exchange column (negatively charged), also at pH 6.1. Predict the order of elution of the amino acids from this column. Propose a strategy to separate the amino acids using one or both columns. Explain your rationale. (Assume only ionic interactions between the amino acids and the ion exchange resin.)

12. For the following peptide sequence, determine the products resulting from the following treatments: (a) Trypsin digestion. (b) Treatment of the peptide with succinic anhydride, then trypsin. (c) Reaction with ethyleneimine followed by trypsin. (d) Chymotrypsin. (e) Cyanogen bromide (CNBr).

Ala-Glu-Lys-Phe-Val-Cys-Tyr-Met-Gly-Phe

13. You have a peptide that is a potent inhibitor of nerve conduction and you wish to obtain its primary sequence. Amino acid analysis reveals the composition to be Ala (5); Lys; Phe. Reaction of the intact peptide with FDNB releases free DNP alanine upon acid hydrolysis. ε-DNP-lysine (but not α-DNP-lysine) is also found. Tryptic digestion gives a tripeptide (composition Lys, Ala_2) and a tetrapeptide (composition Ala_3, Phe). Chymotryptic digestion of the intact peptide releases a hexapeptide and free alanine. Derive the peptide sequence.

14. Performic acid oxidation followed by acid hydrolysis of a decapeptide yielded the following amino acid composition: Ala(1); Asp(2); cysteic acid(2); Gly(2); methionine sulfone(1); Phe(1); Val(1). In addition, one mole of ammonium ion was detected per mole of peptide hydrolyzed. The following results were obtained.

(i) Carboxypeptidase A released Asn, then Ala.

(ii) Two cycles of sequential Edman degradation released the phenylthiohydantoins of Asp and Val, in that order.

(iii) Chymotrypsin released two peptides whose compositions were determined after acid hydrolysis: CHT A (Ala, Asp, Cys, Gly, Met); and CHT B (Asp, Cys, Gly, Phe, Val).

(iv) Treatment of the original peptide with 2-bromoethylamine and then with trypsin released three peptides whose compositions were determined after acid hydrolysis: T-1 (Ala, Asp); T-2 (Asp, modified Cys, Gly, Val); and T-3 (modified Cys, Gly, Met, Phe).

Deduce the primary sequence of the peptide. If there is a portion of the sequence whose assignment is ambiguous or if there is a disulfide bond, so indicate. If there is an ambiguity, what other single cleavage might you use to assign an unambiguous sequence?

15. You have isolated from a rare fungus an octapeptide that prevents baldness and you wish to determine the peptide sequence. The amino acid composition is Lys_2, Asp, Tyr, Phe, Gly, Ser, Ala. Reaction of the intact peptide with FDNB yields DNP-alanine plus 2 moles of ε-DNP-lysine upon acid hydrolysis. Cleavage with trypsin yields peptides whose compositions are: (Lys, Ala, Ser) and (Gly, Phe, Lys) plus a dipeptide. Reaction with chymotrypsin releases free aspartic acid, a tetrapeptide with the composition (Lys, Ser, Phe, Ala), and a tripeptide whose composition following acid hydrolysis is (Gly, Lys, Tyr). What is the sequence?

Solutions

1a. The weak acid will dissociate to establish the equilibrium

$$HA \rightleftharpoons H^+ + A^-$$

$$K_a = \frac{[H^+]\,[A^-]}{[HA]}$$

The concentrations of $[H^+]$ and $[A^-]$ will be equal. The $[H^+]$ concentration at pH 3.00 is 1.00×10^{-3}. HA concentration is the value at equilibrium and will be the initial concentration minus the amount dissociated: 10 mM – 1 mM = 9 mM.

Therefore,

$$K_a = \frac{(1.00 \times 10^{-3})(1.00 \times 10^{-3})}{(9.00 \times 10^{-3})}$$

$K_a = 1.11 \times 10^{-4}$

$pK_a = -\log K_a$

$pK_a = -\log(1.11 \times 10^{-4}) = +4 - \log 1.11$

$pK_a = 3.95$

1b. Applying the Henderson-Hasselbach equation relating pH to pK_a plus ratio of conjugate base to nondissociated acid

$$pH = pK_a + \log\left[\frac{(A^-)}{(HA)}\right]$$

One liter of a 10 mM solution contains 10 millimoles of total HA (undissociated HA and A^-). It was established in (1a) that (HA) was 9 mM and A^- was 1 mM. Hence in one liter of solution there are 9.00 millimoles HA and 1.00 millimole A^- initially.

The amount of NaOH added is 0.06 gm/40 gm/mole = 1.5 m mole. The OH^- will remove protons from a stoichiometric amount of HA to form the anion (A^-). Thus, after addition of 1.5 mmoles NaOH, there will be 7.5 mmoles of HA and 2.5 mmoles of A^- in the solution.

$$pH = 3.95 + \log\frac{(2.5\ \text{mmoles/l})}{(7.5\ \text{mmoles/l})}$$

$$pH = 3.95 + \log 0.333$$

$$pH = 3.95 - 0.48 = 3.47$$

2. Glutamic acid and glutamine each have an α-carboxyl and an α-amino group but differ with respect to the R group. Glutamic acid has a γ-carboxyl group that is an ionizable functional group. The side chain of glutamine is an amide which is not ionizable under biological conditions. Titration of the fully protonated glutamic acid would require three equivalents of base, two to titrate the carboxyl groups and one to titrate the α-amino group. The pK_a values of the carboxyl groups (pK_1 2.1, pK_R 4.1) are sufficiently close that they will likely not be distinguishable as discrete titratable groups in the titration curve. The α-amino group will exhibit a pK of 9.5. The titration curve of glutamine will reveal one carboxyl group titrating with pK_a 2.2 and an α-amino group whose pK_a is 9.1.

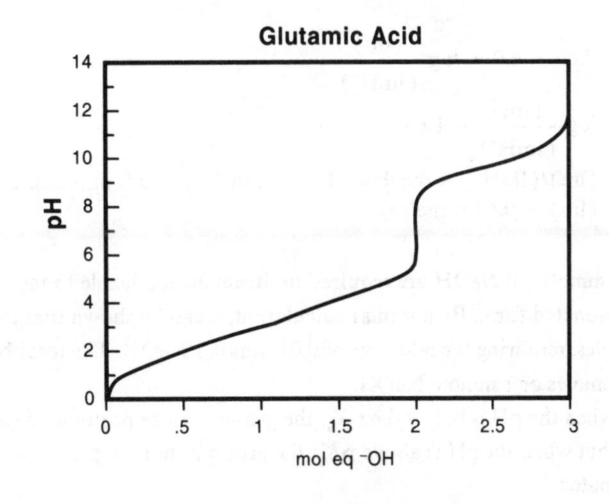

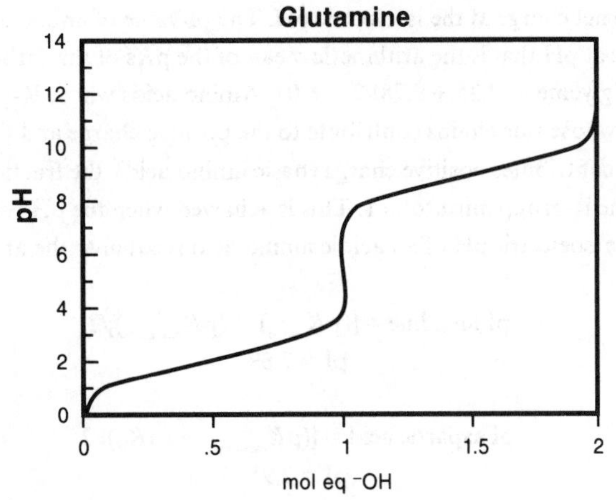

3. The isoelectric point of an amino acid (or protein) is the pH at which the sum of positive and negative charges is zero. Fully protonated histidine will bear a net +2 charge. (α-amino and imidazole group are protonated, each with a positive charge. The α-carboxylate is protonated but is neutral.)

There are two approaches to the problem. By inspection, we note that removal of two protons will decrease the net charge to zero. Thus, 0.05 l × 10 mmoles/l = 0.5 mmoles of histidine. Two mmoles of OH^- are required to titrate each mmole of histidine. Thus 1 mmole of NaOH must be added.

The second approach requires quantifying the final state of protonation of histidine at the isoelectric point. The pI is $(pK_a + pK_R)/2 = 7.69$. At that pH, the α-carboxyl (pK 1.8) will be fully deprotonated and the deprotonation is accomplished by the addition of 0.5 mmole NaOH to the fully protonated molecule.

Apply the Henderson-Hasselbach equation to determine the ratio of unprotonated to protonated imidazole (R) group.

$$7.69 = 6.0 + \log \frac{(\text{Im})}{(\text{ImH}^+)}$$

$$\log \frac{(\text{Im})}{(\text{ImH}^+)} = 1.69$$

$(\text{Im})/(\text{ImH}^+) = 49$; but $(\text{Im}) + (\text{ImH}^+) = 0.5$ mmoles, so
$(\text{Im}) = 0.49$ mmoles

Thus, 0.49 mmoles of NaOH are required to titrate the imidazole to the appropriate ratio of unprotonated to protonated form. By a similar calculation, it can be shown that the amount of unprotonated (α-NH_2) is 0.01 mmoles, requiring the addition of 0.01 mmoles NaOH. The total NaOH required is thus (0.5 + 0.49 + 0.01) mmoles or 1 mmole NaOH.

Note that when the pH is below the pK_a, the group is more protonated (neutral or positively charged) than unprotonated, but when the pH is above pK_a, the group is more unprotonated (neutral or negatively charged) than protonated.

4. The amino acid has no net charge at the isoelectric pH. The pI value of amino acids with side chains that are not ionizable are isoelectric at pH that is the arithmetic mean of the p*K*s of the carboxyl and the α-amino groups. For example, the pI of glycine is (2.35 + 9.78)/2 = 6.07. Amino acids whose R-groups are ionizable may be divided between those whose side chains contribute to the positive charge and those contributing to the negative charge. If the R-group contributes positive charge (basic amino acid), the fractional charge residing on the α-amino group and on the R-group must total 1. This is achieved when the pH is the arithmetic mean of pK_{amino} plus pK_R. Similarly, the isoelectric pH of an acidic amino acid is equal to the arithmetic mean of $pK_{carboxyl}$ and pK_R.

$$\text{pI histidine} = [(pK_{amino}) + (pK_{imidazole})]/2$$
$$\text{pI} = 7.69$$

$$\text{pI aspartic acid} = [(pK_{carboxyl}) + (pK_R)]/2$$
$$\text{pI} = 2.95$$

$$\text{pI arginine} = (pK_{amino} + pK_R)/2$$
$$\text{pI} = 10.74$$

Apply the Henderson-Hasselbach equation to calculate the ratio of unprotonated to protonated groups on aspartate, then calculate the fractional charge on each group. At pH = pI (2.95), the α-amino group will be virtually fully protonated and will contribute one positive charge. The fractional charge on the carboxyl groups are

$$\alpha\text{—COOH: } pK_a = 1.99$$

$$\text{pH} = pK_a + \log\left[\frac{A^-}{HA}\right]$$

$$2.95 = 1.99 + \log\left[\frac{A^-}{HA}\right]$$

$$\log\left[\frac{A^-}{HA}\right] = 0.96$$

$$\left[\frac{A^-}{HA}\right] = 9.1, \text{ and } [A^-] + [HA] = 1$$

$[A^-] = 0.9$ fractional negative charge (90% of the α-carboxylate will be unprotonated at any time)

A similar calculation will reveal that the fractional negative charge on the ß-carboxylate will be significantly less because the pH (2.95) is below the pK_a. Hence,

$[A^-] = 0.1$ negative charge on the ß-carboxylate

Sum positive and negative charge contributions are

α-amino group	α-carboxyl	ß-carboxyl
(+1)	(−0.9)	(−0.1) = 0

These data demonstrate that the net charge on aspartic acid is zero at pH 2.95 and thus verifies 2.95 as the isoelectric pH.

5. The ionizable groups will be more protonated if pH is less than pK_a and will be more unprotonated if pH is greater than pK_a. Examine the table of amino acid pK_a values in the text and determine (a) which of the ionizable R-groups will be protonated at each pH value and (b) whether protonation neutralizes a negative charge or confers a positive charge. Those whose side chains are charged are

(a) pH 2: Arg, His, Lys
(b) pH 7: Arg, Asp, Glu, Lys
(c) pH 12: Arg, Asp, Glu, Tyr, Cys

6. The buffering capacity of an ionizable group is greatest when the pH is equal to pK_a for the group and becomes inconsequential at pH values +/−1 pH unit from pK_a. In principle, we must consider the potential buffering capacity for each ionizable group.

Aspartic acid: pH 2, 4, 10. (The pH range 1 to 5 will be buffered by the α-carboxyl and the ß-carboxyl groups.)

Histidine: pH 2, 9, 6 (imidazole side chain).

Serine: pH 2, 9. (The pK_a of the alcohol is outside the range of pH normally considered for buffers.)

7. The two principal forms differ in degree of protonation of the R-group (epsilon amino-group). Each will have a fully deprotonated carboxyl (net −1) and an unprotonated α-amino group (neutral). The equilibrium mixture will contain 0.028 mmoles protonated and 0.072 mmoles unprotonated epsilon amino-group. These values may be derived by applying calculations demonstrated in problem 4.

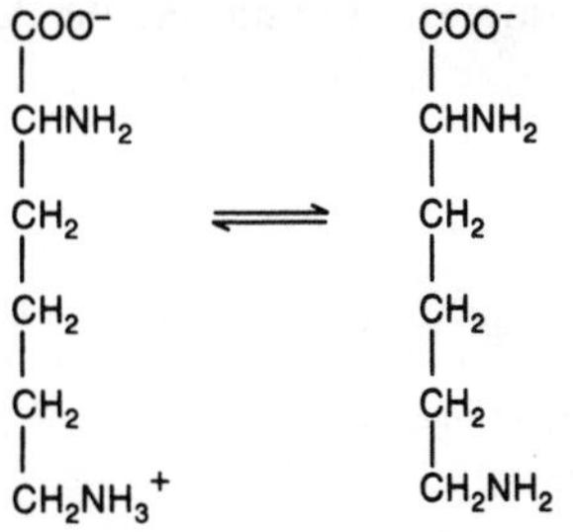

COO^- – $CHNH_2$ – CH_2 – CH_2 – CH_2 – $CH_2NH_3^+$ (0.028 mmoles) ⇌ COO^- – $CHNH_2$ – CH_2 – CH_2 – CH_2 – CH_2NH_2 (0.072 mmoles)

8a. At pH 1, each of the functional groups is protonated, because the pH is well below each pK_a value. The carboxyl groups and the sulfhydryl group are neutral when protonated. The α-amino group from the N-terminal aspartate and the e-amino group of the lysyl residue are each positively charged. The net charge is (+2).

At pH 14, each of the functional groups is unprotonated. The α-amino and e-amino groups are neutral. The carboxylate groups and the thiolate of cysteine are each negatively charged. The net charge is (−3).

8b. The pI will be approximately 6. At this pH, the carboxyl groups will be fully deprotonated (net −2) and the amino groups will be fully protonated (net +2). The cysteine sulfhydryl group will remain protonated and neutral.

9. Charged molecules are more soluble in water than uncharged molecules. The imidazole groups of polyhistidine will have a pK_a of approximately 6. Polyhistidine will bear little positive charge at pH 7.8 and even less at pH 10, causing the polymer to be insoluble. At pH 5.5, the fraction of imidazole groups protonated will be greater and the polymer will be more soluble.

10. At pH 7, the protamine molecules should bear a net positive charge (polycation) and will ionically interact with negatively charged (polyanionic) molecules. Interaction with the nucleic acids, whose phosphodiester linkages are negatively charged at neutral pH, and with negatively charged proteins would be expected. The complexes would precipitate because the large molecular weight complex (e.g., protamine-DNA fragment) has fewer ionic groups to form H-bonds with water than does the uncomplexed molecules.

11. The amino acids will be separated based on molecular charge. Consider the charge of each amino acid at pH 6.1. The isoelectric points of the amino acids in the mixture are: glutamic acid, 3.1; alanine, 6.1; and arginine, 10.74.

Amino acid	pI	net charge at pH 6.1
Glutamic acid	3.1	negative (pH > pI)
Alanine	6.1	neutral (pH = pI)
Arginine	10.74	positive (pH < pI)

At pH 6.1, the negatively charged glutamate will bind to the positively charged, weakly basic resin, but arginine and alanine will pass unretarded through the column. A weakly acid ion exchange resin at pH 6.1, bearing a negative charge, will retain the positively charged arginine, but alanine (0 net charge) and glutamate (net negative charge) will pass unretarded through the column.

A strategy to separate the three amino acids involves operating the columns in series. Each column is buffered at pH 6.1, eliminating complications due to buffer or pH changes. The effluent from the weakly basic ion-exchange column contains arginine and alanine, but arginine will be retained by the acidic column and alanine will be eluted. Washing each column independently with buffer containing increased salt concentration will elute the bound amino acid.

12a. Trypsin, an endopeptidase, cleaves peptide bonds whose carboxyl group is donated by the basic amino acids arginine or lysine. The predicted cleavage pattern is:

(Ala-Glu-Lys) (Phe-Val-Cys-Tyr-Met-Gly-Phe)
T1 T2

12b. Succinic anhydride will react with and succinylate the N-terminal α-amino group and the ϵ-amino group of lysine. Succinylated lysine is not a substrate for trypsin, and the cleavage site is masked. The succinylated peptide is thus resistant to tryptic cleavage.

Reaction of lysine with succinic anhydride

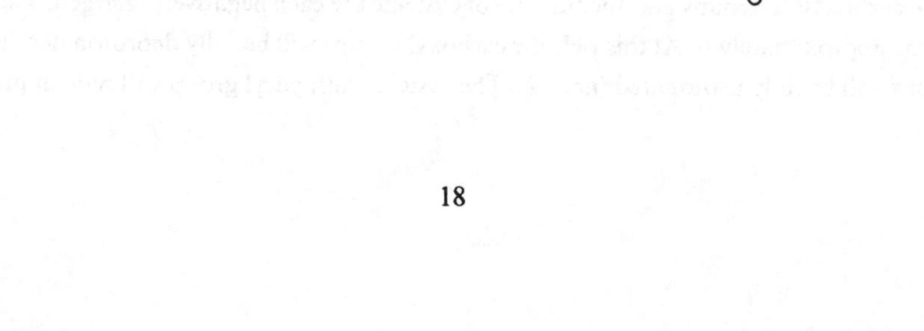

12c. Ethyleneimine reacts with the –SH group of cysteine to form a sulfur-containing homolog of the lysine side chain. The S-(2-aminoethylcysteine) is a substrate for trypsin. Thus, a trypsin cleavage site will be generated at each cysteine, and the cleavage pattern is

(Ala-Glu-Lys) (Phe-Val-modified Cys) (Tyr-Met-Gly-Phe)

Reaction of cysteine –SH with ethyleneimine

$$-CH(R_1)-C(=O)-NH-CH(CH_2SH)-C(=O)-NH\sim + \text{ethyleneimine} \longrightarrow -CH(R_1)-C(=O)-NH-CH(CH_2-S-CH_2-CH_2NH_3^{\oplus})-C(=O)-NH\sim$$

12d. Chymotrypsin, an endopeptidase, most rapidly cleaves peptide bonds whose carboxyl is donated by the amino acids Phe, Tyr, or Trp. Peptides released are

(Ala-Glu-Lys-Phe) (Val-Cys-Tyr) (Met-Gly-Phe)

12e. Cyanogen bromide reacts with the methionine thioether and cleaves the peptide chain at the carboxyl side of methionine. The methionine is converted to a homoserine lactone whose ring opens in acid.

(Ala-Glu-Lys-Phe-Val-Cys-Tyr-Homoserine) (Gly-Phe)

13. FDNB reacts with primary amines and is used to label the N-terminal amino acid. Were lysine the N-terminal, the diDNP derivative would have been observed. Alanine is the other amino acid that reacted with FDNB and is the N-terminal. Thus (Ala, Ala_4, Lys, Phe), where (-) indicates the peptide bond between two amino acids in sequence and (,) separates amino acids whose positions in the sequence have not been determined.

Trypsin cleaves the peptide bonds in which lysine or arginine contribute the carboxyl group. The tripeptide derived from trypsin cleavage must have the sequence (Ala-Ala-Lys). The sequence of the tetrapeptide cannot be determined at this point.

Chymotrypsin cleaves most rapidly at the carboxyl side of Phe, Tyr, or Trp residues in peptides. The release of only Ala and an intact hexapeptide is consistent with the bonding (Phe-Ala) on the C-terminal of the peptide.

Based on the data, propose the following sequence and determine if the sequence is consistent with the data.

Ala-Ala-Lys-Ala-Ala-Phe-Ala

FDNB/Hydrolysis	α-DNP-Ala ; ε-DNP-Lys
Trypsin	(Ala-Ala-Lys) (Ala_3, Phe)
Chymotrypsin	(Ala -Ala-Lys-Ala-Ala-Phe) (Ala)

The proposed sequence is consistent with the data.

14. Performic acid treatment oxidizes methionine to the sulfone and cysteine or cystine to cysteic acid. Ammonium ion release is consistent with either Asn in the peptide or with a C-terminal amide. Carboxypeptidase releases C-terminal amino acids and released Asn, then Ala. Edman degradation reveals the order of the N-terminal residues (Asp, then Val). From these data, one may propose the partial sequence

(Asp-Val,______ ______ ______ ______ ______ ______ -Ala-Asn)

Consider the chymotryptic peptides.

CHT A contains the sole Ala, which is residue 9 in the sequence. Thus (Cys, Gly, Met, Ala-Asn).

CHT B must have a C-terminal Phe and contains the N-terminal Asp and Val. Thus (NH_2-Asp-Val, Cys, Gly,-Phe). We now know that (Asp-Val, {Cys, Gly}- Phe- {Cys, Gly, Met}-Ala-Asn).

The original peptide, lacking either Lys or Arg, is not a substrate for trypsin. Reaction of the peptide with 2-bromoethylamine creates trypsin-sensitive sites at cysteine residues. Moreover, this demonstrates that the two cysteine side chains were not in a disulfide linkage. (These tryptic peptides will have modified cysteine at the Cterminal.)

T-1 contains the C-terminal Ala and As(n) residues but no modified cysteine. T-2 contains the N-terminal sequence and modified cysteine (Asp-Val-Gly-Cys).

T-3 can be ordered (Phe-{Gly,Met}-Cys).

Sequence of the original peptide can be deduced.

(Asp-Val-Gly-Cys-Phe-{Gly, Met}-Cys-Ala-Asn)

The order of Gly and Met cannot be deduced from these data. However, cleavage of the original peptide with cyanogen bromide will sever the peptide at the sole methionine and will establish the order of Gly and Met. If the order is (Gly-Met), Gly will be found in the CNBr peptide, which also contains homoserine. If the order is (Met-Gly), Gly will be found in the cleavage peptide lacking homoserine.

15. DNP-Ala establishes the N-terminal.

(Ala-{Asp,Gly,Lys(2),Phe,Ser,Tyr}

Tryptic peptides are

T-1 (Ala-Ser-Lys) is at the N-terminal of the original peptide

T-2 {Gly,Phe}-Lys

T-3 {Asp,Tyr}

Chymotrypsin treatment yields free Asp, which could only have been released if it were the C-terminal following Phe or Tyr. The tetrapeptide contains the N-terminal sequence, a Lys, and Phe.

(Ala-Ser-Lys-Phe)

The tripeptide contains Tyr as C-terminal and Gly as N-terminal. Therefore (Gly-Lys-Tyr). The original peptide sequence is

(Ala-Ser-Lys-Phe-Gly-Lys-Tyr-Asp).

4 The Three-Dimensional Structure of Proteins

Problems

1. The principal force driving the folding of some proteins is the movement of hydrophobic amino acid side chains out of an aqueous environment. Explain.

2. Outline the hierarchy of protein structural organization.

   Primary structure
   Secondary structure
   Tertiary structure
   Quarternary structure

3. What is the role of loops or short segments of "random" structure in a protein whose structure is primarily α-helix?

   Pack more closely

4. What are some consequences of changing a hydrophilic residue to a hydrophobic residue on the surface of a globular protein? What are the consequences of changing an interior hydrophobic to a hydrophilic residue in the protein?

   Destabilize structure

5. Some proteins are anchored to membranes by insertion of a segment of the N-terminal into the hydrophobic interior of the membrane. Predict (guess) the probable structure of the sequence (Met-Ala-(Leu-Phe-Ala)$_3$-(Leu-Met-Phe)$_3$-Pro-Asn-Gly-Met-Leu-Phe). Why would this sequence be likely to insert into a membrane?

6. Suppose that every other Leu residue in the peptide shown in problem 5 were changed to Asp. Would that necessarily alter the secondary structure? Explain whether insertion into the membrane would be altered.

7. Amino acid side chains coordinate to the metal cofactor in metalloproteins. Examples of these coordination ligands include Asp, Glu, His, and Cys. In most of the proteins studied, the side chains directly surrounding the ligand amino acid are highly conserved among homologous proteins isolated from different organisms, while

nonconservative alterations in amino acid sequence are found at sites distant from the metal binding site. How do these observations fit the argument that biological structure dictates function?

8. Proteins that span biological membranes to provide ion-conducting channels or pores frequently have multiple α-helical segments aligned parallel to each other. Proteins that span the membrane with a single α-helical segment do not allow conduction of ions. Explain why ion conduction through a single α-helical segment does not occur.

9. An investigator purified a protein (protein X) from *E. coli.* She injected protein X into rabbits to generate antibodies that recognize and bind to protein X. Using an electrophoretic technique, she separated the proteins from a crude cell extract of *E. coli* and used the antibody to locate protein X on the gel. To her surprise, the antibody reacted not only with protein X but also with a second, unrelated protein (protein Y). When proteins X and Y were sequenced, she found that the sequence of residues 67–78 in protein X and that of residues 120–131 in protein Y were identical. Help the investigator rationalize the data, recognizing that antigenic determinants (epitopes) of proteins are clusters of amino acids.

10. "Left- and right-handed α helices of polyglycine are equally stable." Defend or refute the statement. (Consider glycine's chirality or lack thereof.)

11. Molecular weight analysis of a protein yields the following information:

Solvent	M_r
Dilute buffer	200,000
6M Guanidinium chloride (GuHCl)	100,000
6M GuHCl + 100 mM 2-mercaptoethanol	75,000 and 25,000

(Guanidinium chloride is a chaotropic (denaturing) reagent and 2-mercaptoethanol can reduce disulfide bonds.) What can you deduce about the protein's quaternary structure?

12. Using the Ramachandran diagram in the text, explain why polypeptides assume only a limited number of regular structures.

13. It might be argued that in protein structure, as in everyday life, it is a "right-handed world." Use examples of protein structure discussed in the chapter to support this contention.

14. Recombinant DNA technology allows the amino acid sequences in protein to be altered. However, not all of these "genetically engineered" proteins yield stable, catalytically active proteins. Why?

15. Write all the quaternary forms possible for a hexamer composed of A and B type subunits. (Homohexamers are allowed.) What forces likely bind the subunits to each other?

Solutions

1. In principle, proteins should assume a conformation yielding the lowest free-energy level. Entropic and enthalpic changes in the system (peptide plus surrounding medium) should sum to a negative value for the folding of the protein. Recall that $\Delta G = \Delta H - T\Delta S$. Thus, interactions that decrease ΔH or increase ΔS contribute to a more negative ΔG. Organized structure in the protein results in a decrease in entropy of the protein. This decrease in entropy must be offset by an increase in entropy in the surroundings. Removing hydrophobic residues from the aqueous interface is thermodynamically favorable and accounts, in part, for the increased entropy in solution surrounding the protein. Water molecules are more highly organized in the space immediately surrounding the hydrophobic residues than in bulk water. Shifting the hydrophobic residues from an aqueous to an anhydrous environment decreases organization of the surrounding water and increases the entropic contribution to folding. In globular proteins dissolved in aqueous media, one finds that in general the hydrophobic residues are exposed on the surface of the protein to interact with water, whereas the hydrophobic residues coalesce in the core or interior of the protein. However, as noted in the text, enthalpically favorable interactions offset the decreased entropy of the folded protein and contributes to stabilization of the folded protein.

2. The ordered sequence of amino acids constitutes the basic or primary structure of the protein. Primary structure dictates organization of the protein into several secondary structures: α-helix, β-sheet, ß-turns, and "random coil." Leszczynski and Rose defined a new class of secondary structure called the omega (Ω) loop. These loops are polypeptide segments of 6 to 16 residues that contain no other regular secondary structure and whose end-to-end width across the loop is less than 10 Å. (Leszczynski, J. F., and G. D. Rose, Loops in globular proteins: A novel category of secondary structure. *Science.* 243:849–855, 1986.)

   The tertiary structure is the arrangement and interaction of the secondary structural elements in a single protein. The interactions occur primarily through the amino acid side chains. In a single polypeptide protein or monomeric protein, tertiary structure is the highest order. Association of the same or different proteins into a multimer constitutes quaternary structure, e.g., dimer or tetramer formation. Further aggregation of multisubunit proteins forms aggregates or supramolecular structures whose molecular weight may exceed 10^7.

These aggregates often are multienzyme complexes involved in primary metabolism, e.g., pyruvate dehydrogenase complex and α-ketoglutarate dehydrogenase complex (see chapter 14 in the text), or are involved in electron transport, e.g., mitochondrial NADH dehydrogenase (see chapter 15 in the text). A compilation of the composition of pyruvate dehydrogenase, isolated from several sources can be found in Patel, M. S., and T. E. Roche, Molecular biology and biochemistry of pyruvate dehydrogenase complexes. *The FASEB Journal.* 4:3224–3233, 1990.

3. The α-helix is a rather rigid, rodlike secondary structural element that cannot easily change direction in space without breaking the helical arrangement. Frequently, ß bends or loops, or in some instances segments of random coil structure, allow the helical elements to change direction and pack into a more compact globular structure. Myoglobin has a high helical content, but the segments of helix fold back on one another and pack into a globular protein.
4. Surface or solvent exposed residues hydrogen bond with water molecules and are important in solubility of the protein in aqueous media. Replacement of a surface hydrophilic with a hydrophobic residue would increase the organization of water molecules around the hydrophobic residue and would be expected to decrease, by a small amount, the overall structural stability of the protein molecule as well as the surrounding solvent. Changing an interior hydrophobic to a hydrophilic residue likely also would destabilize the structure by potentially disrupting an area of hydrophobic interaction. The hydrophilic character of the residue, with or without a formal charge, would preclude close approach of a neighboring hydrophobic residue. Each of these replacements would diminish, to a small degree, the overall structural stability of the protein. Single substitutions would likely be accommodated without major disruption to the overall structure of the protein, but there are exceptions. Substitution of valine for glutamic acid at position 6 of the ß subunit of hemoglobin results in aggregation of the deoxyhemoglobin (sickle-cell hemoglobin). Erythrocyte flexibility is markedly diminished.
5. The sequence given can assume an α-helical arrangement with the hydrophobic side chains located along the outside of the helix and exposed to solvent. The α-helix will be distorted at the proline residue and may enter a β-turn. The arrangement of the hydrophobic residues would likely limit water solubility, because the increased organization of water structure surrounding the hydrophobic side chains decreases the entropic contribution to the stability of the system. However, the segment of hydrophobic helix would be stabilized by insertion into the hydrophobic environment of the membrane. Moving the hydrophobic side chains on the helix from aqueous contact into the lipid bilayer would increase the entropic contribution of the water molecules to the system. Hydrophobic segments are found in proteins that are bound to biological membranes.
6. Replacement of Leu with Asp would not necessarily alter the secondary structure; α-helix formation is not precluded. However, the helix containing the aspartate residues would be considerably more hydrophilic and would be more difficult to insert into the membrane. It would be energetically less favorable to remove an anionic residue from the aqueous environment, where the charge is somewhat shielded by water, and insert the charge into a nonpolar environment. The peptide in which 50% of Leu was replaced with Asp would not be likely to insert into the membrane.
7. Metals are held in proteins by interaction with specific amino acid residues (ligands). The metal is bound to the protein in one of a few specific three-dimensional arrangements of the ligands. Binding of the metal to the protein may require the ligands to be arranged roughly in a square planar, a tetrahedral, or an octahedral orientation. The precise geometry of the metal-ligand complex is required for biological activity. Ligands to the metal do not necessarily arise from the same segment of the protein. For example, the ligands to Fe in the iron-containing superoxide dismutase from *E. coli* are histidyl side chains (His-26, His-73, and His-160) and an aspartyl residue (Asp-156). The numbers represent the position of the amino acid in the primary sequence with the N-terminal as number 1. Within the same family of proteins, the ligands are apparently highly conserved. (See Carlios, A., et al., Iron superoxide dismutase. Nucleotide sequence of the gene from *Escherichia coli* K_{12} and correlation with crystal structure. *J. Biol. Chem.* 263:1555–1562, 1988, and Barra, D., et al., The primary structure of iron superoxide dismutase from *Photobacterium leiognathi*. *J. Biol. Chem.* 262:1001–1009, 1987.)

Changing one or more of the ligands may abolish metal binding to the protein or may alter the spatial arrangement around the metal so that the biological function is abolished. Conservation of residues around the ligand side chains preserves the local environment and conformation of the ligands, allowing the metal to be bound. Nonconservative replacements at sites distant from the metal binding site may have little influence on the metal binding region.

8. Examination of space-filling models reveals that the core or center of the α helix is not a hollow tube; rather, the space is occupied by the atoms comprising the peptide bonds. There is no space through which an ion or water may migrate. Thus, single-span α-helices do not allow ion conduction across membranes. Multiple-span helical segments may (and do) form roughly cylindrical channels, or pores, in which the α-helices are on the periphery of the cylinder. Hydrophobic side chains interact at the lipid bilayer-protein interface, whereas hydrophilic residues protrude toward the center of the cylinder. In that way, a hydrophilic channel, capable of transporting ions, may form in the membrane.

9. Antibodies are formed in the host animal in response to a foreign protein (in this case). Epitopes are groups or clusters of amino acids, recognized as foreign, to which antibodies are made. One might argue that the common sequence shared by the proteins X and Y are common epitopes. The antibody to protein X binds to the common epitope of protein Y. If this is true, the apparently unrelated proteins may react with the same antibody.

10. The small size of the glycine side chain (H–) and the lack of chirality allow glycine to assume conformations not possible with other amino acid residues. Examination of the Ramachandran plot reveals that there are two areas defined by (Phi) and (Psi) that "allow" glycine to form right-handed or left-handed helices. There is no apparent reason to suggest that one helical arrangement of glycine is more stable than the other.

11. The protein (200K M_r) is dissociated into two units of identical molecular weight (100K M_r) when denatured with guanidinium chloride in the absence of reductant. Thus, these 100K M_r units are not linked by disulfide bonds. However, each 100K M_r unit is comprised of two subunits (75K and 25K) that are linked by disulfide bonds. Treatment of the protein with the denaturant under reducing conditions yields 75k and 25k M_r proteins. The structure can be proposed schematically:

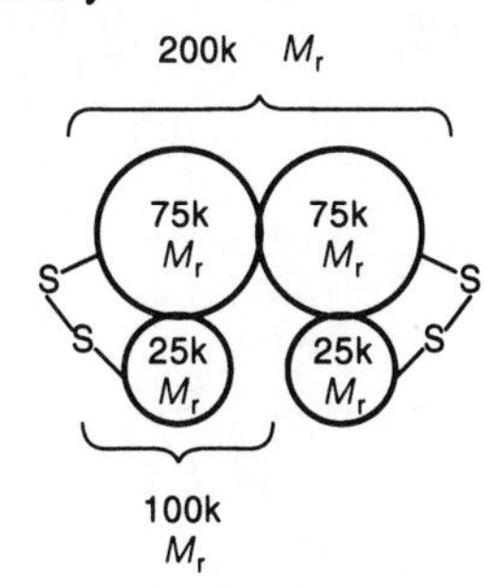

12. Each peptide bond may be considered to be an approximately planar structure which rotates as a single unit around the peptide C_α—N and the C_α-carbonyl bonds. The planar structure of the peptide bond removes a number of degrees of freedom of rotation. Moreover, rotation of the peptide bond around the C_α—N and C_α-carbonyl carbon bonds is further restricted because of unfavorable interactions between atoms comprising the peptide bonds or between atoms on the peptide and the side chains. The sterically allowed rotational angles that minimize energetically unfavorable interactions between peptide bond substituents accommodate only a limited number of regular structures. (Examine the Ramachandran plot in the text to identify these allowed conformations.)

13. Examples are right-handed helices, the right-hand twist of beta sheets, and right-hand crossovers. (See figs. 4.22 and 4.23 in text.)

14. From the dictum that primary structure ultimately dictates protein function, it follows that replacement of amino acid residues within the protein may alter protein function. Conservative replacement (glu→asp, leu→ile) may have little if any discernable effect on structure and function. Nonconservative replacements at sites distant from the active site may also have little effect on biological activity. However, there are residues at the active site

of enzymes that perform catalytic roles. Replacement of these residues may have a profound effect on enzyme activity. Wholesale substitutions also are likely to disrupt the function of the protein. Replacement of hydrophobic with hydrophilic residues, or *vice versa,* may alter the manner in which the protein folds or the manner in which subunits interact. Insertion of proline may distort an α-helical segment critical to protein function.

15. Consider all the possible combinations of the subunits to form a hexamer: A_6, A_5B_9, A_4B_2, A_3B_3, A_2B_4, AB_5, B_6. There are seven possible forms of the protein (isoforms or isoenzymes). Not all of the forms may occur and, in principle, tissue specificity may be required for the formation of a specific isoform. Consider the tetrameric lactate dehydrogenase, composed of H (heart) and M (skeletal muscle) derived subunits. One finds virtually all H_4 forms in cardiac muscle and M_4 in skeletal muscle. Various ratios of the other isoenzymic forms are observed in other tissues. Not only do the isoenzymes differ structurally, they differ in kinetic characteristics that will be discussed in later chapters.

    Subunits may be joined noncovalently by ionic or hydrophobic interactions of amino acid side chains on adjacent subunits or covalently by intersubunit disulfide bonds. The noncovalent forces may be disrupted and the subunits separated under conditions of high ionic strength or with chaotropic reagents. The intersubunit disulfide bonds can be cleaved by reducing agents such as dithiothreitol or 2-mercaptoethanol.

5 Functional Diversity of Proteins

Problems

Step	Volume (ml)	Total Protein (mg)	Total Units	Specific Activity (U/mg)	Yield (%)	Purification (*n*-fold)
Cell extract	2,800	70,000	2,700			
$((NH_4)_2SO_4)$ fractionation	3,000	25,400	2,300			
Heat treatment	3,000	16,500	1,980			
DEAE chromatography	80	390	1,680			
CM-cellulose	50	47	1,350			
Bio-Gel A	7	35	1,120			

1. **A method for the purification of 6-phosphogluconate dehydrogenase from *E. coli* is summarized in the table. For each step, calculate the specific activity, percentage yield, and degree of purification (*n*-fold). Indicate which step results in the greatest purification. Assume that the protein is pure after gel (Bio-Gel A) exclusion chromatography. What percentage of the initial crude cell extract protein was 6-phosphogluconate dehydrogenase?**

2. **Although used effectively in the 6-phosphogluconate dehydrogenase isolation procedure, heat treatment cannot be used in the isolation of all enzymes. Explain.**

3. **Assume that the isoelectric point (pI) of the 6-phosphogluconate dehydrogenase is 6. Explain why the buffer used in the DEAE cellulose chromatography must have a pH greater than 6 but less than 9 in order for the enzyme to bind to the DEAE resin.**

4. Will the 6-phosphogluconate dehydrogenase bind to the CM-cellulose in the same buffer pH range used with the DEAE-cellulose? Explain. In what pH range might you expect the dehydrogenase to bind to CM-cellulose? Explain.

5. Examine the isolation procedure shown in problem 1 and explain why gel exclusion chromatography is used as the final step rather than as the step following the heat treatment.

6. A student isolated an enzyme from anaerobic bacteria and subjected a sample of the protein to SDS polyacrylamide gel electrophoresis. A single band was observed upon staining the gel for protein. His adviser was excited about the result, but suggested that the protein be subjected to electrophoresis under nondenaturing (native) conditions. Electrophoresis under nondenaturing conditions revealed two bands after the gel was stained for protein. Assuming the sample had not been mishandled, offer an explanation for the observations.

7. A salt-precipitated fraction of ribonuclease contained two contaminating protein bands in addition to the ribonuclease. Further studies showed that one contaminant had a molecular weight of about 13,000 (similar to ribonuclease) but an isoelectric point 4 pH units more acidic than the pI of ribonuclease. The second contaminant had an isoelectric point similar to ribonuclease but had a molecular weight of 75,000. Suggest an efficient protocol for the separation of the ribonuclease from the contaminating proteins.

8. You have a mixture of proteins with the following properties:

Protein 1: M_r 12,000, pI = 10
Protein 2: M_r 62,000, pI = 4
Protein 3: M_r 28,000, pI = 8
Protein 4: M_r 9,000, pI = 5

Predict the order of emergence of these proteins when a mixture of the four is chromatographed in the following systems:

(a) DEAE-cellulose at pH 7, with a linear salt gradient elution.

(b) CM-cellulose at pH 7, with a linear salt gradient elution.

(c) A gel exclusion column with a fractionation range of 1,000–30,000 M_r, at pH 7.

9. The absorbance at 280 nm of an 0.5 mg ml^{-1} solution of protein A was 0.75. In addition, a shoulder on the absorbance spectrum at 288 nm was observed. An absorbance of 0.2 at 280 nm was measured for protein B in the same buffer and at the same concentration as protein A, but no spectral feature at 288 nm was observed. Amino acid analysis revealed that each protein contained approximately the same molar quantities of Tyr and Phe. Suggest a reason for the differences in absorbance at 280 nm and 288 nm.

10. The absorbance of a protein solution measured at 278 nm was 0.846. The protein content of that solution, calculated from quantitative amino acid analysis, was 460 µg/ml. Calculate the extinction coefficient of the protein in units of ml mg^{-1} cm^{-1}. Assume the molecular weight of the protein to be 42,000. Calculate the molar extinction coefficient.

11. You have isolated a manganese-containing pyrophosphatase and wish to determine its molecular weight. A Sephadex G-100 gel exclusion column was calibrated using proteins whose molecular weight had been established, with the following results.

Protein	Molecular Weight	Elution Volume (ml)
Serum albumin	69,000	84
Ovalbumin	43,000	92
Carbonic anhydrase	29,000	100
Chymotrypsinogen	23,000	104
Myoglobin	17,000	110
Blue dextran	2,000,000	60

The pyrophosphatase eluted from the calibrated column in 94 ml. Determine the molecular weight of the enzyme.

12. Individual proteins of known subunit molecular weight and the pyrophosphatase whose *native* molecular weight had been determined (problem 11) were denatured in buffer containing both SDS and 2-mercaptoethanol. A sample of each protein was electrophoretically separated by SDS-polyacrylamide gel electrophoresis. The marker dye migrated 12 cm. Staining the gel for protein revealed the following:

Protein	Molecular Weight (subunit)	Distance Migrated (cm)
Serum albumin	69,000	1.6
Catalase	60,000	2.2
Ovalbumin	43,000	3.5
Carbonic anhydrase	29,000	5.2
Myoglobin	17,000	7.4

The pyrophosphatase migrated 6.7 cm. However, when the 2-mercaptoethanol was omitted from the sample buffer, the pyrophosphatase migrated 3.8 cm. Determine the subunit molecular weight and comment on the quaternary structure of the enzyme.

13. A mutant form of alkaline phosphatase was focused to its isoelectric point and was found to differ from the native (nonmutant) alkaline phosphatase by an amount corresponding to one additional positive charge relative to the native enzyme. Assuming that the mutant enzyme's isoelectric point reflects the substitution of a single amino acid residue, what are the possibilities?

14. You wish to purify an ATP-binding enzyme from a crude extract that contains several contaminating proteins. In order to purify the enzyme rapidly and to the highest purity, you must consider some sophisticated strategies, among them affinity chromatography. Explain how affinity chromatography may be applied to this separation and explain the physical basis of the separation.

15. You have isolated a protein complex that sediments in the ultracentrifuge similarly to a hemoglobin marker. If the ultracentrifugation is repeated under identical conditions except that 2 M NaCl is added to the dilute buffer, the protein sediments similarly to a myoglobin marker. What conclusion can you reach about the properties of the protein complex?

16. Predict the effect on O_2 transport of a mutant form of hemoglobin with a markedly decreased affinity for glycerate-2,3-bisphosphate.

17. The concentration of glycerate-2,3-bisphosphate (GBP) is approximately 5 mM in the erythrocyte. Calculate the concentration of hemoglobin in the erythrocyte and compare it with that of GBP. Assume that the hemoglobin content of blood is 14 gm/100 ml and that the erythrocytes occupy 45% of whole blood volume. For the calculations, assume that hemoglobin (M_r = 64,500) occupies 100% of the erythrocyte volume.

Solutions

Step	Volume (ml)	Total Protein (mg)	Total Units	Specific Activity (U/mg)	Yield (%)	Purification (*n*-fold)
Cell extract	2,800	70,000	2,700	0.039	100	1
$((NH_4)_2SO_4)$ fractionation	3,000	25,400	2,300	0.091	85	2.3
Heat treatment	3,000	16,500	1,980	0.12	73	3.1
DEAE chromatography	80	390	1,680	4.3	62	110
CM-cellulose	50	47	1,350	29	50	740
Bio-Gel A	7	35	1,120	32	41	820

1. Sample calculation (DEAE cellulose step)
Specific activity is defined as units of enzymatic activity per mg protein.

$$\text{Specific activity} = \frac{1{,}680 \text{ U}}{390 \text{ mg}}$$
$$\text{Specific activity} = 4.3 \text{ U/mg}$$

Yield is calculated as the percentage of the initial activity units remaining after the specified isolation step.

$$\text{Yield} = \frac{1{,}680 \text{ U}}{2{,}700 \text{ U}} \times 100$$
$$\text{Yield} = 62\%$$

N-fold purification is the ratio of the specific activity at a given step in the purification to the specific activity of the enzyme in the cell extract.

$$n\text{-fold purification} = \frac{4.3 \text{ U mg}^{-1}}{3.9 \times 10^{-2} \text{ U mg}^{-1}}$$
$$n\text{-fold purification} = 110$$

The DEAE chromatography yields the greatest purification. The specific activity of the enzyme following the DEAE chromatography was increased 36-fold compared with the specific activity of the previous step. The ammonium sulfate fractionation caused a 2.3-fold increase; the heat step, 1.4-fold; CM-cellulose, 6.6-fold; and gel exclusion chromatography, 1.1-fold.

The specific activity of the pure 6-phosphogluconate dehydrogenase was increased 820-fold compared to the cell extract. The enzyme must have been present as 1/820 of the total protein. The initial percentage was therefore

$$\frac{1}{820} = 1.2 \times 10^{-3} = 0.12\%$$

2. Heat treatment of protein solutions denatures and precipitates some of the proteins, while others remain both soluble and stable. Thermal lability is determined empirically for each enzyme or protein of interest. The enzyme or protein of interest may be rapidly denatured by the thermal treatment, necessitating alternative strategies for isolation.
3. The protein is predicted to adhere to DEAE if the net charge on the protein is negative and that of the diethylaminoethyl exchange group is positive. The DEAE exchanger group is protonated and positively charged at pH up to approximately 9 but is unprotonated and neutral above pH 9, assuming a pK_a of about 8.5 for the DEAE exchanger. The protein will be negatively charged at pH more basic than the isoelectric point (pI = 6.0). Thus, buffer in the pH range just above 6 to just below 9 will ensure that the exchanger group is positively charged and the protein is negatively charged, and ionic interaction between the two would likely occur.
4. The ion-exchanger group on the CM-cellulose is a weak carboxylic acid whose pK is around 4. Above pH 4, the carboxylic acid is unprotonated and negatively charged. However, in the buffer system given in problem 3, the protein is negatively charged and will not adhere to the negatively charged CM-cellulose. The protein is positively charged at pH more acidic than the pI. Therefore the protein will probably adhere to (bind to) the CM-cellulose if the pH is between 4 and 6.
5. Gel exclusion chromatography is generally of limited value when large volumes of protein solution are used. The most effective separations occur when the smallest possible volume of protein solution is loaded onto the gel exclusion column, avoiding the broad, diffuse, overlapping protein bands that usually result when large volumes are loaded onto the column. The 3-liter protein pool after heat treatment is hardly conducive to gel exclusion chromatography on less than industrial scale. For example, the volume loaded onto gel exclusion columns is ideally 5% of the column volume. Were this criterion applied to chromatography of the 3-liter sample, a column containing 60 liters of gel exclusion resin would be required. In addition, the less highly purified sample is more likely to have a contaminating protein that would coelute with the desired protein.
6. Proteins are separated by SDS-PAGE on the basis of molecular weight of the denatured protein. The molecular weight is the minimum value or subunit molecular weight. Two proteins that differ in some properties but have the same subunit molecular weight will likely appear as a single protein band after SDS-PAGE. Thus, a 40,000 M_r dimer composed of 2 × 20,000 M_r subunits will not be separated from an 80,000 M_r tetramer composed of 4 × 20,000 M_r subunits. Native, or nondenaturing, electrophoresis separates proteins based on mass/charge characteristics. The student's "pure" protein contains at least two components separable by the criterion of mass/charge but which share a common subunit molecular weight. It is also possible that the multiple protein bands appearing in the nondenaturing gel arose through deamidation of glutamine or asparagine residue side chains.
7. Contaminant B (75,000 M_r) can be easily separated from both ribonuclease and contaminant A by gel exclusion chromatography but not by ion-exchange chromatography. The similarity of pI values for the ribonuclease and contaminant B suggests that separation based on charge likely would be of little value. Contaminant A may be separated from ribonuclease by ion-exchange chromatography (pI values differ markedly) but not by gel exclusion chromatography (molecular weights are similar).

   Ion-exchange chromatography is suited not only for separation of proteins but also for concentration of the proteins in dilute solution. Assume the volume of the contaminated ribonuclease solution is too large for gel exclusion chromatography. The solution is dialyzed in buffer whose pH is more acidic than the pI of ribonuclease but more basic than the pI of contaminant A. Under these conditions, the ribonuclease and contaminant B will be positively charged, but contaminant A will be negatively charged. A CM-cellulose ion exchange column in the same buffer as the proteins will bear a negative charge on the carboxymethyl groups. Contaminant A will elute in the low ionic strength buffer used to load the column, while ribonuclease and contaminant B are predicted to adhere to the column but will likely be eluted at approximately the same point in a linear KCl gradient. The pooled fraction from the CM-cellulose (containing ribonuclease and contaminant B) in a smaller volume than the original sample may now be chromatographed on a gel exclusion resin that excludes 50,000 M_r globular proteins. Contaminant B will elute in the void volume, but ribonuclease will be retarded in the included volume.

8a. DEAE cellulose will have a net positive charge at pH 7 due to protonation of the weakly basic tertiary amine (pK_a approximately 8.5). The exchanger will bind or retard negatively charged proteins. The charge on each protein may be estimated from the pI. At pH values greater than pI, the protein will be negatively charged. At pH below pI, the protein will be positively charged.

Protein	pI	Charge at pH 7
Protein 1	10	(+)
Protein 2	4	(-)
Protein 3	8	(+)
Protein 4	5	(-)

Proteins 1 and 3 should elute in the initial wash buffer, but proteins 2 and 4 are predicted to bind to the column. Based solely on isoelectric point, one might predict that protein 4, then protein 2, would be eluted in the salt gradient.

8b. The carboxyl group of the CM-cellulose will be negatively charged at pH 7 because of deprotonation of the weak acid (pK_a approximately 4). The charge on each protein was established in the previous section.

Proteins 2 and 4 would be eluted in the initial wash buffer from the column, whereas proteins 1 and 3 would be predicted to adhere to the column. Based solely on pI values, protein 3 would be predicted to elute prior to protein 1 in the KCl gradient. In each separation of proteins based on ion exchange (e.g., problems 8a and 8b), the actual elution order must be determined experimentally. Proteins having the same or similar pI but differing in the absolute number of charged residues will likely differ in their elution characteristics.

8c. Gel exclusion chromatography separates globular proteins on the basis of relative size, which is a function of molecular weight. Larger proteins (higher molecular weight) elute before smaller (lower molecular weight) proteins. The gel exclusion resin proposed to separate the four proteins (8a) has an exclusion limit of 30,000. Proteins with molecular weight greater than the limit (protein 2, 62,000 M_r) are excluded from entry into the gel and elute in the void volume (V_o). The other proteins in the solution will elute in the order protein 3, protein 1, protein 4.

9. The absorbance of light in the 270 nm to 290 nm wavelength range is due primarily to the content of tyrosine and tryptophan. Proteins whose amino acid content includes tryptophan absorb light at 280 nm and frequently have a discernable absorbance "shoulder" at 288 nm. Moreover, the molar extinction of tryptophan at 280 nm is approximately fivefold that of tyrosine. The characteristics of the absorbance spectrum of protein A and the absorbance value at 280 nm are consistent with tryptophan as a component of the amino acid composition. Protein B has little if any tryptophan in the amino acid composition, a prediction consistent with the low absorbance at 280 nm.

10. From the Beer-Lambert relationship,

$$A = \epsilon b c$$

where

A is the absorbance of the solution at a discrete wavelength
b is the pathlength of light through the solution
c is the concentration of the absorbing species in solution and
ϵ is the extinction coefficient relating concentration of absorbing species to absorbance at a discrete wavelength.

$$\epsilon = \frac{A}{(b)(c)} = \frac{(0.846)}{(1\ \text{cm})\ (0.46\ \text{mg ml}^{-1})}$$
$$\epsilon = 1.84\ \text{ml mg}^{-1}\ \text{cm}^{-1}$$

The molar concentration of protein is

$$\frac{(0.46\ \text{mg ml}^{-1})}{(42{,}000\ \text{mg mmole}^{-1})} = \frac{(0.46\ \text{mg ml}^{-1})}{(4.2 \times 10^4\ \text{mg mmole}^{-1})}$$

$$(C) = 1.1 \times 10^{-5}\ \text{mmole ml}^{-1}$$
$$(C) = 1.1 \times 10^{-5}\ \text{M}$$

Now,

$$\epsilon = \frac{(0.846)}{(1\ \text{cm})\ (1.1 \times 10^{-5}\ \text{M})}$$
$$\epsilon = 7.7 \times 10^4\ \text{M}^{-1}\ \text{cm}^{-1}$$

11. The molecular weight of the pyrophosphatase can be estimated from a calibrated gel exclusion column. A standard curve is constructed by graphing the partition function (V_e/V_o) versus the logarithm of the known molecular weights.

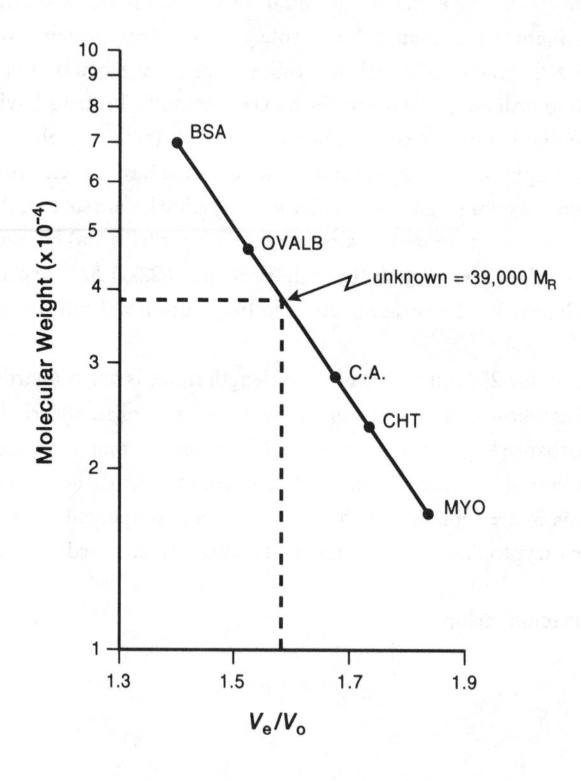

V_e is the elution volume, determined for each protein, and V_o is the column void volume, measured by the elution of Blue dextran. The molecular weight of the pyrophosphatase may then be determined from its partition function (V_e / V_o).

Protein	Molecular Weight	V_e / V_o
Serum albumin	69,000	1.40
Ovalbumin	43,000	1.54
Carbonic anhydrase	29,000	1.68
Chymotrypsinogen	23,000	1.73
Myoglobin	17,000	1.83

The (V_e / V_o) calculated for the pyrophosphatase is 1.57, and the molecular weight is approximately 39,000. Note that the molecular weight could have been determined by graphing log M_r versus elution volume (V_e) or versus the function

$$K_{ave} = \frac{V_e - V_o}{V_t - V_o}$$

where V_t is the total volume of the hydrated resin in the column.

12. Calculate an R_m value for each molecular weight standard based on the distance the protein standard migrated divided by the distance the tracking dye migrated. These values graphed versus the logarithm of the respective molecular weights establish a standard curve from which the subunit molecular weight of the pyrophosphatase can be estimated.

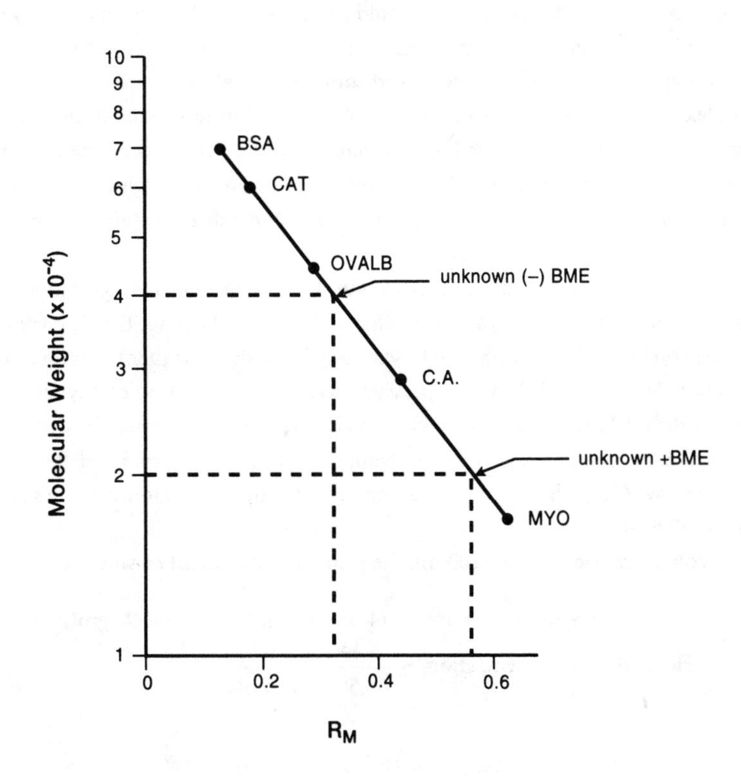

Protein	Subunit Molecular Weight	R_f
Serum albumin	69,000	0.13
Catalase	60,000	0.18
Ovalbumin	43,000	0.29
Carbonic anhydrase	29,000	0.43
Myoglobin	17,000	0.62

Pyrophosphatase (+ 2-mercaptoethanol) 0.56 (20,000 M_r)
Pyrophosphatase (- 2-mercaptoethanol) 0.32 (40,000 M_r)

Native (nondenatured) molecular weight determined with the calibrated gel exclusion chromatography is 39,000. SDS-PAGE of pyrophosphatase that had been denatured and reduced with 2-mercaptoethanol yielded a 20,000 M_r peptide, whereas omission of the reductant yielded a 40,000 M_r peptide. The pyrophosphatase is thus a dimer of 20,000 molecular weight subunits joined by intersubunit disulfide bonds.

13. Increasing the net positive charge on the protein by a single amino acid substitution may be accomplished by
    (a) substitution of an acidic amino acid (Asp or Glu) by a neutral amino acid (e.g., Asn or Gln) or
    (b) substitution of a basic amino acid residue (Lys or Arg) for a neutral amino acid residue.

14. Affinity chromatography is an elegant, rather specific method to separate a protein or enzyme from a mixture. The technique is based on the specific, strong binding of a substrate, substrate analog, inhibitor, or antibody to a protein. An initial strategy to isolate the ATP-binding protein involves constructing an affinity resin by linking ATP to an insoluble support through a spacer side chain of appropriate length. The spacer allows the protein to bind the affinity probe with minimal steric or chemical interference from the insoluble support. However, the protein of interest, or another protein in the crude extract, may hydrolyze ATP and destroy the affinity probe. As an alternative approach, construct the affinity probe using a nonhydrolyzable analog of ATP (e.g., imido- or methylene analog). The ATP-binding protein(s) would be predicted to bind to the affinity probe. The contaminating proteins are washed from the column and the bound protein eluted with ATP, increased ionic strength, pH alteration, or an empirically determined combination of these.

15. The protein complex sediments similarly to the 64,000 M_r marker protein hemoglobin but sediments similarly to the 17,000 M_r myoglobin marker when 2 M NaCl is included in the buffer. The structure predicted for the complex is a tetramer of approximately 16,000 M_r subunits associated through ionic or electrostatic interactions but not by covalent bonds. The high ionic strength of the 2 M NaCl disrupts the ionic interactions, causing dissociation of the tetramer to the monomers.

16. Glycerate-2,3-bisphosphate (GBP) binds in the pocket in the center of the hemoglobin $\alpha_2 \beta_2$ tetramer. 2,3-GBP binds more tightly to deoxyhemoglobin than to oxyhemoglobin. In the lung, the O_2 partial pressure is sufficiently large to drive the deoxyHb/oxyHb equilibrium toward oxyHb with subsequent decrease in GBP affinity. In the O_2-consuming tissues, the decreased O_2 partial pressure favors deoxygenation of oxyhemoglobin. The deoxyHb form is stabilized by the GBP, shifting the deoxyHb/oxyHb equilibrium toward the deoxy form. Thus, more O_2 is released to the tissues in the presence of GBP than would occur in its absence. If GBP were not present, or if hemoglobin failed to bind GBP, there would be a diminished supply of O_2 to the tissues and that, if severe, could lead to anoxia and cell death.

17. If the hemoglobin concentration is 14 gm/100 ml, the hemoglobin content of whole blood may be estimated.

$$\text{Hemoglobin content} = 14 \text{ gm/100 ml} \times 10 = 140 \text{ gm/l}$$

$$\text{Hemoglobin concentration} = \frac{140 \text{ gm l}^{-1}}{64{,}500 \text{ gm/mole}}$$

$$[\text{Hb}] = 2.17 \times 10^{-3} \text{ M } (2.17 \text{ mM})$$

in whole blood, but the erythrocytes occupy 45% of the whole blood volume.

$$\text{Therefore, [Hb] in erythrocytes} = \frac{2.17 \text{ mM}}{0.45}$$

[Hb] = 4.8 mM and is approximately equal to the concentration of GBP.

6 Carbohydrates, Glycoproteins, and Cell Walls

Problems

1. Compare the Haworth projections of D-glucose, D-mannose, and D-galactose. Indicate the differences between the structures.

2. Indicate the chiral carbons in D-glucose and determine the number of possible stereoisomers.

3. Is it possible that carbohydrates could have produced the diversity required to catalyze the myriad cellular reactions now relegated primarily to proteins?

4. In solution, D-glucose has a specific rotation of $[\alpha]_o^{20}D = +52.7°$. The specific rotation of pure ß-D-glucose is +18.7° and that of pure α-D-glucose is +112.2°. Calculate the fraction of the α and ß anomers in solution.

5. If either pure crystalline α- or ß-D-glucose is dissolved in water, the final solution will contain the same fraction of α and ß anomers as determined in problem 4. Explain chemically how this mutarotation occurs.

6. You are given two containers of polysaccharide by a colleague who has labeled one container as cellulose and one as glycogen. In his haste he may have mislabeled them and has asked you to verify which is which by methylation analysis. Indicate what product(s) you would expect from exhaustive methylation and mild acid hydrolysis of each polysaccharide and how the products could be used to differentiate the samples.

7. (a) Glycogen, starch, and cellulose are all polymers of glucose. Suggest reasons, based on structure, that the physical form of each is appropriate to its role in nature. Why are the polymer forms of starch and glycogen much more desirable than an equivalent amount of free glucose in the cell?

7. (b) Suggest how a cell might selectively synthesize starch but not cellulose.

8. Consider the packing of lipid triglycerides in adipocytes and of glycogen granules in the liver. Comment on the feasibility of using only glycogen, rather than lipid, as sole energy reserve. (See the section in chapter 7 titled "Some fatty acids are stored as an energy reserve in triglycerides.")

9. Given the trisaccharide D-mannose-ß(1,3) D-glucose-α(1,6) D-galactose, draw the structure of the trisaccharide using Haworth projections. Name and draw the structures of the products of exhaustive methylation with dimethyl sulfate and mild acid hydrolysis of the trisaccharide.

10. Chemical degradation of glycosaminoglycans causes reduced viscosity of the synovial fluid and subsequent damage to joints. Explain.

11. A tetrasaccharide has the following composition: D-Man (2), D-Gal (1), D-Glc(1). The tetrasaccharide gave a positive reducing sugar test in which the glucose residue was oxidized. Exhaustive methylation and mild acid hydrolysis released 2,3,4,6-tetra-*O*-methylmannose, 2,3-di-*O*-methylgalactose, and 2,3,6-tri-*O*-methylglucose. Treatment of the tetrasaccharide with an α-mannosidase released mannose and a trisaccharide that yielded the following methylation products: 2,3,4,6-tetra-*O*-methylmannose, 2,3,6-tri-*O*-methylgalactose, and 2,3,6-tri-*O*-methylglucose. Deduce the sequence and specificity of anomeric linkages and indicate any ambiguity.

12. Which enzymes can be used to ascertain the presence of sialic acid, galactose, *N*-acetylglucosamine, and mannose in a complex carbohydrate?

13. How might a cell synthesizing an N-linked glycoprotein specifically glycosylate only two of ten asparagine residues in the protein?

14. What structural features of oligosaccharides complicate the determination of their sequence as compared with the sequence determination of proteins?

Solutions

1. D-Mannose is the 2-epimer and D-galactose is the 4-epimer of D-glucose as shown in the figure.

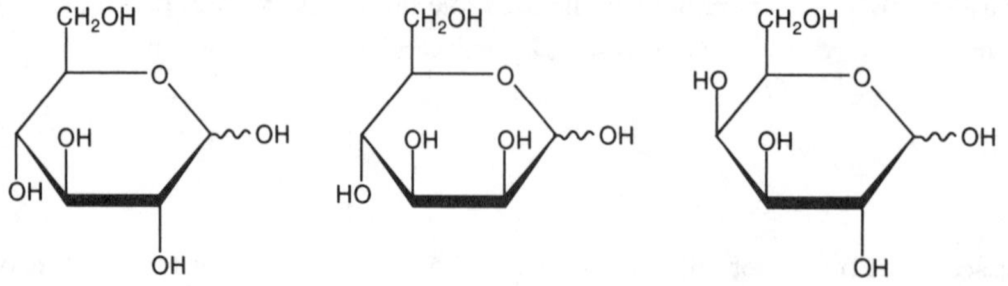

D-glucose | D-mannose (2-epimer of glucose) | D-galactose (4-epimer of glucose)

2. The chiral carbons in D-glucose, shown as a Fischer projection, are indicated by asterisks.

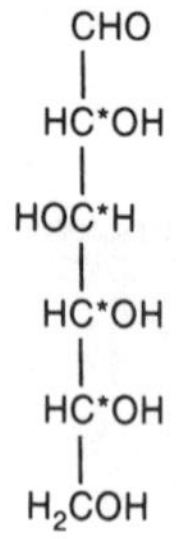

CHO
|
HC*OH
|
HOC*H
|
HC*OH
|
HC*OH
|
H_2COH

There are four asymmetric (chiral) carbons. The number of possible stereoisomers is $2^4 = 16$. If the pyranose ring form is considered, there are two additional stereoisomeric forms represented by the α and ß anomers. The pyranose ring is generated by formation of the hemiacetal between the C - 5 OH group and the C-1 aldehyde and creates a fifth asymmetric center, yielding thirty-two stereoisomeric forms. Thus, α- and β-D-glucose are two of the 32 possible stereoisomeric forms of the aldohetoses.

3. Polysaccharides exist, in some instances, as highly branched structures. The number of different structural arrangements possible in polysaccharides is staggering. When one considers higher order polymers of the various sugars, the diversity of structures may exceed the possible arrangement of linear amino acids in polypeptides. However, the carbohydrates may be unable to provide the diversity of chemical reactivity or binding characteristics that can be provided by the amino acids. Amino acids provide diverse functional side chains (strongly basic, acidic, and hydrophobic entities) that carbohydrates lack. In addition, one would have to consider information transfer to allow specific synthesis of catalytic carbohydrates. Amino acids are linked into linear polypeptide chains under the direction of a linear sequence of nucleotides. Construction of branched carbohydrates from a linear template would be difficult.

4. The fraction of each anomeric form may be calculated using two linear equations. Let X = fraction of ß anomer and Y the fraction of α anomer. The fractions equal 1 (or 100%).

$$X + Y = 1$$

The specific rotation contributed to the mixture is a function of the amount of each anomer.

$$18.7X + 112.2Y = 52.7$$

Solving the two linear equations reveals that the mixture contains 0.364 (36.4%) α anomer and 0.636 (63.6%) β anomer.

5. Solutions of pure α or β anomer are thermodynamically less stable than the final mixture. The anomers will mutarotate until a lower energy final steady-state mixture is achieved. Each anomeric form undergoes reversible opening and reclosing of the pyranose ring, with the equilibrium position strongly favoring ring closure. In the ring-open configuration, the aldehyde group may rotate so that either side (face) of the carbonyl is available for readdition of the -OH group from C-5, thus generating the mixture of anomers.

6. Cellulose is a polymer of glucose linked in a linear chain through β(1,4) glycosidic bonds. Exhaustive methylation and mild acid hydrolysis should yield a preponderance of 2,3,6-tri-*O*-methylglucose, derived from the glucosyl residues in the chain, and a small fraction of 2,3,4,6-tetra-*O*-methylglucose from the nonreducing terminal residues.

    Glycogen is a highly branched polymer of glucose molecules linked via α(1,4) glycosidic bonds with α(1,6)-linked branches. Methylation and acid hydrolysis will yield the following mixture: 2,3,4,6-tetra-*O*-methylglucose, derived from the nonreducing terminal glucose residues; 2,3,6-tri-*O*-methylglucose, from glucosyl residues in the chain; 2,3-di-*O*-methylglucose from the glucose residues at the branch point. One can differentiate glycogen from cellulose by the presence of the 2,3-di-*O*-methylglucose. One would also expect the ratio of the 2,3,4,6-tetra-*O*-methylglucose to 2,3,6-tri-*O*-methylglucose to be larger in glycogen than in cellulose because of the larger number of branches in glycogen, each contributing a nonreducing glucosyl residue.

7a. Glycogen and starch are polymers of glucose linked through α-glycosidic bonds. Glycogen is more highly branched than is starch. Each form assumes an open, helical coiled form that hydrogen-bonds with water. The highly branched structure of glycogen provides numerous nonreducing termini from which glucose-1-phosphate is released by the catalytic action of glycogen phosphorylase. The rapid release of glucose is required to provide energy to the skeletal muscle during the adrenalin-stimulated "fight or flight" response and as a source of plasma glucose provided by the liver for short-term blood glucose homeostasis.

    Cellulose is a linear polymer of β(1,4)-linked glucose molecules. The glucosyl chains associate through interchain hydrogen bonds. Virtually all the potential H bonds are consumed in this manner, yielding a rigid, water-insoluble structure suitable for its structural role in plant cell walls.

    Glycogen or starch exhibit lower osmotic pressure than would an equivalent amount of free glucose, allowing the cell to store carbohydrate without danger of osmotically induced swelling and lysis of the cell.

7b. Starch and cellulose are each secondary gene products whose synthesis depends on the appropriate catalytic activities. Proteins (enzymes) are primary gene products whose synthesis is regulated within the cells. Hence, cells that express only the enzymes catalyzing starch formation selectively form starch but not cellulose. Cellulose synthesis requires a specific set of biosynthetic enzymes that are expressed only in cells that synthesize cellulose. Synthesis of starch requires enzymatic activities to form the α-glycosidic bonds and to form the branches. Cellulose synthesis requires formation of the β-glycosidic bond.

8. Lipids are packed as hydrophobic droplets in specific cells called adipocytes (see chapter 7 in the text). The hydrophobic lipids provide a compact source of energy. Glycogen is hydrated, and the water of hydration contributes to the mass of the stored carbohydrate. Even if lipid and carbohydrate supplied the same metabolic energy per gram, the hydrophobic lipid would be a preferred storage form. However, as we will see in chapter 17, the energy available from lipid is approximately 2.5-fold that of carbohydrates. To provide comparable energy would require a mass of carbohydrate 2.5 times that of triglyceride. The actual stored mass of carbohydrate would, in reality, be significantly greater due to the water of hydration of the hydrophilic carbohydrate.

9. **Exhaustive methylation and mild acid hydrolysis yield: 2,3,4,6-tetra-*O*-methyl D-mannose; 2,4,6-tri-*O*-methyl D-glucose; and 2,3,4-tri-*O*-methyl D-galactose. The anomeric configuration of the reducing terminal galactose was not specified and will be a mixture of the α and ß anomers. The structures of the methylated products are:**

Methylated products

2, 3, 4, 6-tetra-O-methyl D-mannose:

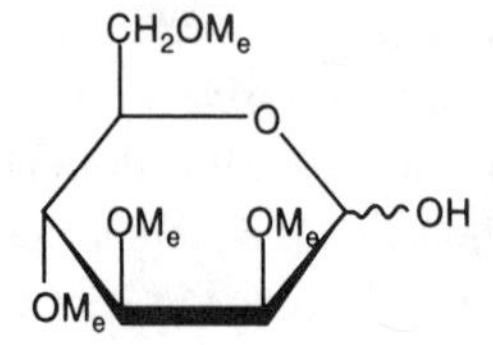
CH2OMe
O
OMe
OMe
OH
OMe

2, 4, 6-tri-O-methyl D-glucose:

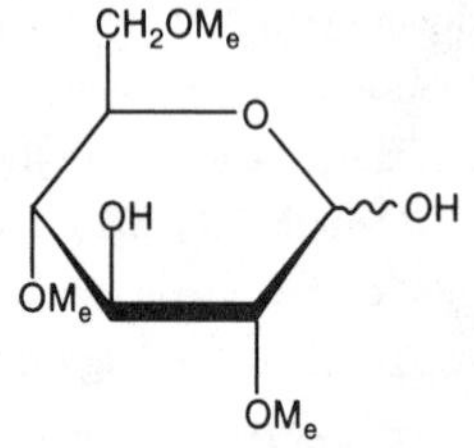
CH2OMe
O
OH
OH
OMe
OMe

2, 3, 4-tri-O-methyl D-galactose:

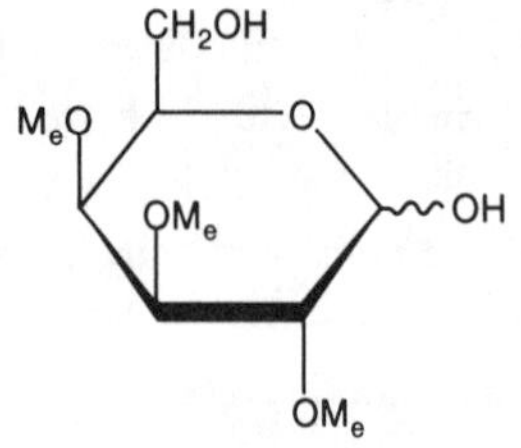
CH2OH
MeO
O
OMe
OH
OMe

Structure of trisaccharide (D-mannose-β-(1→3) - D-glucose-α(1→6) - D-galactose)

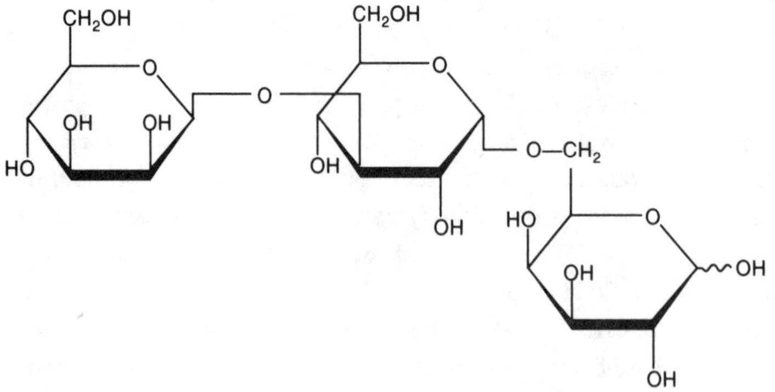
CH2OH
O
OH
OH
HO
O
CH2OH
O
OH
OH
O—CH2
HO
O
OH
OH
OH

10. Glycosaminoglycans are linear polysaccharide chains with a high sulfate and carboxylate content and form a viscous, mucoid solution. The lubricating properties of the viscous solution eases movement of the bones within joints and cushions the impact of joint motion. Chemical degradation of the components within the synovial fluid may lead to decreased viscosity with resulting diminution of lubricating properties. Eventually, debilitating damage to the joint may occur.

11. Glucose is the reducing terminal sugar and has its 4-OH group in a glycosidic linkage. The mannose residues are nonreducing termini of the branched structure and yield a single methylated product. The structure of the tetrasaccharide consistent with these data is

D-man (1, 6) ╲
 D-gal-(1,4) D-Glc
D-man (1, 4) ╱

Treatment of the tetrasaccharide with α-mannosidase released free mannose and a trisaccharide. Methylation analysis of the trisaccharide yielded 2,3,4,6-tetra-*O*-methyl D-mannose, 2,3,6-tri-*O*-methylgalactose, and 2,3,6-tri-*O*-methyl D-glucose. However, the availability of the C-6 OH group of D-galactose for methylation after the α-mannosidase treatment shows that one mannose residue was linked α (1 → 6) to the galactose. The other mannose, resistant to the α-mannosidase treatment, must be linked ß (1 → 4). None of the data reveal the anomeric specificity of the D-Gal-D-Glc linkage. The structural information is consistent with

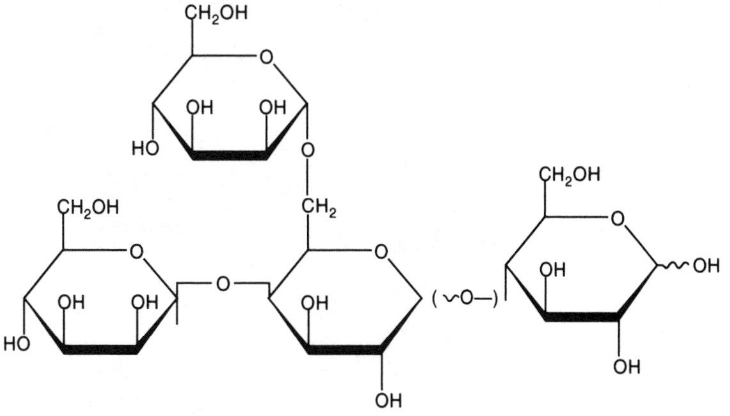

D-Man α(1→ 6) ╲
 D-Gal (1→ 4) D-Glc
D-Man β(1→ 4) ╱

12. Exoglycosidases catalyze the sugar-, anomer-, and linkage-specific cleavage of monosaccharides from the nonreducing termini of oligosaccharides. Treatment of the oligosaccharide with a specific exoglycosidase may release a specific monosaccharide if the sugar is present in the correct anomeric configuration. Thus, the following exoglycosidases may be used to release these sugars.
 (a) sialic acid: *N*-acetylneuraminidases
 (b) galactose: α or ß galactosidases
 (c) *N*-acetylglucosamine: *N*-acetylglucosaminidases
 (d) mannose: α or ß mannosidases

13. The preformed complex carbohydrate core is transferred to specific asparagine residues in the protein sequence. Asparagines bearing the carbohydrate are found in the amino acid sequence Asn-X-Ser(Thr) in which a variety of amino acids except proline have been found at position X. Asparagine in this particular sequence of amino acids is a substrate for the transfer of the glycosyl unit, whereas other asparagines, with different neighboring amino acids, are not. Asparagines accepting oligosaccharide are on the surface of the protein and are frequently in a ß turn or 100P in the protein. (See Bause, E., and G. Legler, *Biochem. J.* 195: 639–644, 1981, and Bause, E., *Biochem. J.* 209:331–336, 1983.)

14. Proteins are linear sequences of amino acids linked by peptide bonds. The linkages are always formed between the α-carboxyl group of an amino acid and the α-amino group of the neighboring residue. Oligosaccharides, in contrast, frequently are highly branched structures. The sugars are linked through glycosidic bonds to any one of several hydroxyl groups on the adjacent sugar (positional isomers) and in one of two anomeric configurations. Complete elucidation of the oligosaccharide structure requires knowledge not only of the identity of the monosaccharides, but also of the linkage position and anomeric configuration of each monosaccharide in the structure. When the complex carbohydrate is attached to protein, as is often the case, it is also important to determine the amino acid to which the carbohydrate is attached.

7 Lipids and Membranes

Problems

1. "Individuals whose diet is devoid of plant tissue may develop essential fatty acid deficiency." Defend or refute this statement.

2. Compare the relative efficiency of extraction of free fatty acid, fatty acid methyl ester, and triacylglycerol into organic solvent. Which component(s) have a pH dependency of extraction? Why?

3. Both triacylglycerol and phospholipids have fatty acid ester components, but only one can be considered amphipathic. Indicate which is amphipathic and explain why.

4. The term phosphatidylcholine (PC) defines a class of phospholipids. What portion of the phosphatidylcholine molecule is common to all members of the class? What portion of the molecule is variable among the PCs?

5. Use Haworth formulas to represent the structure of the glycosphingolipid referred to as Trihexosylceramide (D-Gal-α-1,4-D-Gal-ß-1,4-D-Glc-ß-1,1-ceramide). What type of linkage bonds the carbohydrate to the ceramide?

6. If phosphatidylcholine were dispersed in an aqueous buffer, what types of molecular associations would yield the most stable suspension?

7. Triton X-100 (M_r = 625, CMC = 0.24 mM) and solid deoxycholate (M_r = 414, CMC = 4 mM) are each used to solubilize proteins from membranes before application of further isolation techniques.

(a) Which of these detergents would be more easily removed by dialysis from an 0.1% (wt/vol) aqueous solution?

(b) In which of these detergents would ion-exchange chromatography be more likely successful for protein purification?

(c) In which of the detergents would gel exclusion chromatography be likely to approximate the true molecular weight of an integral membrane protein in 0.1% aqueous solution of the detergent? Why?

8. Quinones that are structurally similar to vitamins K_1 and K_2 (see fig. 11.24) are associated with the membranes of mitochondria and chloroplasts. How does the structure represented by vitamin K_2 explain the molecule's association with a hydrophobic membrane?

9. The relative orientation of polar and nonpolar amino acid side chains in integral membrane proteins is "inside out" relative to that of the amino acid side chains of water-soluble globular proteins. Explain.

10. What physical properties are conferred on biological membranes by phospholipids? How could the charge characteristics of the phospholipids affect binding of peripheral proteins to the membrane? What role might divalent metal ions play in the interaction of peripheral membrane proteins with phospholipids?

11. Predict the effects of the following on the phase transition temperature (T_m) and/or phospholipid mobility in sure dipalmitoylphosphatidylcholine vesicles (T_m = 41°C).

(a) Raising the temperature from 30° to 50°C.

(b) Introducing dipalmitoleoylphosphatidylcholine into the vesicles.

(c) Introducing dimyristoylphosphatidylcholine into the vesicles.

(d) Incorporating integral membrane proteins into the vesicles.

12. Differentiate between peripheral and integral membrane proteins with respect to location, orientation, and interactions that bind the protein to the membrane. What are some strategies used to differentiate peripheral and integral proteins by means of detergents or chelating agents?

13. Frequently, integral membrane proteins are glycosylated with complex carbohydrate arrays. Explain how glycosylation further enhances the asymmetric orientation of integral proteins.

14. In examining a cell extract from a culture of Gram negative bacteria, you identified a protein suspected to be periplasmic. Upon performing the osmotic shock experiment described in the text, you find a small amount (approximately 10%) of the total enzymatic activity in the extracellular fraction. What control or other experiments might you do to convince yourself that the protein is periplasmic? If you had only data from the experiment described, what other possibilities could explain your results?

15. You are characterizing a protein in a membrane fraction that was dissolved in octylglucoside. You have estimated the molecular weight to be approximately 60,000. However, upon exhaustive treatment to remove most of the detergent, the protein elutes from a 100,000 M_r cutoff gel exclusion column in the void (excluded) volume. What can you conclude from these data?

16. Integral transmembrane proteins often contain helical segments of the appropriate length to span the membrane. These helices are composed of hydrophobic amino acid residues. In transmembranous proteins with multiple segments that span the membrane, you may find some hydrophilic residue side chains. Why are hydrophilic side chains not favored in single-span membrane proteins? How may the hydrophilic side chains be accommodated in multiple-span proteins?

Solutions

1. Mammals are unable to synthesize the unsaturated fatty acids, linoleic (C18:2 $^{cis\Delta 9,12}$) and linolenic (C18:3$^{cis\Delta 9,12,15}$), that are required for synthesis of phospholipids and of other polyunsaturated fatty acids. The essential fatty acids must therefore be supplied in the diet, and individuals whose diets are devoid of these fatty acids could develop a deficiency. Plant tissues have the enzymes required to synthesize these polyunsaturated fatty acids and are a dietary source of linolenic and linoleic acids. However, meat or milk from herbivores contain these essential fatty acids, albeit in lower concentration than in plant tissue, and could supply the amounts of linoleic and linolenic acids required for adequate nutrition. Hence the postulate is not true.
2. Extraction of solute from an aqueous to a nonaqueous solution depends on its relative solubility in each of the solvents. The solubility will depend to a large degree on the relative hydrophobicity or hydrophilicity of the solute. Hydrophilic molecules readily dissolve in aqueous media, whereas hydrophobic molecules do not. Triacylglycerides and the fatty acid methyl ester are hydrophobic and will readily partition into the nonaqueous solvent. Neither of these molecules has an ionizable group; their extraction into a nonpolar solvent is therefore not expected to be pH-dependent. The free (unesterified) fatty acid is an alkyl carboxylic acid whose alkyl chain is hydrophobic but whose carboxylate is hydrophilic.

   The pK_a value for these carboxylic acids is approximately 5 and the carboxyl group will exist as an anion at pH > 5. Hence, extraction of the fatty acid anion into nonaqueous media is retarded by the energetically unfavorable transfer of a charged group into a nonpolar environment where solvation and H-bonding are precluded. Protonation of the carboxylate (pH < pK_a) abolishes the charge on the molecule, diminishes the hydrophilicity, and increases the ease of extraction into organic solvent.
3. Triacylglycerides are neutral lipids, readily soluble in organic but not aqueous media, and are not amphipathic. Triacylglycerides are stored in adipocytes in a hydrophobic environment, providing a ready source of metabolic energy. (See chapters 14 and 17 of the text.) Phospholipids are amphipathic because the polar phosphoglyceryl portion of the molecule is soluble in aqueous media, whereas the fatty acids esterified to the glyceryl portion are hydrophobic. Moreover, the phosphate group may be esterified with choline, ethanolamine, or serine, adding to the polar nature of that portion of the molecule. Phospholipids in biological membranes associate in bilayers in which the hydrophobic fatty acyl "tails" form the interior of the bilayer while the polar phosphoryl groups interact with the aqueous medium on either side of the bilayer.
4. Phosphatidylcholine describes a class of molecules consisting of *sn*-glycerol-3-phosphate linked through a phosphodiester bond to the -OH group of choline (see the structure shown here). This portion of the molecule is common to all the phosphatidylcholines. The fatty acids esterified to the C-1 and C-2 OH groups of the glyceryl portion of the molecule (R_1 and R_2 in the structure) provide the diversity of phosphatidylcholine structures.

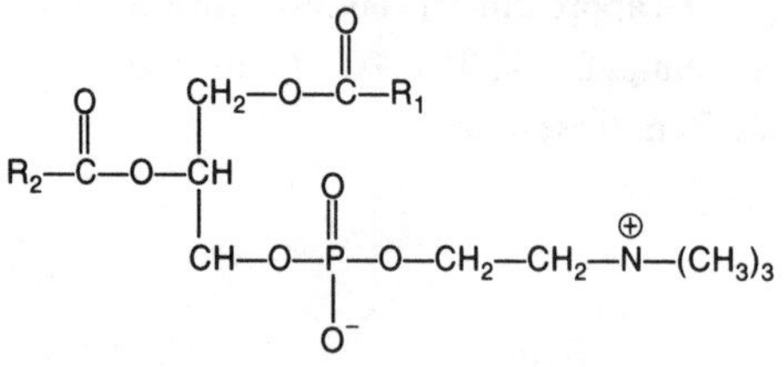

$$
\begin{array}{l}
\qquad\qquad CH_2{-}O{-}C(=O){-}R_1 \\
R_2{-}C(=O){-}O{-}CH \\
\qquad\qquad CH{-}O{-}P(=O)(O^-){-}O{-}CH_2{-}CH_2{-}\overset{\oplus}{N}{-}(CH_3)_3
\end{array}
$$

5. Review Haworth representations of galactose and of glucose. The linkage to ceramide is a beta glycosidic linkage and the structure is shown here.

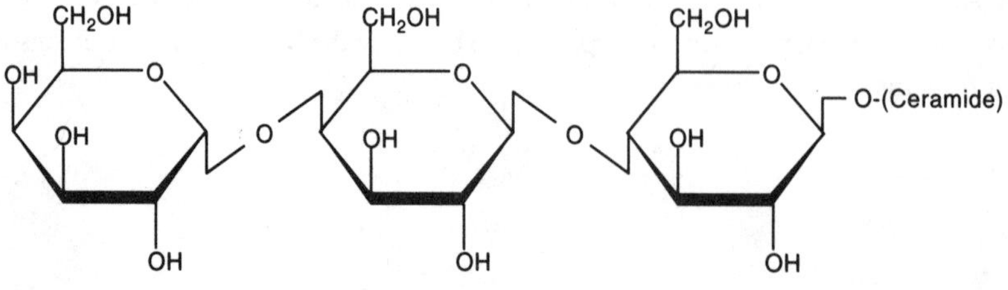

D-Gal-α-(1→4-) - D-Gal-β (1→4) - D - Glc-β (1→1) Ceramide

6. Phosphatidylcholine is an amphipathic molecule whose nonpolar fatty acyl groups would be most stable in an environment that excludes water but whose phosphorylcholine group would be most stable solvated in water. The formation of a multilammelar bilayer vesicle filled with aqueous buffer would provide a more stable environment. In the bilayer, as is the case in the membrane bilayer, the polar groups are in aqueous contact, whereas the nonpolar acyl groups face the interior of the bilayer.

7. An 0.1% (wt/vol) solution of each detergent will have the following concentration:

$$\text{Triton concentration} = \frac{10 \times 10^{-1} \text{ mg ml}^{-1}}{6.25 \times 10^{2} \text{ mg mmole}}$$

$$= 1.6 \times 10^{-3} \text{ mmole ml}^{-1} = 1.6 \text{ mM}$$

The Triton concentration (1.6 mM) is greater than the critical micellar concentration (0.24 mM) and it will exist in solution as micelles.

$$\text{Sodium deoxycholate concentration} = \frac{10 \times 10^{-1} \text{ mg ml}^{-1}}{4.14 \times 10^{2} \text{ mg mmole}^{-1}}$$

$$= 2.4 \times 10^{-3} \text{ mmole ml}^{-1} = 2.4 \text{ mM}$$

The concentration of sodium deoxycholate (2.4 mM) is well below the critical micellar concentration (4 mM) and it will not form micelles.

(a) Triton X-100 is present above the critical micellar concentration and will likely not pass through dialysis membrane because the micelles are larger than the dialysis membrane pores. If the dialysis membrane retained 12,000 M_r globular proteins, micelles with as few as twenty-five to thirty Triton molecules would, in principle, be too large to traverse the membrane. Sodium deoxycholate would freely diffuse across the dialysis membrane because there are no micelles in the solution.

(b) Triton is neutral and would impose no charge characteristics of the detergent on the protein of interest and could be used in the ion-exchange chromatography. Sodium deoxycholate would impose a negative charge upon association with the protein but in some instances might still be effectively used in ion-exchange chromatography.

(c) Proteins dissolved in 0.1% (wt/vol) triton would be incorporated into the micelles and would assume, to a great extent, size characteristics of the micelle. Artifactually large molecular weight values would therefore be measured on gel exclusion chromatography. Assuming that the molecules of deoxycholate that are associated with the protein add only minimal mass, the molecular weight of the protein measured in the presence of sodium deoxycholate should be more accurate.

8. Napthoquinones structurally similar to vitamin K_2 are lipophilic molecules having a naphthoquinone ring system attached to a hydrophobic polyprenyl side chain. The quinone is dissolved in the membrane in part because of the hydrophobic nature of the polyprenyl chain. Ordering of water molecules around the polyprenyl side chain in an aqueous environment is thermodynamically less favorable than if the polyprenyl side chain were in a nonaqueous environment because these water molecules would be less well ordered. Once dissolved into the hydrophobic membrane bilayer, the quinones rapidly diffuse laterally within the membrane. Ubiquinones (UQ) are benzoquinones substituted with hydrophobic polyprenyl side chains comprised of up to 10 prenyl units. Ubiquinone molecules diffuse among the primary dehydrogenases and the cyt bc_1 complexes to facilitate electron transport in mitochondria (see chapter 15 in the text).

9. Globular proteins found in the cytosol have tertiary structures that allow the hydrophobic amino acid side chains to assume a more thermodynamically favorable association in the interior of the protein because water is excluded. The polar side chains are most frequently found on the surface of the protein in contact with the

aqueous environment. Hydration of the polar groups and the diminished organization of water upon removal of the nonpolar side chains from aqueous contact stabilize the water-soluble protein.

Integral membrane proteins have a large portion of the surface "dissolved" in the hydrophobic lipid bilayer of the membrane. In this case, exterior polar groups would be less stable than exterior hydrophobic groups.

The hydrophobic amino acid side chains on the exterior of the integral membrane protein are stable in the water-free environment of the hydrophobic membrane interior. The polar side chains must be inside the protein, associated with water or in ionic bonds with other polar side chains. The portion of the membrane protein protruding from the membrane and in contact with the aqueous environment is composed primarily of polar amino acids, similar to soluble proteins.

10. Phospholipids are amphipathic molecules that associate as a lipid bilayer to form membranes. The hydrophobic side chains of the esterified fatty acyl groups are oriented toward the center of the membrane, whereas the phosphate and any substituent esterified to the phosphate face the aqueous phase. The phospholipids impose a charged hydrophilic surface to the membrane while maintaining a hydrophobic bilayer that is in general impermeable to ionic species. In addition, the nature of the esterified fatty acyl group affects the fluidity of the membrane.

    Peripheral proteins are associated with the exterior or interior face of the phospholipid bilayer, whereas the integral proteins are partially dissolved in the hydrophobic membrane. Peripheral proteins bind to the membrane primarily through weak ionic interactions, either directly or through metal ions. In the latter case, divalent metal ions may interact with the negatively charged phosphate or the carboxyl group of phosphatidyl serine of the phospholipid and with negatively charged amino acid side chains of the peripheral protein.

11. Membrane phase transition occurs as the membrane is heated because the components of the membrane become increasingly disordered. The phase transition temperature is dependent on the substituents of the phospholipids and the amounts of proteins and other components of the membrane. Components that increase organization of the membrane components will increase the T_m, whereas the opposite is true for components that tend to disrupt membrane organization.
    (a) The temperature shift from 30°C, which is below the T_m, to 50°C, a temperature above the T_m, will increase the motion of the membrane components and increase the relative fluidity of the membrane.
    (b) Addition of the phospholipid containing the unsaturated fatty acid disrupts the organization (packing) of the vesicles because of the bend in the fatty acyl chain imposed by the *cis*-double bond. The phase transition temperature will decrease.
    (c) Addition of the phospholipid with shorter fatty acyl groups (lower melting temperature) will increase the fluidity of the vesicle and decrease the phase transition temperature (T_m).
    (d) Integral membrane proteins interact with the phospholipid acyl chains in the membrane, causing an increase in organization of the membrane phospholipids adjacent to the protein. The T_m is thus increased and the transition occurs over a broader temperature range than that of the pure phospholipid vesicle.

12. Peripheral proteins are bound to the inner or outer aspects of the membrane through weak ionic interactions that include association with phospholipid head groups, by electrostatic or ionic interaction with a hydrophilic region of an integral membrane protein or through divalent metal ion bridging to the membrane surface. Integral proteins are dissolved into the lipid bilayer of the membrane through interactions of the hydrophobic amino acid side chains and fatty acyl groups of phospholipids. These interactions exclude the hydrophobic residues from aqueous contact (see fig. 7.25 in the text).

    Peripherally bound proteins may be released without disrupting the membrane. Thus, increased salt concentration shields ionic charges and weakens the charge–charge interactions between the peripheral protein and the membrane components. Although increased ionic strength may weaken the salt bridges contributed by divalent metal ions, application of a chelator to sequester the divalent ions may promote release of the peripheral protein. Changing the relative proportion of protonated/deprotonated groups by adjusting pH would, in principle, affect binding of peripheral proteins to the membrane.

None of the conditions described would be expected to release integral membrane proteins. For example, high ionic strength fosters, rather than weakens, hydrophobic interactions that bind integral proteins to the membrane. In order to remove integral membrane proteins, the membrane must be disrupted by addition of detergents or other chaotropic reagents to solubilize the protein and to prevent aggregation and precipitation of the hydrophobic proteins upon their removal from the membrane. For example, succinate dehydrogenase, a membrane-bound primary dehydrogenase, is an integral protein in the inner mitochondrial membrane and is removed only upon dissolution of the membrane with chaotropic reagents.

13. Integral membrane proteins have negligible rates of transverse motion (flip) across the membrane presumably because there is a high energy barrier to the movement of the solvent-exposed hydrophilic portion of the protein through the hydrophobic interior of the membrane. The highly hydrated oligosaccharide associated with some integral protein contributes considerable hydrophilic character to the protein at the aqueous interface and contributes to the energy barrier to transverse motion of the protein across the membrane.

14. The small amount of enzymatic activity released by the osmotic shock could result from the lysis of some of the cells. One could assay the fraction released by osmotic shock for enzymatic activities of proteins known to be either in the cytoplasm or in the periplasmic space. If the cytoplasmic marker activity were absent, there likely was no rupture of the cells and the protein of interest might be periplasmic. Another approach is to use a protein modifying reagent unable to penetrate the cytoplasmic membrane. If the protein of interest reacts with the reagent, there is further evidence of a periplasmic location.

    In the absence of data from the controls, there is no compelling reason to suggest a periplasmic location of the protein.

15. The protein dissolved in octylglucoside is most likely an integral membrane protein whose surface amino acid side chains are hydrophobic. The detergent must be present to prevent aggregation of protein molecules. The aggregate is more stable in the absence of detergent because aggregation excludes the water that would surround the hydrophobic amino acid side chains. The data are consistent with aggregation of the protein upon removal of detergent. The aggregate is at least a dimer (>100,000 M_r).

16. The amino acid side chains in an α-helix protrude from the axis of the helix and interact either with solvent or with other amino acid side chains (in a folded protein). In the α-helix of single-span integral protein, each amino acid side chain would interact with the hydrophobic interior of the membrane. In principle, it is energetically unfavorable to place a hydrophilic residue in a nonaqueous environment. In integral proteins with multiple α helices that span the membrane, hydrophilic side chains from different helical segments may interact and in some cases form a channel through which ions may diffuse. Portions of the helical segments exposed to the lipid will contain primarily hydrophobic amino acid residues.

8 Enzyme Kinetics

Problems

1. Explain what is meant by the order of a reaction, using the reaction below as an example. What is the reaction order for each reactant? (Consider the forward and reverse reaction.) For the overall reaction?

$$A + B \leftrightarrow 2C$$

2. In a first order reaction a substrate is converted to product so that 87% of the substrate is converted in 7 min. Calculate the first-order rate constant. In what time would 50% of the substrate be converted to product?

3. K_m is frequently equated with K_s, the [ES] dissociation constant. However, there is usually a disparity between those values. Why? Under what conditions are K_m and K_s equivalent?

4. Differentiate between the enzyme-substrate complex and the transition-state intermediate in an enzymatic reaction.

5. An enzyme was assayed with substrate concentration of twice the K_m value. The progress curve of the enzyme (product produced per minute) is shown here. Give two possible reasons why the progress curve becomes nonlinear.

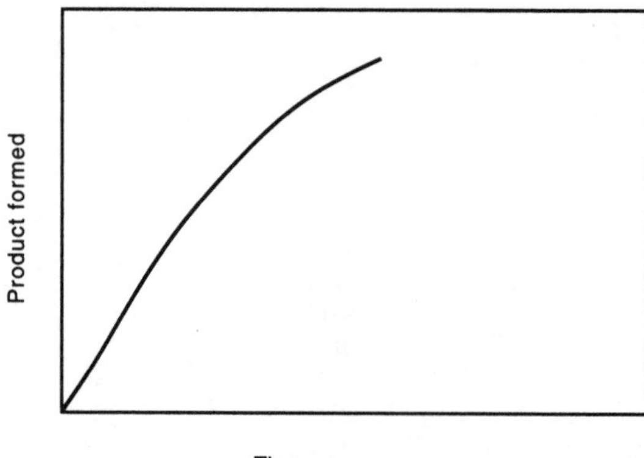

6. What is the steady-state approximation and under what conditions is it valid?

7. Assume that an enzyme-catalyzed reaction follows Michaelis-Menten kinetics with a K_m of 1 μM. The initial velocity is 0.1 μM/min at 10 mM substrate. Calculate the initial velocity at 1 mM, 10 μM, and 1 μM substrate. If the substrate concentration were increased to 20 mM, would the initial velocity double? Why or why not?

8. If the K_m for an enzyme is 1.0×10^{-5} M and the K_i of a competitive inhibitor of the enzyme is 1.0×10^{-6} M, what concentration of inhibitor is necessary to lower the reaction rate by a factor of 10 when the substrate concentration is 1.0×10^{-3} M? 1.0×10^{-5} M? 1.0×10^{-6} M?

9. Assume that an enzyme-catalyzed reaction follows the scheme shown:

$$E + S \underset{k_2}{\overset{k_1}{\rightleftharpoons}} ES \underset{k_4}{\overset{k_3}{\rightleftharpoons}} E + P$$

Where $k_1 = 10^9\ M^{-1}\ s^{-1}$, $k_2 = 10^5\ s^{-1}$, $k_3 = 10^2\ s^{-1}$, $k_4 = 10^7\ M^{-1}\ s^{-1}$, and $[E_T]$ is 0.1 nM. Determine the value of each of the following.

(a) K_m

(b) V_{max}

(c) Turnover number

(d) Initial velocity when $[S]_o$ is 20 μM.

10. A colleague has measured the enzymatic activity as a function of reaction temperature and obtained the data shown in this graph. He insists on labeling point A as the "temperature optimum" for the enzyme. Try, tactfully, to point out the fallacy of that interpretation.

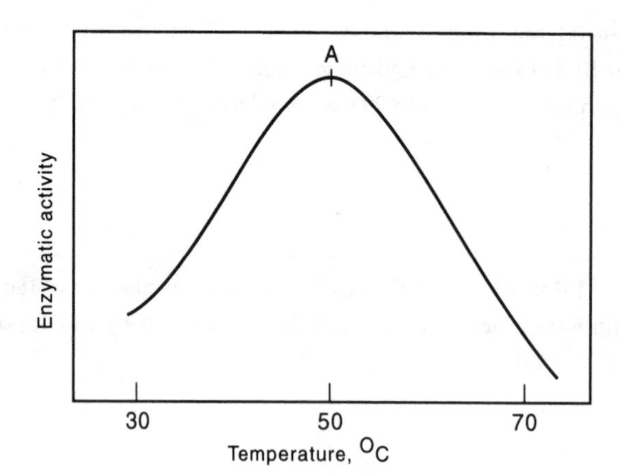

11. You have isolated an enzyme that catalyzes a bimolecular reaction:

$$A + B \leftrightarrow P + Q$$

The initial velocity data yielded intersecting double-reciprocal plots with [A] varied at fixed [B] or [B] varied at fixed [A]. Which kinetic pattern—sequential (ordered or random), or Ping-Pong—might you rule out?

12. You have isolated a tetrameric NAD^+-dependent dehydrogenase. You incubate this enzyme with iodoacetamide in the absence or presence of NADH (at ten times the K_m concentration) and you periodically remove aliquots of the enzyme for activity measurements and amino acid composition analysis. The results of the analyses are shown.

	(No NADH present)			**(NADH Present)**		
Time (min)	Activity (U/mg)	His (Residues/mole)	Cys	Activity (U/mg)	His (Residues/mole)	Cys
0	1,000	20	12	1,000	20	12
15	560	18.2	11.4	975	20	11.4
30	320	17.3	10.8	950	20	10.8
45	180	16.7	10.4	925	19.8	10.4
60	100	16.4	10.0	900	19.6	10.0

(a) What can you conclude about the reactivities of the cysteinyl and histidyl residues of the protein?

(b) Which residue could you implicate in the catalytic active site? On what do you base the choice? Are the data conclusive concerning the assignment of a residue to the active site? Why or why not?

(c) After 1 h you dilute the enzyme incubated with iodoacetamide but no NADH. Would you expect the enzyme activity to be restored? Explain.

13. The following initial velocity data were obtained for an enzyme.

[S] (mM)	Velocity (Ms^{-1}) × 10^7
0.10	0.96
0.125	1.12
0.167	1.35
0.250	1.66
0.50	2.22
1.0	2.63

Each assay at the indicated substrate concentration was initiated by adding enzyme to a final concentration of 0.01 nM. Derive K_m, V_{max}, k_{cat}, and the specificity constant.

14. You have measured the initial velocity of an enzyme in the absence of inhibitor and with inhibitor A or inhibitor B. In each case, the inhibitor is present at 10 μM. The data are shown in the table.

[S] (mM)	Velocity (Ms^{-1}) x 10^7 Uninhibited	Velocity (Ms^{-1}) x 10^7 Inhibitor A	Velocity (Ms^{-1}) x 10^7 Inhibitor B
0.333	1.65	1.05	0.794
0.40	1.86	1.21	0.893
0.50	2.13	1.43	1.02
0.666	2.49	1.74	1.19
1.0	2.99	2.22	1.43
2.0	3.72	3.08	1.79

(a) Determine K_m and V_{max} of the enzyme.

(b) Determine the type of inhibition imposed by inhibitor A and calculate K_i(s).

(c) Determine the type of inhibition imposed by inhibitor B and calculate K_i(s).

15. Irreversible inactivation of an enzyme by a compound may be confused with noncompetitive inhibition. Why? How could you distinguish between a reversible noncompetitive inhibitor and an irreversible inactivator? Enzyme supply is not limiting.

Solutions

1. Reaction order is the power to which a reactant concentration is raised in defining the rate equation. The rate equation describing the forward reaction in the example is

$$v_f = k(A)(B)$$

where (A) and (B) are expressed as molar concentration and k has the units $M^{-1}\ s^{-1}$. The units of v_f are $M\ s^{-1}$. The concentration of substrates A and B are each raised to the first power in the rate equation. The reaction is first order in A and in B. The overall reaction order is the sum of the order contributions of each reactant and is second order.

The rate equation for the reverse reaction is

$$v_r = k_2\ (C)^2$$

where (C) is expressed as M and k has the dimensions $M^{-1}\ s^{-1}$. The concentration term of reactant (C) in the rate equation is raised to the second power and the reaction is second order.

2. Let (s_0) be the concentration of substrate present before the reaction is initiated (t_0) and let (S_x) represent the concentration of substrate remaining after time (T_x). After 7 min, 87% of the substrate was consumed and 13% remains. Rearranging the equation for first-order reaction yields

$$2.3 \log (S_x)/(S_o) = -kt_x \text{ and}$$
$$2.3 \log (S_o)/(S_x) = kt_x$$

where (S_o) = 100% and S_x is 13% after 7 min reaction time.

$$2.3\log\ (100)/(13) = k(7\ \text{min})$$
$$k = \frac{2.3\ \log(7.7)}{(7\ \text{min})}$$
$$k = 0.29\ \text{min}^{-1}$$

The value of k now may be used to calculate the time required to convert 50% of the substrate to product.

$$2.3 \log(100)/(50) = (0.29\ \text{min}^{-1})\ (t)$$
$$t = 2.3\ (0.3)/(0.29)$$
$$t = 2.4\ \text{min}$$

For a first-order reaction, the time required for conversion of 50% substrate to product is called the half-time ($T_{0.5}$) and is independent of the initial concentration of substrate.

3. The Michaelis-Menten equation for the reaction

$$E + S \underset{k_2}{\overset{k_1}{\rightleftharpoons}} ES \overset{k_3}{\rightarrow} E + P$$

defines K_m in terms of the individual rate constants: $K_m = (k_2 + k_3)/k_1$

K_s is the dissociation constant for ES → E + S and in terms of the individual rate constants is k_2/k_1. K_m is equal to the [ES] dissociation constant [K_s] when the value of k_3 is much smaller than the value of k_2.

4. The enzyme-substrate complex forms upon binding of substrate at the enzyme-active site, the initial step in catalysis. The substrate at this point is chemically unchanged but is physically restrained and correctly oriented for subsequent reaction. The enzyme-substrate complex may dissociate to regenerate free substrate and enzyme or may continue along the catalytic pathway to an activated enzyme-substrate complex ($ES^{\#}$), the transition state. The transition state is the chemical intermediate on the reaction path between substrate and product and is the form present at the highest energy point along the reaction pathway. Although the enzyme-bound transition-state complex is the highest energy intermediate, the energy maximum is lower than that of the uncatalyzed reaction. The transition state intermediate is stabilized by binding to the active site and represents the chemical form of the complex capable of proceeding to product or returning to substrate, with equal probability of either event.

5. To assess the maximal activity for an enzyme (V_{max}), the rate is measured under saturating substrate concentration ([S] at least 10 × K_m) so that the observed rate depends only on enzyme concentration. At substrate concentrations lower than saturation, the observed rate depends on concentration of enzyme as well as the actual concentration of substrate relative to the K_m value. The initial rate of the reaction is measured to minimize error introduced by the reverse reaction, the conversion of product to substrate, that occurs as the product accumulates in the reaction mixture. Nonlinearity of the rate of product formation may be caused by:
   (a) Consumption of substrate. Since the initial substrate concentration was only 2 × K_m, as the substrate concentration decreases, so does the observed rate.
   (b) Accumulation of product during the course of the reaction. The product inhibits the reaction, or the reverse reaction becomes significant.

6. Steady-state approximation is based on the concept that the formation of [ES] complex by binding of substrate to free enzyme, and breakdown of [ES] to form product plus free enzyme occur at equal rates. A graphical representation of the relative concentrations of free enzyme, substrate, enzyme-substrate complex, and product is shown in figure 8.9 in the text. Note that [S] is initially much larger than [E_T]. If we assume

$$E + S \underset{k_2}{\overset{k_1}{\rightleftharpoons}} ES \xrightarrow{k_3} E + P$$

and that [P] is zero, the rate of formation of [ES] is expressed as

$$v_f = k_1 [E][S]$$

and the rate of breakdown is expressed as

$$v_r = k_2 [ES] + k_3 [ES]$$
$$v_r = (k_2 + k_3) [ES]$$

At steady state, the two velocities v_f and v_r are equal. One can apply the distribution (or conservation) expression (E_t = [E] + [ES]) and the kinetic constants to arrive at the Michaelis-Menten expression, an expression derived using the steady-state assumption.

$$v = \frac{V_m[S]}{K_m + [S]}$$

where $V_m = k_3 [E_t]$ and $K_m = (k_2 + k_3)/(k_1)$.
Steady-state approximation may be assumed until the substrate concentration is depleted with a concomitant decrease in the concentration of [ES].

7. The reaction velocity at each substrate concentration is determined using the Michaelis-Menten expression.

$$v = \frac{V_m[S]}{K_m + [S]}$$

In this example, K_m is 1 μM. The initial velocity is equal to V_m when [S] is 10 mM. In this instance, (K_m + S) is ($10^{-6} + 10^{-2}$)M or (10^{-2} M).
Therefore, when [S] is 10^{-2} M, $V_o = V_m$ and has the value 0.1 μM min^{-1}. The velocity at each of the other substrate concentrations may now be calculated. For example, at 1 μM substrate,

$$v = \frac{(10^{-7}\ M\ min^{-1})(10^{-6}\ M)}{(10^{-6}\ M + 10^{-6}\ M)}$$
$$v = \frac{(10^{-13})\ M^2\ min^{-1}}{(2 \times 10^{-6}\ M)}$$
$$v = 5 \times 10^{-8}\ M\ min^{-1}$$

when [S] is 10^{-3} M, $v_o = V_m$ (1×10^{-7} M min^{-1})
[S] is 10^{-5} M, $v_o = 9 \times 10^{-8}$ M min^{-1} (90% V_m)
[S] is 10^{-6} M, $v_o = 5 \times 10^{-8}$ M min^{-1} (50% V_m)

Since the initial velocity is maximal velocity when the substrate concentration is 10mM (1,000 × K_m), doubling the substrate concentration will not measurably increase the initial velocity.

8. Enzyme velocity in the presence of a competitive inhibitor is expressed as

$$v_i = \frac{V_m\ [S]}{K_m[(1 + (I/K_i)] + [S]}$$

where K_i is the inhibitor dissociation constant. Let v be the velocity of the reaction in the absence of the inhibitor. If v is 100%, the inhibitor concentration yielding v_i = 10% v is sought. Now,

$$\frac{v}{v_i} = \frac{V_m\ [S]}{K_m + [S]} \times \frac{K_m(1 + [I/K_i]) + [S]}{V_m\ [S]}$$

$v/v_i = 10$ and the expression simplifies:

$$10[K_m + S] = K_m + K_m/K_i\ [I] + [S]$$
$$9(K_m + [S] = K_m/K_i\ [I]$$

Substitute the known values and derive the expression

$$9(10^{-5}M + S) = 10^{-5}/10^{-6}(I)$$
$$(I) = 0.9(10^{-5}M + S)$$

The amount of inhibitor required to inhibit the reaction 90% may now be calculated.

When $[S] = 1.0 \times 10^{-3}$ M, $[I] = 9.1 \times 10^{-4}$ M
$[S] = 1.0 \times 10^{-5}$ M, $[I] = 1.8 \times 10^{-5}$ M
$[S] = 1.0 \times 10^{-6}$M, $[I] = 9.9 \times 10^{-6}$ M

Thus, increasing the concentration of substrate at a fixed inhibitor concentration yields less inhibition.

9a.

$$K_m = (k_2 + k_3)/k_1$$
$$K_m = \frac{(10^5 \text{ s}^{-1} + 10^2 \text{ s}^{-1})}{(10^9 \text{ M}^{-1} \text{ s}^{-1})}$$
$$K_m = 10^{-4} \text{ M}$$

9b.

$$V_m = k_3 (E_T) \text{ where } E_T = 10^{-10} \text{ M.}$$
$$V_m = 10^{-8} \text{ M s}^{-1}$$

9c. Turnover number is moles of substrate converted to product per time unit per mole of enzyme.

$$\text{Tn} = V_m/E_T = k_3 = k_{cat}$$
$$\text{Turnover number} = 100 \text{ s}^{-1}$$

9d.

$$v = \frac{V_m (S)}{K_m + (S)}$$

Substitute the kinetic constants derived in parts (9a) and (9b) and substrate concentration (2×10^{-5} M).

$$v = \frac{(10^{-8} \text{ M s}^{-1})(2 \times 10^{-5} \text{ M})}{(10^{-4} \text{ M} + 2 \times 10^{-5}\text{M})}$$
$$v = 1.72 \times 10^{9} \text{ M s}^{-1}$$

10. The rate of a chemical reaction increases with temperature as defined by the Arrhenius expression (see chapter 8 in the text). Since an enzyme is a catalyst for a chemical reaction, the rate of an enzyme catalyzed reaction increases with increased temperature. However, the catalyst, a protein, is structurally labile and is inactivated (denatured) at elevated temperatures. The precise temperature at which the enzyme is inactivated varies with the specific enzyme. The figure in problem 10 illustrates the expected increase in reaction rate with increased temperature until the temperature at point A is reached. The temperature at point A roughly approximates the maximum temperature at which the catalyst (enzyme) is stable. Denaturation or inactivation removes the catalyst from the reaction and the reaction rate decreases because the observed velocity is dependent on the concentration of enzyme. There is no "temperature optimum" for a catalyst (enzyme).

11. Intersecting initial velocity plots are consistent with ordered sequential or random sequential but not Ping-Pong mechanisms. Initial velocity plots that are parallel are consistent with Ping-Pong mechanism. A Ping-Pong kinetic mechanism can be ruled out.

Product inhibition studies should differentiate random from ordered kinetic mechanisms. In the ordered reaction mechanism, the first substrate to bind and the last product to dissociate from free enzyme are competitive with each other because they bind to the same enzyme form (E) and other combinations are not competitive. Consider the enzyme reaction diagram now termed Cleland diagram. (See Cleland, W. W., in *The Enzymes, Student Edition* Vol. II, ed. P. D. Boyer, New York: Academic Press.)

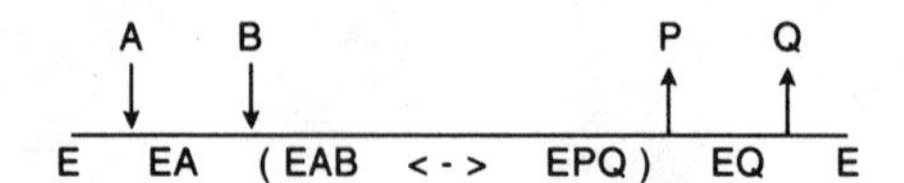

Product Q will competitively inhibit the reaction when substrate A is the varied substrate. In principle, no other combinations exhibit competitive inhibition.

For a random reaction mechanism, theoretically, each product would be competitive with its corresponding substrate. Consider the diagram

A B P Q
EA EQ
E— EAB $\rightleftharpoons$ EPQ —E
EB EP
B A Q P

Those combinations (Q versus A, P versus B) will, in theory, exhibit competitive inhibition.

12a. Iodoacetamide reacts with cysteinyl sulfyhdryl groups, the imidazole group of histidine residues, and α- ϵ-amino groups. The data are consistent with the reaction of iodoacetamide with some, but not all, cysteine and histidine groups in the protein in the time required to inactivate 90% of the enzyme. Addition of NADH to the inactivation mixture protected the enzyme from inactivation and prevented reaction of histidine, but not cysteine residues, with iodoacetamide.

12b. Histidine residues may be implicated at the active site based on the following information.

(1) Addition of NADH, a coenzyme that binds at the active site, prevents inactivation of the enzyme and also inhibits the reaction of histidine, but not cysteine, with iodoacetamide (IAM).

(2) There is a linear relationship between loss of catalytic activity and the number of histidyl residues that are modified by IAM, shown in the figure.

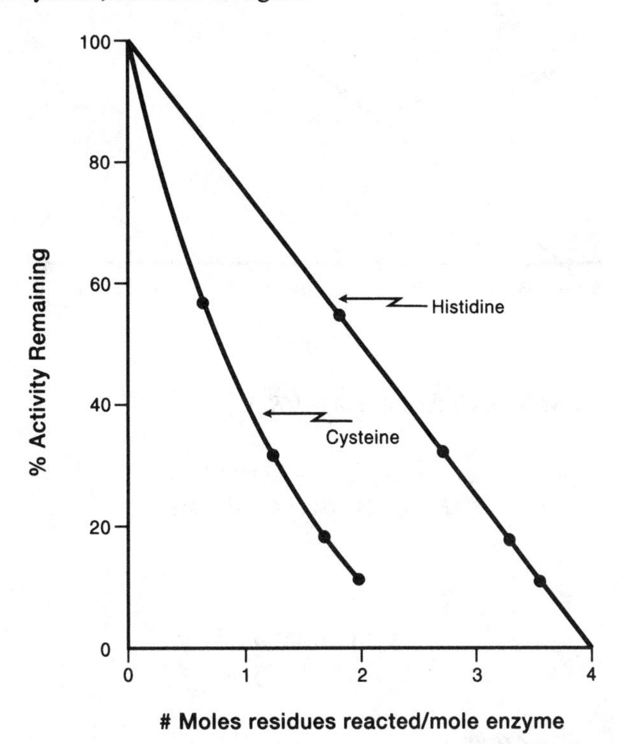

Extrapolation of the line to 100% inactivation of the enzyme reveals that four histidine residues would react with IAM. The relationship of cysteine residues reacting as a function of enzyme inactivation is nonlinear. Although these data are consistent with one reactive histidyl side chain per active site, it is possible that the histidine may be just outside the active site, and reaction with IAM causes occlusion of the active site. In the final analysis, x-ray crystallographic analysis should reveal the positions of these reactive histidine residues.

12c. The reaction of iodoacetamide with the amino acid side chain forms a covalent bond that is not reversible upon dilution. The iodoacetamide-treated enzyme would remain inactive upon dilution or during exhaustive dialysis. Were the inhibitor noncovalently bound, dilution of the enzyme-inhibitor complex would be expected to reverse the inhibition. This demonstrates the difference between reversible inhibitors used in kinetic analysis and irreversible inactivation.

13. The kinetic data are analyzed by graphing the reciprocal of the velocity as a function of the reciprocal of the substrate concentration. The kinetic constants V_m and K_m are obtained from the resulting Lineweaver Burk plot.

Substrate (mM^{-1})	Velocity ($sec\ M^{-1} \times 10^{-7}$)
10	1.04
8	0.89
6	0.74
4	0.60
2	0.45
1	0.38

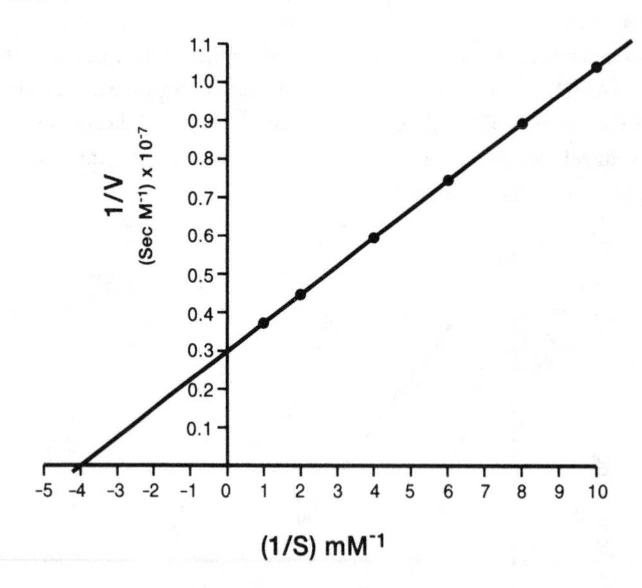

The intercept on the negative X axis is the value of $(-1/K_m)$.

$$-(K_m)^{-1} = -4.17\ mM^{-1}$$
$$K_m = 0.24\ mM\ (2.4 \times 10^{-4}\ M)$$

The Y intercept is $(V_m)^{-1}$.

$$(V_m^{-1}) = 0.305 \times 10^{7}$$
$$V_m = 3.3 \times 10^{-7}\ M\ s^{-1}$$

Turnover number (k_{cat} or k_3) = $V_m/(E_T)$

$$k_{cat} = (3.3 \times 10^{-7}\ M\ s^{-1})/(10^{-11} M) = 3.3 \times 10^{4}\ s^{-1}$$

Specificity constant = k_{cat}/K_m

$$= (3.3 \times 10^4)/(2.4 \times 10^{-4})$$
$$= 1.4 \times 10^8 \text{ M}^{-1} \text{ s}^{-1}$$

Enzymes operating near catalytic perfection have k_{cat}/K_m values approaching the limit set by diffusion (approximately 10^8 M^{-1} s^{-1}). This enzyme meets the criterion and operates near catalytic perfection.

14. The type of inhibition imposed by each inhibitor can be determined graphically using the Lineweaver-Burk plot. The reciprocal of the velocity measured in the absence and in the presence of each inhibitor is graphed versus the reciprocal of the corresponding substrate concentration.

1/(S) (mM^{-1})	1/Velocity s M^{-1} (× 10^{-7}) Uninhibited	1/Velocity s M^{-1} (× 10^{-7}) Inhibitor A	1/Velocity s M^{-1} (× 10^{-7}) Inhibitor B
3.0	0.606	0.952	1.26
2.5	0.538	0.826	1.12
2.0	0.469	0.699	0.980
1.5	0.400	0.575	0.840
1.0	0.334	0.450	0.699
0.5	0.269	0.325	0.559

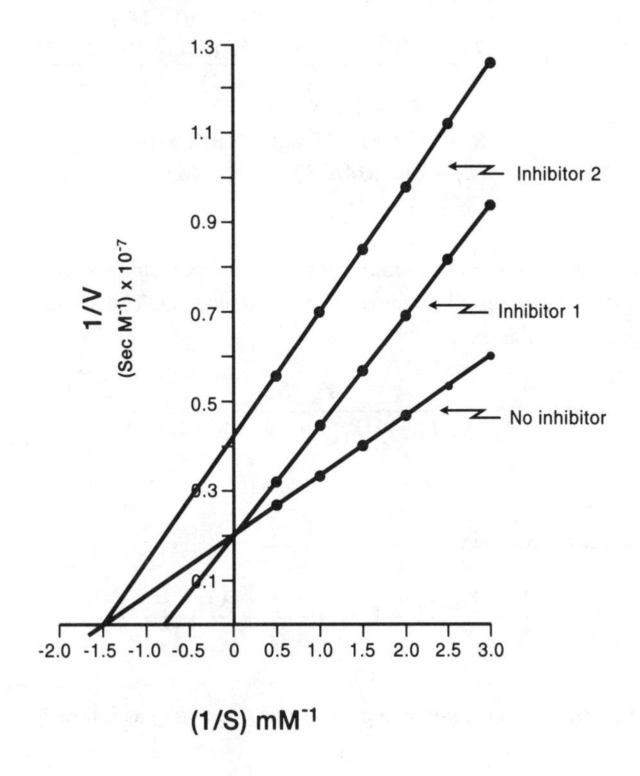

Inhibitor A has an effect on the slope but not on the y-intercept of the double reciprocal plot when compared with the data from the uninhibited enzyme. Infinite substrate concentration abolishes the effect of inhibitor A and is diagnostic for competitive inhibition. Therefore, inhibitor A is competitive with the substrate for the enzyme active site. The kinetic constants (K_m and V_m) are only defined for the uninhibited enzyme and are derived graphically as shown in problem 13.

V_m is 5.0×10^{-7} M s^{-1} and K_m is 0.67 mM.

The effect of a competitive inhibitor on the velocity of the enzyme-catalyzed reaction is given by

$$v_i = \frac{V_m\,(S)}{K_m\,(1 + I/K_i) + S}$$

The reciprocal of this expression, in Lineweaver-Burk form,

$$\frac{1}{v_i} = \frac{K_m}{V_m}(1 + I/K_i)\,\frac{(1)}{(S)} + \frac{1}{V_m}$$

reveals that the competitive inhibitor affects the slope term of the kinetic expression but not the Y-intercept term (V_m). The value of K_i can be calculated from the slope of the double reciprocal plot.

$$\text{slope} = 2.51 \times 10^6 (\text{s M}^{-1}/\text{mM}^{-1})$$

$$\text{slope} = \frac{K_m\,(1 + I/K_i)}{V_m}$$

$$(2.51 \times 10^6 \text{ s M}^{-1}/\text{mM}^{-1}) = \frac{0.67 \text{ mM } (1 + I/K_i)}{5.0 \times 10^{-7} \text{ M s}^{-1}}$$

$$(1 + I/K_i) = \frac{2.51 \times 10^6 \text{ s M}^{-1}/\text{mM}^{-1})\,(5.0 \times 10^{-7} \text{ M s}^{-1})}{(0.67 \text{ mM})}$$

$$(1 + I/K_i) = 1.87$$

$$I/K_i = 0.87 \text{ but [I] was 10 } \mu\text{M so:}$$

$$K_i = (10\ \mu\text{M}/0.87) = 11\ \mu\text{M}$$

Inhibitor B affects both the slope and the Y-intercept of the double reciprocal plot compared to the uninhibited enzyme, a pattern consistent with mixed or noncompetitive inhibition. The Michaelis-Menten expression, derived for noncompetitive inhibition, is

$$v_i = \frac{V_m\,(S)}{K_m\,(1 + I/K_{i\,s}) + (S)\,(1 + I/K_{i\,i})}$$

The reciprocal form of the expression is

$$\frac{1}{v_i} = \frac{K_m}{V_m}(1 + I/K_{i\,s})\,\frac{(1)}{(S)} + \frac{(1 + I/K_{i\,i})}{V_m}$$

K_i values can now be derived from the y-intercept ($K_{i\,i}$) and slope ($K_{i\,s}$) as follows:

(A) K_{ii}

$$\text{Y-intercept} = \frac{(1 + I/K_{ii})}{V_m}$$

$$0.42 \times 10^7\ M^{-1}\ s = \frac{(1 + I/K_{ii})}{(5.0 \times 10^{-7}\ M\ s^{-1})}$$

$$(1 + I/K_{ii}) = (0.42 \times 10^7\ M^{-1}\ s)\ (5.0 \times 10^{-7}\ M\ s^{-1})$$

$$(1 + I/K_{ii}) = 2.1$$

$$I/K_{ii} = 1.1 \text{ but [I] was } 10\ \mu M, \text{ so}$$

$$K_{ii} = 9.1\ \mu M$$

(B) K_{is}

$$\text{slope} = \frac{K_m\ (1 + I/K_{is})}{V_m}$$

$$\text{slope} = 2.8 \times 10^6\ s\ M^{-1}\ /\ mM^{-1}$$

$$2.8 \times 10^6\ s\ M^{-1}/mM^{-1} = \frac{0.67\ mM^{-1}\ (1 + I/K_{is})}{5.0 \times 10^{-7}\ M\ s^{-1}}$$

$$(1 + I/K_{is} = (2.8 \times 10^6\ s\ M^{-1}/mM^{-1},\ (5.0 \times 10^{-7}\ M\ s^{-1})$$

$$1 + I/K_{is} = 2.1$$

$$I/K_{is} = 1.1 \text{ but [I] was } 10\ \mu M \text{ so}$$

$$K_{is} = 9.1\ \mu M$$

In this example, K_{ii} and K_{is} are equal, but that is not usually the case.

15. Irreversible inactivation may be confused with noncompetitive inhibition because the inactivating agent decreases the amount of active enzyme, yielding a slope and y-intercept effect if the data were analyzed by a Lineweaver-Burk plot. The V_m would apparently decrease but the K_m would be unchanged.

Reversible inhibitors usually inhibit the enzyme instantaneously upon combination of enzyme and inhibitor. Dilution of the inhibitor-enzyme complex or application of techniques to separate large and small molecules will diminish or reverse the inhibition. Reversible inhibitors thus interact with the enzyme at a specific site and are bound through *noncovalent* interactions.

Irreversible inactivators covalently modify amino acid side chain(s) on the enzyme and inactivate the enzyme by chemically altering an amino acid side chain in the active site, by promoting a conformational change in the enzyme, or by occluding the active site. Covalent interactions are not reversed by dilution or by application of other separation techniques.

Structural analogs of the substrate often compete with the true substrate for the active site and impose reversible competitive inhibition. Substrate analogs having chemically reactive groups may bind at the active site, allowing the reactive side chain to covalently modify, and thus irreversibly inactivate the enzyme.

9 Mechanisms of Enzyme Catalysis

Problems

1. (a) In what ways are the mechanistic features of chymotrypsin, trypsin, and elastase similar?

   (b) If the mechanisms of these enzymes are similar, what features of the enzyme active site dictate substrate specificity?

2. (a) If you monitor the lactate dehydrogenase reaction by the formation of NADH (increase in absorbance at 340 nm), should increasing pH make it easier to measure the dehydrogenation of lactate to pyruvate?

   (b) Would a chemical trapping agent for pyruvate serve the same purpose at lower pH values? Why?

   (c) Dehydrogenase activity may be measured by reoxidizing the NADH and reducing a tetrazolium dye. The reduced dye is intensely colored. What are the advantages of measuring the LDH reaction by means of the tetrazolium dye system?

3. For many enzymes, V_{max} is dependent on pH. At what pH would you expect V_{max} of RNase to be optimal? Why?

4. If a lysine were substituted for the aspartate in the trypsin side chain binding crevice, would you expect the enzyme to be functional? If it were functional, what effect would you predict the substitution to have on substrate specificity?

5. Carboxypeptidase A preferentially cleaves C-terminal aromatic residues from proteins. When the aromatic substrate side chain is bound, water is expelled from the active site. How does the release of water stabilize binding of substrate in the active site?

6. You have isolated a metalloenzyme that preferentially cleaves basic amino acids from the carboxyl terminal of proteins. Would you expect the enzyme to retain an arginine in the active site as does carboxypeptidase A? Why or why not? What other residues would you predict to be in the substrate binding site for the new enzyme? How would these residues dictate cleavage specificity?

7. RNase can be completely denatured by boiling or by treatment with chaotropic agents (e.g. urea), yet can refold to its fully active form upon cooling or removal of the denaturant. By contrast, when enzymes of the trypsin family and carboxypeptidase A are denatured, they do not regain full activity upon renaturation. What aspects of trypsin and carboxypeptidase A structure preclude their renaturation to the fully active form?

8. (a) The amino acids in the active site of the protease papain are shown. Predict a feasible reaction mechanism for papain.

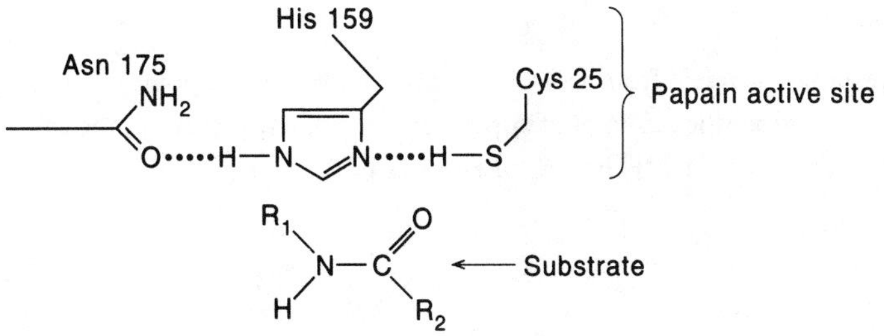

His 159

Asn 175

NH_2

Cys 25

Papain active site

O•••••H—N N•••••H—S

R_1 O

N—C ← Substrate

H R_2

(b) *N*-ethylmaleimide (NEM) reacts rapidly with cysteine thiolate anion via a Michael addition. What is the product of the reaction between NEM and cysteine? Would you expect the rate of R-SH reaction with *N*-ethylmaleimide to be more rapid at pH 5 or pH 7.5? Why?

(c) Would you expect cysteine 25 (see part a) to be more reactive with *N*-ethylmaleimide than any of the other cysteine residues in the protein? Explain.

9. Why do structural analogs of the transition-state intermediate of an enzyme inhibit the enzyme competitively and with low K_i values?

10. Transition-state analogs of a specific chemical reaction have been used to elicit antibodies with catalytic activity. These catalytic antibodies have great promise as experimental tools as well as having commercial value. Why is it reasonable to assume that the binding site for the transition-state analog on the antibody would mimic the enzyme active site? What difficulties might be encountered if a catalytic antibody were sought for a reaction requiring a cofactor (coenzyme)?

11. Superoxide dismutases catalyze the reaction

$$O_2^- + O_2^- + 2H^+ \rightarrow O_2 + H_2O_2$$

The catalytic mechanism of the superoxide dismutases involves active site transition metals (Cu, Fe, Mn) that undergo valence changes in the catalytic cycle. Write equations that represent the catalytic cycle of each of these transition metals. (Remember, in each case you must finish with the catalyst in the same state as when you began.) (See Fridovich, I., *Adv. Enzymol.* 58:61–97, 1986.)

12. The superoxide dismutase isolated from most sources has an isoelectric point around 5.

(a) What problem does the acidic pI of superoxide dismutase present to the catalytic disproportionation of superoxide anion at pH 7? (The pK_a of $HO_2\bullet$ is 4.8.)

(b) How might strategically placed basic residues assist in catalysis?

(c) Would the acidic pI of the enzyme or presence of basic residues at the active site impede release of product from the active site? Why?

13. Using site-directed mutagenesis techniques, you have isolated a series of recombinant enzymes in which specific lysine residues were replaced with aspartate residues. The enzymatic assay revealed the following.

Enzyme Form	Activity (U/mg)
Native enzyme	1,000
Recombinant Lys 21 → Asp 21	970
Recombinant Lys 86 → Asp 86	100
Recombinant Lys 101 → Asp 101	970

(a) What might you infer about the role(s) of Lys 21, 86, and 101 in the catalytic mechanism of the native enzyme?

(b) Speculate on the location of Lys 21 and Lys 101. Would you expect these residues to be conserved in an evolutionary sense?

(c) Would you expect Lys 86 to be evolutionarily conserved? Why or why not?

14. You have isolated a microorganism capable of metabolizing the deoxysugar shown below. You find that the compound is phosphorylated by a specific kinase on the C-1 OH group and then the bond between C-3—C-4 is cleaved, yielding dihydroxyacetone phosphate and acetaldehyde. You also find that the cleavage is catalyzed by a zinc-containing enzyme.

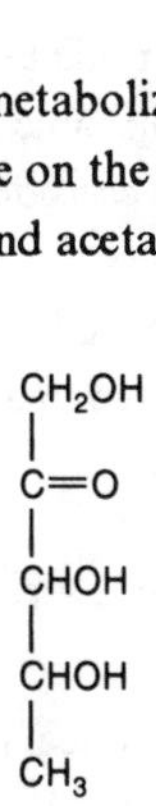

$$\begin{array}{c} CH_2OH \\ | \\ C{=}O \\ | \\ CHOH \\ | \\ CHOH \\ | \\ CH_3 \end{array}$$

(a) The mechanism of cleavage of the sugar includes generation of a carbanion on C-3. Explain how zinc might stabilize the carbanion. (Consider the role of zinc in the carboxypeptidase A mechanism.)

(b) Would you expect the nonphosphorylated deoxysugar to be a substrate? Why or why not?

15. During catalysis, covalent chemical bonds are broken and formed. In that sense, almost all enzymes perform covalent catalysis. However, the terms "covalent" and "noncovalent" have particular meanings in an enzyme mechanism. Define the difference in these terms as they apply to catalysis.

Solutions

1a. The proteolytic enzymes trypsin, chymotrypsin, and elastase share the following mechanistic features.

(a) *Each enzyme has the active site "catalytic triad" of aspartate, histidine, and serine.* In the currently proposed model for the serine protease mechanism, the aspartate is hydrogen-bonded to the imidazole group of histidine. The histidine accepts the proton from the serine hydroxyl group during nucleophilic addition of the serine —OH to the substrate peptidyl bond. The active-site aspartate and the oxyanion hole contribute to the electrostatic stabilization of the tetrahedral transition state intermediate. (See Warshel, A. et al., *Biochemistry.* 28:3629–3637, 1989.)

(b) *Each forms an acyl-enzyme intermediate during peptide bond cleavage.* The nucleophilic OH group of serine adds to the carboxyl of the substrate peptide bond, and an acyl ester is formed between the serine OH and the N-terminal portion of the peptide substrate. The C-terminal portion of the substrate diffuses from the active site.

(c) *The acyl-enzyme intermediate is deacylated by hydrolysis.* Addition of water hydrolyzes the acyl ester and restores the active site residues to their original chemical state. See figure 9.11 in text.

1b. The specificity of each enzyme is dictated by the size and relative hydrophobicity of the substrate binding crevice that is adjacent to the catalytically active serine residue.

Trypsin cleaves peptide bonds on the carboxyl side of internal lysine or arginine residues. The charged, polar side chains of those amino acids are bound in a crevice that has an acidic residue, aspartate, located at its base. Electrostatic interaction occurs between the aspartate and the positively charged side chains of the Lys or Arg residues.

Chymotrypsin preferentially cleaves peptide bonds on the carboxyl side of internal Phe, Tyr, or Trp residues. These amino acid side chains are hydrophobic aromatic (Phe and Trp) or polar but not charged (Tyr). The active site crevice that binds these residues is considerably more hydrophobic than is the binding pocket in trypsin.

Elastase cleaves on the carboxyl side of the less bulky, uncharged residues, such as alanine or valine. The smaller size of the side chains is accommodated in a shallower, active site crevice.

2a. The reaction catalyzed by lactate dehydrogenase

$$CH_3\text{—}CHOH\text{—}COO^- + NAD^+ \rightarrow CH_3\text{—}CO\text{—}COO^- + NADH + H^+$$

is shifted toward product formation, as is any chemical reaction, either by decreasing product concentration, increasing reactant concentration, or both. A proton is released as a product of the lactate dehydrogenase reaction, and increasing pH lowers the H^+ concentration with a shift of the reaction toward pyruvate formation. An increase of 1 pH unit lowers proton concentration by a factor of 10, but there are limits to the amount that pH can effectively be increased in an enzyme-catalyzed reaction, because many enzymes are inactivated at high pH.

2b. Chemically trapping the pyruvate will have the same effect as increasing the pH of the lactate dehydrogenase reaction. Including a chemical trap to react with pyruvate removes the product, resulting in a shift toward product production. However, NADH, the other product, will continue to accumulate.

2c. Reoxidation of the NADH by the nonenzymatic reaction

$$\text{NADH} + \text{tetrazolium}_{ox}\ (\text{yellow}) \rightarrow \text{NAD}^+ + \text{formazan (purple)}$$

has the advantages of:

(a) removing the product NADH from the equilibrium;

(b) replenishing the reactant NAD^+, providing a longer linear initial rate; and

(c) increasing the sensitivity of the lactate dehydrogenase activity measurement. NADH absorbs light at 340 nm with an extinction coefficient of 6.22 $mM^{-1}\ cm^{-1}$. The formazan absorbs at 578 nm with an extinction decrease of 15 $mM^{-1}\ cm^{-1}$. The increased sensitivity of the formazan chromophore thus decreases the amount of enzyme required in each assay. (See Abdallah, M. A., and J.-F. Biellmann, *Eur. J. Biochem. 112:* 331–33, 1980, for details of the tetrazolium-coupled assay.)

3. RNase activity depends on concerted acid–base reaction catalyzed by two histidyl residues at the active site. Removal of the C′–2 OH proton is best accomplished by an unprotonated imidazole (pH > pK). Removal of the proton allows the alkoxide to add to the phosphorus of the phosphodiester, forming a cyclic 2′–3′ phosphodiester with expulsion of the C–5′ oxygen on the adjacent ribose ring. Addition of a proton to the C–5′ alkoxide ($-O^-$) of the leaving-group ribose is best accomplished by a protonated imidazole (pH < pK). The best compromise is pH at pK of histidine (pH 6), where each of the histidines exists with the maximal fraction of each imidazole in the appropriate degree of protonation. The actual pH optimum is 7, suggesting that the pK_a values of the imidazole side chains in the active site of RNase are shifted relative to the imidazole side chain in free histidine (see chapter 9 in the text).

4. Trypsin cleaves peptide bonds on the carboxyl side of arginine or lysine residues. This specificity is achieved through specific interaction of the positively charged side chains with an aspartic acid residue located at the base of the substrate side chain binding pocket. The aspartate in the binding crevice is not to be confused with Asp 102, a component of the catalytic triad. Substitution of lysine for the aspartate in the substrate binding pocket should have no effect on the ability of the enzyme to hydrolyze peptide bonds. The catalytic triad (Asp-His-Ser) would remain functional assuming that substitution caused minimal disruption of the protein conformation. Specificity of bond cleavage would most certainly change. The favorable electrostatic interaction between the Lys or Arg side chain of the peptide substrate and the aspartate in the binding pocket would be replaced by electrostatic repulsion upon substitution of the lysine residue. Thus, selective binding of Arg or Lys to the substrate pocket would be precluded. The modified trypsin might therefore (a) cleave peptide bonds randomly, but exclude bonds on the carboxyl side of Lys or Arg residues, or (b) exhibit specificity for peptide cleavage on the carboxyl side of acidic amino acids (Asp, Glu). Side chains of these amino acids may fit well into the substrate binding pocket and have favorable electrostatic interaction with the lysine therein.

5. Water structure and its thermodynamic contribution to biochemical reactions must be considered in understanding reaction mechanisms. Water molecules that surround hydrophobic residues or solvate ionic amino acid side chains in the enzyme active site are less randomly distributed than are water molecules in bulk solution (see chapter 2 in the text). Substrate binding to the enzyme is frequently accompanied by a conformational change of the protein concomitant with expulsion of water from the active site. The conformational change fosters hydrophobic and ionic interactions between the enzyme active site and the substrate and ultimately between enzyme and transition-state intermediate. The decreased organization of the water molecules expelled from the active site increases the entropy (ΔS) of the system and decreases ΔG. Moreover, the pK_a values of acidic or basic functional groups at the active site may differ significantly from those of the free amino acids because removal of water favors the uncharged forms.

6. The active site could retain the arginine residue to provide electrostatic interaction and stabilization of the substrate carboxyl terminus as is the case in carboxypeptidase A. The substrate binding site in the enzyme would be predicted to contain acidic and polar side chains to bind the basic side chain of the residue to be cleaved. Mechanistic features including polarization of the carbonyl by metal (possibly zinc) and the electrostatic stabilization of the tetrahedral anionic intermediate by the active site transition metal may be similar to those of carboxypeptidase A.

7. Proteins are assembled from amino acids into the primary sequence based on information encoded in the gene sequence. The primary sequence dictates the formation of secondary structures, and these structures fold and interact to form the tertiary structure. Renaturation of denatured protein is also dictated by the primary structure of the protein.

The trypsin family of enzymes and carboxypeptidase A are each synthesized as proenzymes that are proteolytically activated after synthesis. The proteolyzed, active enzymes have primary structures that differ from the gene product (proenzyme) and would not be expected to refold into a conformation equivalent to that of the proenzyme. It is not surprising that the renatured enzyme is not catalytically active. In addition, zinc is a cofactor required for carboxypeptidase A activity and must be available during the renaturation process. Addition of Zn^{2+} to the renaturing proenzyme would likely result in the correct ligand attachment to the metal. The correct placement of the ligands may not occur in the refolding of denatured, proteolytically activated enzymes.

8a. The active center amino acid residues shown in the diagram form a catalytic triad similar to the Asp-His-Ser arrangement found in the trypsin family of proteases, except for the substitution of the thiol (–SH) for the alcohol (–OH).

The nucleophilicity of the cysteine sulfhydryl is enhanced by polarization of the sulfur-hydrogen bond by the neighboring imidazole group from histidine. The thiolate adds to the carbonyl, forming an anionic tetrahedral intermediate presumably stabilized by electrostatic interactions with adjacent groups of peptide bond hydrogens. The release of the C-terminal portion of the peptide substrate leaves a thioester as the covalent adduct at the active site. Subsequent hydrolysis of the thioester thus releases the N-terminal portion of the substrate. A mechanism illustrating these fundamentals is shown.

8b. The sulfhydryl group of cysteine adds to the activated double bond of *N*-ethylmaleimide to form the S-(*N*-ethyl) succinimido thioether adduct whose structure is shown.

Cysteine

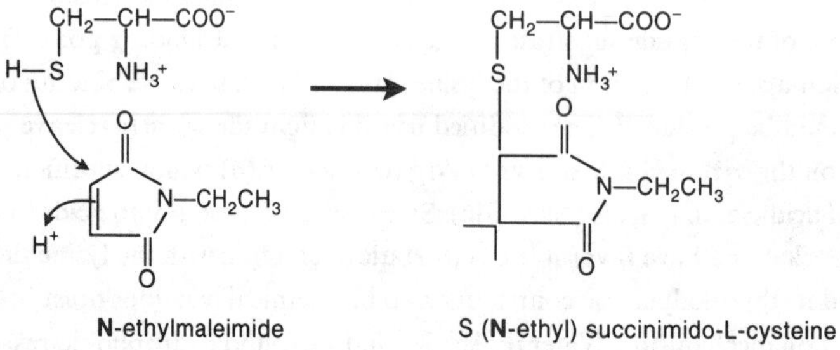

CH_2—CH—COO^- ; H—S ; NH_3^+ ; O ; N—CH_2CH_3 ; H^+ ; O

CH_2—CH—COO^- ; S ; NH_3^+ ; O ; N—CH_2CH_3 ; O

N-ethylmaleimide

S (N-ethyl) succinimido-L-cysteine

One would predict the reactivity of the cysteine SH group with *N*-ethylmaleimide (NEM) to increase with increasing pH if the thiolate anion reacts more rapidly with NEM than does the protonated SH group. The pK_a of the cysteine –SH group is 8.3. The fraction of the cysteine existing as the thiolate anion can be calculated using the Henderson-Hasselbach expression.

$$pH = pK_a + \log (RS^-)/(RSH)$$

where RS^- and RSH represent the fraction of sulfhydryl group in the thiolate and protonated forms, respectively.

At ph 5, the ratio of $(RS^-)/(RSH)$ is 5.0×10^{-4}. However, at pH 7.5, the ratio $(RS^-)/(RSH)$ is 1.6×10^{-1}, a value about 320-fold larger than the ratio at pH 5. The reaction of cysteine with NEM at pH 7.5 is therefore faster than at pH 5.5 because of the larger fraction of the thiolate anion.

8c. The cysteinyl SH group in the active center of papain would be predicted to react more rapidly with NEM than would any of the other cysteines in the protein because of the increased nucleophilic character of the active site sulfhydryl group. The increased nucleophilicity is a result of the polarization of the proton from the SH group afforded by the adjacent imidazole.

9. The enzyme active site stabilizes the transition-state intermediate formed during catalysis with a resulting decrease in the activation energy of the reaction. The structure of the transition-state intermediate is complementary to the structure of the active site and is thus bound much more tightly than are either substrates or products. Transition-state analogs, having structural features only slightly different from the transition-state intermediate, would be expected to bind very tightly to the active site and thus have a low K_i. The analog would compete with substrate or product for the active site and would exhibit competitive inhibition.

10. The transition-state intermediate is bound more tightly to the enzyme active site than are either substrates or products and presumably fits well into the three-dimensional structure of the active site. Transition-state analogs are chemically similar to the transition-state intermediate and should also be tightly bound by the enzyme active site. The transition-state analogs, used as antigens, should elicit amino acid residues in the antibody combining site that are structurally complementary to the enzyme active site. Hence, rate enhancement and substrate specificity similar to the enzyme could be mimicked. Antibodies catalyzing reactions requiring a cofactor, or specific coenzyme functional residue, pose the added difficulty of choosing and synthesizing the appropriate transition-state analog, or attachment of a cofactor by chemical means after the catalytic antibody is formed. (See Schultz, P. G., R. A. Lerner, and S. J. Benkovic, *Chem Eng. News.* May 28, 1990, 26–40, and Benkoric, S. J., *Annu. Rev. Biochem.* 61:29–54, 1992, for reviews.)

11. Superoxide dismutases (SOD) are metalloproteins that catalyze the disproportionation of superoxide molecules to O_2 and H_2O_2. In the disproportionation process, one of the superoxide molecules is oxidized and the metal cofactor at the SOD active site is reduced. The catalytic cycle is completed upon reduction of a second superoxide molecule and reoxidation of the active site metal ion. The two partial reactions catalyzed by superoxide dismutase are schematically illustrated using the iron-dependent form of SOD as an example.

(a) $$Fe^{3+} + O_2^- \rightarrow Fe^{2+} + O_2$$

(b) $$Fe^{2+} + O_2^- + 2\,H^+ \rightarrow Fe^{3+} + H_2O_2$$

The overall reaction is

$$2\,O_2^- + 2\,H^+ \rightarrow O_2 + H_2O_2$$

12a. At pH 7 the superoxide dismutase (pI about 5) will have a net negative charge, as will the superoxide anion. The collision of the negatively charged substrate with the anionic enzyme molecule would be impeded by electrostatic repulsion if the anionic charge were equally distributed over the enzyme surface.

12b. Basic residues that are positively charged at pH 7 and placed strategically in the active site could steer and propel the substrate toward the catalytically active metal ion. The net charge on the superoxide dismutase remains unchanged, but the charge distribution is not uniform over the enzyme surface. (See Cudd, A., and I. Fridovich, *J. Biol. Chem. 257:*11443–447, 1982.)

12c. Neither a negatively charged enzyme nor the basic residues should affect diffusion of the uncharged products (O_2 and H_2O_2) from the active site.

13a. Recombinant enzyme in which lysine residues at positions 21 and 101 were replaced with aspartic acid residues (Lys 21 → Asp 21 and Lys 101 → Asp 101) exhibited only 3% less maximal catalytic activity than did the nonrecombinant (native) enzyme. Therefore, these residues were neither functional active-site residues, nor were they critical for maintaining a catalytically competent conformation of the enzyme. Recombinant enzyme in which Lys 86 was replaced with Asp 86 exhibited catalytic activity that was only 10% that of the nonrecombinant enzyme, suggesting that Lys 86 is a critical residue either at or near the catalytic active center or is essential for maintaining a catalytically competent conformation of the enzyme.

13b. Lysins 21 and 101 are likely outside the catalytic site and may not be evolutionarily conserved. However, that assessment is based solely on measurement of catalytic activity *in vitro.* Amino acid residues on the surface of proteins, a location reasonably assumed for these residues, have other roles that may be physiologically relevant: (a) as part of structural motifs recognized by regulatory enzymes, (b) as one of a cluster of charged amino acid side chains important in assembly of multienzyme complexes, or (c) in interaction of the enzyme with the surface of the membrane.

13c. Lysine 86 is required for maximum enzymatic activity and would likely be conserved. Amino acids whose side chains are involved either with the binding of substrate or cofactors, chelating transition metals, as reactants in the catalytic chemistry, or as essential structural determinants are usually strongly conserved among groups of enzymes.

14a. Phosphorylated sugar is bound to the enzyme active site through interactions that include electrostatic association of the carbonyl oxygen and phosphate oxygen with zinc. A base group in the active site removes a proton from the sugar C-4 OH, yielding an oxyanion. The oxyanion rearranges to form a carbonyl at C-4 with cleavage of the bond between C-3 and C-4. The carbanion formed on C-3 is in resonance with a zinc-stabilized endate.

The substrate is the phosphorylated sugar:

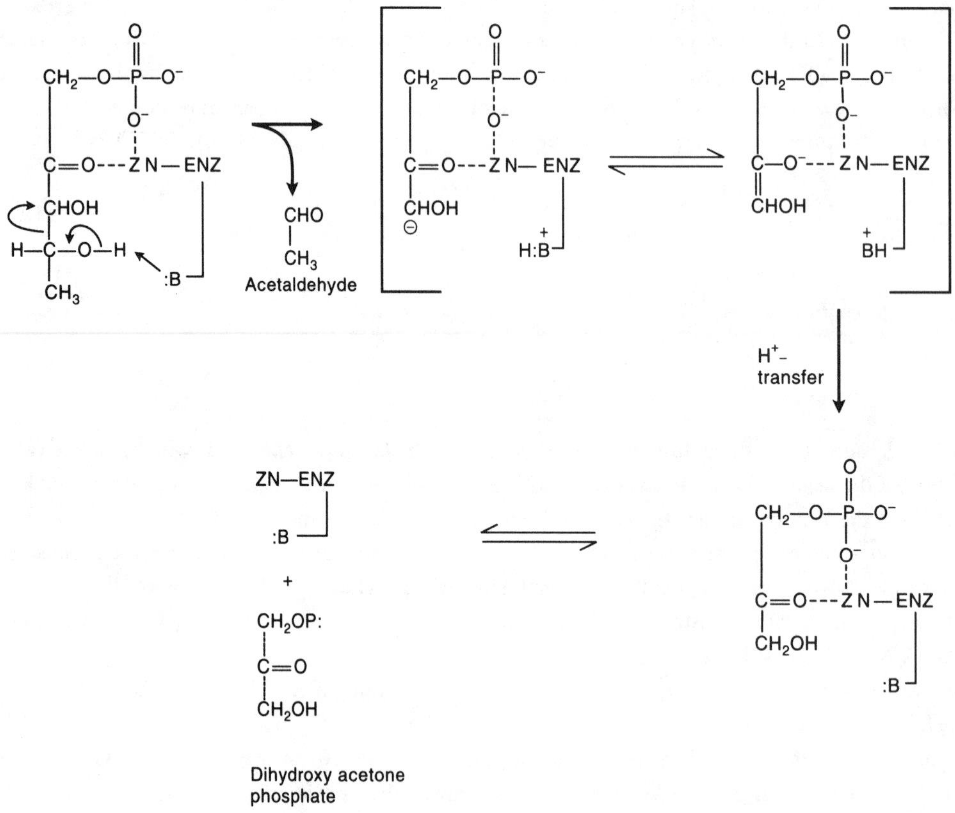

Acetaldehyde

H^+- transfer

Dihydroxy acetone phosphate

14b. The substrate may be anchored to a positively charged area in the active site by electrostatic interaction with the anionic phosphate group. The nonphosphorylated sugar, lacking this means of interaction at the active site, would likely be a poor substrate.

15. An intermediate that is covalently bound to an active site amino acid residue is an essential part of the mechanism in covalent catalysis. Examples of enzymes whose mechanism involves covalent catalysis include liver fructose-1,6-bisphosphate aldolase, amino transferases, the trypsin family of proteases, papain, and pyruvate dehydrogenase. In noncovalent catalysis, the intermediates are bound through ionic or hydrogen bonds or through hydrophobic interaction. Examples of enzymes employing that mechanism include lactate dehydrogenase, hexokinase, triosephosphate isomerase, and aconitase.

10 Regulation of Enzyme Activities

Problems

1. Show by writing the reactions how the combination of a protein kinase and the corresponding phosphoprotein phosphatase, if unregulated, theoretically forms a futile cycle.

2. The cAMP-dependent protein kinases phosphorylate specific Ser (Thr) residues on target proteins. Given the availability of serine and threonine residues on the surface of globular proteins, how might a protein kinase select the "correct" residues to phosphorylate?

3. Assume that the flow diagram shown represents an amino acid biosynthetic pathway where G, J, and H are amino acids and A is a common precursor. Products G, H, and J are required by the cell. Enzymes catalyzing the steps are numbered. Suggest a plausible scheme for the regulation of specific enzymes by their products.

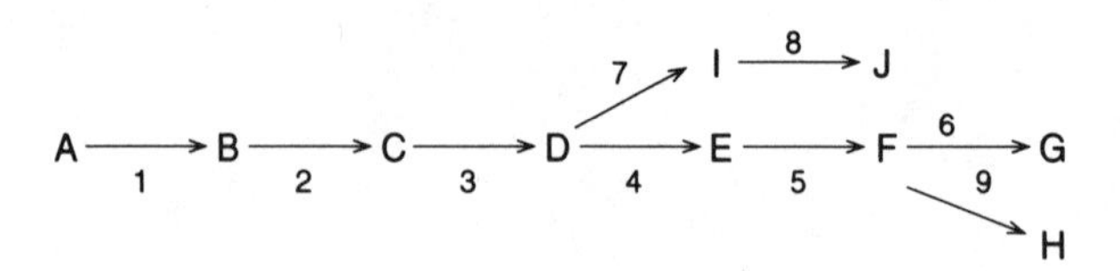

4. Aspartate carbamoyltransferase is an allosteric enzyme in which the active sites and the allosteric effector binding sites are on different subunits. Explain how it might be possible for an allosteric enzyme to have both kinds of sites on the same subunit.

5. ATP is both a substrate and an inhibitor of the enzyme phosphofructokinase (PFK). Although the substrate fructose-6-phosphate binds cooperatively to the active site, ATP does not bind cooperatively. Explain how ATP may be both a substrate and an inhibitor of PFK.

6. Calculate the substrate concentration [S] is terms of K_m for a hyperbolically responding enzyme when the velocity is 10% V_{max} or 90% V_{max}. What is the ratio of substrate concentrations that affect this nine-fold velocity change ($S_{0.9}/S_{0.1}$)? How would the ($S_{0.9}/S_{0.1}$) ratio differ for an allosterically responding enzyme?

7. The substrate concentration yielding half-maximal velocity is equal to K_m for hyperbolically responding enzymes. Is this relationship true for allosterically or sigmoidally responding enzymes? How do positive or negative allosteric effectors change the substrate concentration required for half-maximal velocity?

8. Examine the relationship of aspartate carbamoyltransferase (ACTase) activity to aspartate concentration shown in the figure. Estimate the ($S_{0.9}/S_{0.1}$) ratio for the reaction under the following conditions: (a) normal curve, (b) plus 0.2-mM CTP, (c) plus 0.8-mM ATP. Do these ratios differ significantly? Explain.

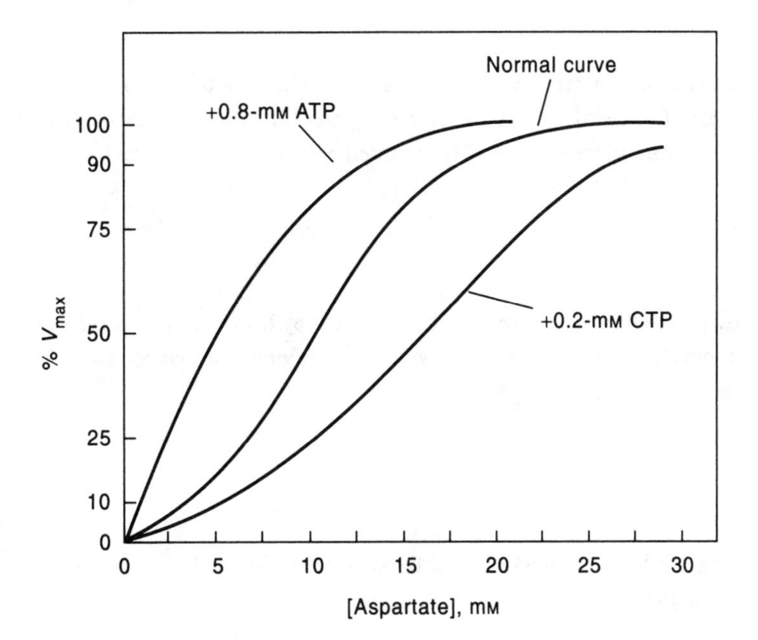

9. If you separated hemoglobin into dimers of (α + β) subunits, would you expect the dimers to bind more or less O_2 at low O_2 tensions? Explain. What effect would 2,3-bisphosphoglycerate have on oxygen binding? (Review O_2 binding to hemoglobin in chapter 5 of text.)

10. Treatment of ACTase with mercurials causes loss of allosteric regulation by ATP and CTP and eliminates positive cooperativity with aspartate. Can you suggest a strategy to "lock" an allosterically regulated enzyme into the active form without dissociating the subunits?

11. NAD^+ binding to the dimeric liver alcohol dehydrogenase (LADH) causes rearrangement of the active site residues to foster binding of ethanol, yet LADH shows no cooperativity in binding either NAD^+ or ethanol. What element of the conformational change is lacking in the LADH that is present in allosteric enzymes?

12. Light-dependent activation of key enzymes in the photosynthetic CO_2 fixation involves activation by thioredoxin–mediated reduction of critical disulfides on the enzymes. Write a reaction linking the reductant ferredoxin (a single-electron donor) to the reduction of protein disulfides using thioredoxin as an intermediate.

13. Assuming that the thioredoxin-dependent activation of proteins resulting from disulfide reduction is reversible, what possibilities exist for the "deactivation" of the enzymes? What are the metabolic and energetic ramifications to the cell in utilizing irreversibly activated or inactivated enzymes?

14. In some instances, protein kinases are activated or inhibited by low-molecular-weight modifiers. Explain how a metabolite may more effectively regulate an enzyme by modifying a protein kinase rather than directly inhibiting the target enzyme.

15. Cite some advantages in having calmodulin tightly associated with the Ca^{2+}-regulated enzyme rather than in an uncomplexed form in the cell.

Solutions

1. In principle, futile cycles may occur if two enzymes (or two metabolic pathways) catalyze reactions that are the reverse of each other. The result of a futile cycle is generally the hydrolysis of high energy intermediates, nonproductive use of reductant, or both. Futile cycling may be minimized in the cell by regulation of key enzymes in the opposing pathways or by physically separating the opposing pathways in different subcellular compartments. Each of those strategies has been demonstrated to occur in the cell.

   Protein kinase and phosphoprotein phosphatase catalyze opposing reactions. Consider the reactions

$$\text{Protein (Ser)} + \text{ATP} \rightarrow \text{protein (Ser-P}_i\text{)} + \text{ADP} \quad (1)$$

$$\text{Protein (Ser-P}_i\text{)} + \text{H}_2\text{0} \rightarrow \text{protein (Ser)} + \text{P}_i \quad (2)$$

$$\text{Sum: ATP} + \text{H}_2\text{O} \rightarrow \text{ADP} + \text{P}_i$$

   The enzymes are (1) protein kinase and (2) phosphoprotein phosphatase. The (Ser) represents a serine residue, in this example, that is reversibly phosphorylated. However, threonine and tyrosine side chains are also phosphorylated by protein kinases.

2. Phosphorylation of serine or threonine (and possibly tyrosine) on the target protein may be largely influenced by the amino acid sequence around these residues. These amino acid sequences may define a specific motif or recognition site for the protein kinase. Kemp (*TIBS.* 15:342–346 (1990)) and Kennelly and Krebs (*J. Biol. Chem.* 266:15555–58 (1991)) have reviewed possible recognition sequences (motifs) on the substrates for a number of protein kinases. The authors caution that not all serine (threonine) residues in a specific motif are necessarily phosphorylated. Serine (or threonine) residues may not be phosphorylated because of other factors, such as topographical features on the target protein, that prevent binding of the target protein to the protein kinase active site.

3. Metabolic pathways are frequently regulated at the step that commits a metabolite to the pathway, at branches in the pathway, and at the enzyme that catalyzes the final step in a branch of the pathway. In the example, assume that products G, H, and J are essential end products. Each would likely inhibit its own production without inhibiting production of the others. Thus, product J could feedback-inhibit enzyme 7 and inhibit enzyme 8 by product inhibition, G inhibits enzyme 6, and H inhibits enzyme 9. Products G and H could cumulatively inhibit enzyme 4, assuming it were the committed step for that branch of the pathway. The committed step (enzyme 1) may be cumulatively inhibited by each of the products. Each product alone may cause only minimal inhibition.

4. Allosteric regulation of an enzyme having a binding site for a regulatory molecule and an active site on the same subunit is not uncommon. Regulatory molecules bind at sites separate from the active site and induce a conformational change that affects substrate binding to the active site on the same as well as adjacent subunits.

5. Phosphofructokinase (PFK) catalyzes the transfer of phosphate from ATP to C-1 of fructose -6-phosphate. Hence ATP is a substrate for the enzyme. PFK is the committed step in the glycolytic pathway whose product is pyruvate, a metabolite whose oxidation drives ADP phosphorylation in the mitochondria. Activity of PFK is therefore responsive to the relative amount of ATP in the cytosol. Two possible mechanisms by which PFK is inhibited by APT include (1) substrate inhibition by ATP or (2) interaction of ATP at a regulatory site different from the active site. Substrate inhibition occurs upon binding of the substrate (ATP) to the enzyme-product complex, forming a dead-end (catalytically inactive) ternary complex. Thus the complex (PFK-fructose-1,6-bisphosphate-ATP) would be catalytically inactive. PFK is inhibited upon binding ATP to a regulatory site that

is different from the active site. ATP thus plays a dual role as a substrate and as a ligand that binds to a regulatory site (see fig. 10.6 in the text).

6. Assume a single-substrate enzyme whose substrate-velocity profile is a rectangular hyperbola (i.e., obeys Michaelis-Menten kinetics). The velocity expression for the enzyme is

$$v = \frac{V_m[S]}{K_m + [S]}$$

where V_m is the maximal velocity, K_m is the Michaelis constant, and [S] is substrate concentration. Let $v = 0.1V_m$.

$$0.1V_m = \frac{V_m\,[S]}{K_m + [S]}$$

$$K_m + [S] = \frac{V_m\,[S]}{0.1\,V_m}$$

$$K_m = 9[S]$$

$$[S] = 1/9(K_m) \text{ when } v = 0.1\,V_m$$

Let $v = 0.9\,V_m$

$$0.9V_m = \frac{V_m\,[S]}{K_m + [S]}$$

$$K_m + [S] = \frac{V_m\,[S]}{0.9\,V_m}$$

$$9[K_m + [S]] = 10[S]$$

$$[S] = 9\,K_m \text{ when } v = 0.9\,V_m$$

Therefore, the ratio of [S] required to change a hyperbolically responding enzyme from 10% V_m to 90% V_m is 81. For an allosterically responding enzyme, the ratio would be < 81 and may be significantly lower (e.g., see the figure for problem 8. The value varies from 5 to 15 for ACTase, depending on the presence of other modifiers). (Atkinson, D. E., in *Cellular Energy Metabolism and Its Regulation.* New York: Academic Press, 1977, 126–135.)

7. The substrate concentration required for half-maximal activity ($S_{0.5}$) of an allosterically regulated enzyme will depend on the relationship between enzyme reaction velocity and substrate concentration as affected by activators and inhibitors, if present. Hence ($S_{0.5}$) may be decreased with allosteric activators and may be increased with allosteric inhibitors. At a given substrate concentration, the activity of an allosterically regulated enzyme may vary widely and depend on the presence of other modifiers (activators or inhibitors). This feature makes these enzymes well suited for their role in metabolic regulation.

8. Based on the data in the figure, the values for the ratio ($S_{0.9}/S_{0.1}$) are approximately 5 for the normal, nonmodified reaction; 4 to 5 in the presence of 0.2-mM CTP; and 13 in the presence of 0.8-mM ATP. The cooperativity, based on the ratio of substrate required to increase activity approximately nine-fold is not significantly different with or without 0.2-mM CTP. However, the absolute amount of substrate that yields 90% V_m is 19 mM in the absence of CTP but 29 mM in the presence of 0.2-mM CTP. Binding of ATP decreases the cooperativity and the [v] versus [aspartate] curve is shifted to a more hyperbolic form. The low substrate ratio required to cause large increases in velocity allows allosterically regulated enzymes to function efficiently as metabolic switches for the cell. Consider the activity of ATCase with 10-mM aspartate and the allosteric modifiers given. The ACTase activity with 10-mM aspartate varies with the presence or absence of allosteric

modifiers. With no allosteric modifiers (normal curve), v = app. 50% V_m; with ATP, v = app. 75% V_m; with 0.8-mM CTP, v = app. 20% V_m.

9. Binding of O_2 to the iron of hemoglobin causes the iron to shift to a low-spin *d*-electron configuration from a high-spin configuration. The low-spin Fe^{2+} has a smaller ionic radius compared to the high-spin Fe^{2+}, allowing the iron to move approximately 0.7 Å into the plane of the heme ring. The imidazole group liganded to the heme iron also moves. This motion is transmitted to the other subunits in the tetrameric hemoglobin through the interaction of the two α and two β subunits. Separation of the tetramer into two (α + β) dimers would destroy the cooperativity exhibited by the tetramer. There could be some cooperativity exhibited by the (α + β) dimer, but not likely on the scale exhibited by the hemoglobin tetramer. Thus, there should be a greater degree of saturation with O_2 at lower O_2 tensions to the dimer than to the tetramer. The O_2-binding curve for the (α and β) dimer would likely resemble the O_2-binding curve of myoglobin.

   The 2,3-bisphosphoglycerate (2,3-BPG) binds to deoxyhemoglobin and stabilizes the conformation of the deoxyhemoglobin relative to the oxyform. The 2,3-BPG binds in a crevice formed at the intersection of the subunits of the tetramer. Were the tetramer dissociated to dimers, the crevice would be destroyed, precluding the binding and subsequent stabilizing effect of 2,3-BPG. Therefore, 2,3-BPG would not be expected to alter the O_2 affinity of the (α & β) dimers.

10. The active form of an allosterically activated enzyme could be "locked" into the active conformation by using a chemically reactive analog that covalently binds at the regulatory site or by directly chemically modifying a residue at the allosteric site so that the desired enzyme conformation predominates.

11. Examination of NAD^+ binding to the dimeric horse liver alcohol dehydrogenase reveals that the enzyme active site shifts from an open to a closed form concomitant with large conformational rearrangements. (Eklund, H., J.-P. Samama, and T. A. Jones, *Biochemistry*. 23:5982–5996 (1984)). These rearrangements cause the enzyme to close around the extended form of NAD^+ and establish the catalytically competent active site. Each subunit of the dimer apparently undergoes the same structural reorganization. However, the conformational change caused by NAD^+ binding does not differentially affect binding of substrate or stabilization of the transition complex at one subunit compared with the other. If there is no differential binding of substrate to the subunits, there is no cooperativity. Thus, not all conformational changes are associated with cooperative kinetic behavior.

12. Thioredoxin alters the activity of target proteins by reducing specific disulfide bonds to dithiols.

$$E(\text{–S–S–}) + Tr(SH)_2 \rightarrow E(SH)_2 + Tr(\text{–S–S–})$$

where E(–S–S–) and Tr(–S–S–) are the oxidized forms and $E(SH_2)$ and $Tr(SH)_2$ are reduced forms of the target protein or target enzyme and thioredoxin, respectively. Oxidized thioredoxin is reduced in the plant cell by reduced ferredoxin.

$$2\ FD_{red} + Tr(\text{–S–S–}) \rightarrow 2\ Fd_{ox} + Tr(SH)_2.$$

It should be noted that two moles of ferredoxin are required per mole of oxidized thioredoxin reduced because ferredoxin is a one electron-equivalent redox reagent, whereas reduction of the disulfide requires two electrons.

13. Thioredoxin reduces specific disulfide bonds on target proteins as a means of regulating enzymatic activity. The sulfhydryl groups on the activated enzymes must be reoxidized to the disulfide form for this regulation to be reversible. Whether the disulfide is formed by an enzymatically catalyzed process or occurs spontaneously via low molecular weight oxidants is not known.

   Irreversibly modified enzymes in principle must be degraded to allow recycling of the amino acids. Although the cell can reclaim the amino acids, there is still a large metabolic expense in *de novo* synthesis of the protein. Reversible modification is therefore more energetically economical for the cell. (Cseke, C., and B. B. Buchanan, *Biochim. et Biophys. Acta*. 853:43–63 (1986).)

14. Inhibition of a target enzyme through direct interaction of a low-molecular-weight modifier eliminates the amplification or cascade factor afforded by the protein kinase. Direct inhibition is a stoichiometric process

(E + I → EI). The amount of inhibitor required to inhibit the enzyme will be a function of K_i (the inhibitor constant). Activation of a protein kinase causes the catalytic alteration of the target proteins. Hence, the effect of a small concentration of I can be amplified through the catalytic activity of the protein kinase.

15. In principle, the activation of tightly associated calmodulin by Ca^{2+} should be kinetically less complex than if the calmodulin were free. Free calmodulin would respond to a Ca^{2+} signal by forming a Ca^{2+}-calmodulin complex which then must collide with and bind to a target protein. The activation of the target protein thus requires the binding of three to four Ca^{2+} atoms per molecule of calmodulin and then a bimolecular reaction of Ca^{2+}-calmodulin with the target protein. Calmodulin complexed to the target protein is a single entity which must collide with Ca^{2+} atoms. The latter reaction should be kinetically less complex and would occur with a higher probability.

11 Vitamins, Coenzymes, and Metal Cofactors

Problems

1. What structural features of biotin and lipoic acid allow these cofactors to be covalently bound to a specific protein in a multienzyme complex yet participate in reactions at active sites on other enzymes of the complex?

2. The following reactions are catalyzed by pyridoxal-5′-phosphate-dependent enzymes. Write a reaction mechanism for each, showing how pyridoxal-5′-phosphate is involved in catalysis.

(a)

$$CH_3-\overset{\overset{O}{\|}}{C}-COO^- + R-\underset{\underset{NH_3^+}{|}}{CH}-COO^- \dashrightarrow CH_3-\underset{\underset{NH_3^+}{|}}{CH}-COO^- + R-\overset{\overset{O}{\|}}{C}-COO^-$$

(b)

$$H_3^+N-(CH_2)_4-\underset{\underset{NH_3^+}{|}}{CH}-CO_2^- \rightarrow CO_2 + H_3^+N-(CH_2)_5-NH_3^+$$

(c)

$$^-O_2C-CH_2-\underset{\underset{NH_3^+}{|}}{CH}-CO_2^- \rightarrow CO_2 + CH_3-\underset{\underset{NH_3^+}{|}}{CH}-CO_2^-$$

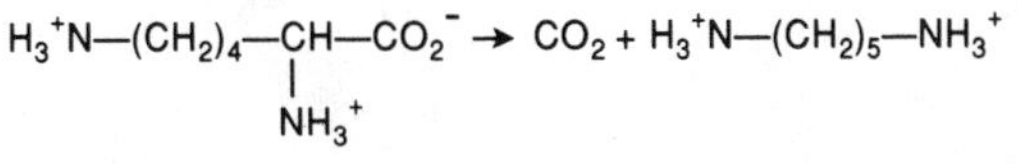

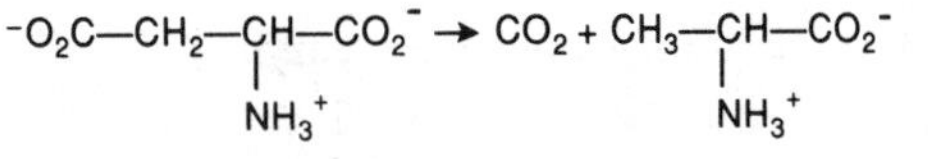

3. $NADP^+$ differs from NAD^+ only by phosphorylation of the C-2′ OH group on the adenosyl moiety. The redox potentials differ only by about 5 mV. Why do you suppose it is necessary for the cell to employ two such similar redox cofactors?

4. Thiamine-pyrophosphate-dependent enzymes catalyze the reactions shown below. Write a chemical mechanism that shows the catalytic role of the coenzyme.

(a)

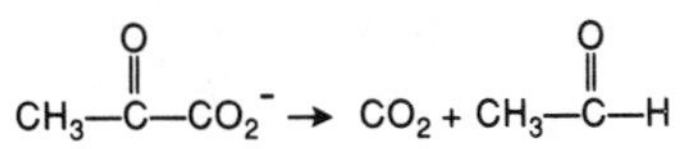

$$CH_3-\overset{\overset{O}{\|}}{C}-CO_2^- \rightarrow CO_2 + CH_3-\overset{\overset{O}{\|}}{C}-H$$

(b)

$$ⓅOCH_2-(CHOH)_2-\overset{\overset{OH}{|}}{CH}-\overset{\overset{O}{\|}}{C}-CH_2OH + HOPO_3^{2-} \rightarrow$$

$$ⓅO-CH_2-(CHOH)_2-CHO + CH_3-\overset{\overset{O}{\|}}{C}-OPO_3^{2-} + H_2O$$

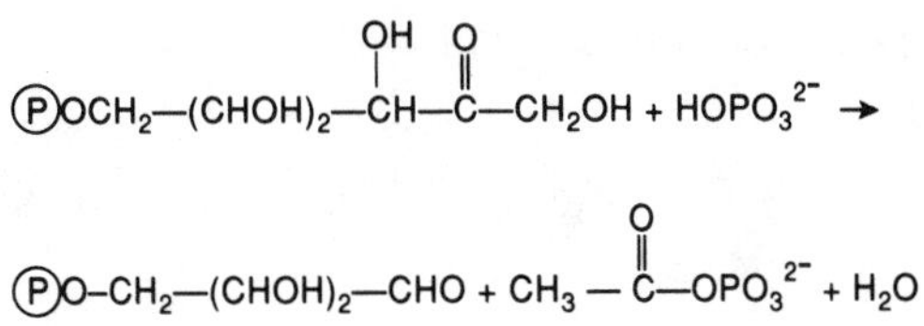

5. Malate synthase catalyzes the condensation of acetyl-CoA with glyoxalate to form L-malate.

(a) Write a chemical reaction illustrating the reaction catalyzed by malate synthase.

(b) Explain the chemical basis for using acetyl-CoA rather than acetate in the condensation reaction.

(c) The product released from the enzyme is L-malate rather than malyl-CoA. Explain the role of thioester hydrolysis in the condensation reaction.

6. Write the mechanism showing how NAD(P)$^+$ is involved in the following reactions.

(a) Malate dehydrogenase

$$\begin{array}{c} COO^- \\ | \\ CHOH \\ | \\ CH_2 \\ | \\ COO^- \end{array} + NAD^+ \dashrightarrow \begin{array}{c} COO^- \\ | \\ C{=}O \\ | \\ CH_2 \\ | \\ COO^- \end{array} + NADH + H^+$$

(b) Malate dehydrogenase (decarboxylating; malic enzyme)

$$\begin{array}{c} COO^- \\ | \\ CHOH \\ | \\ CH_2 \\ | \\ COO^- \end{array} + NADP^+ \dashrightarrow \begin{array}{c} COO^- \\ | \\ C{=}O \\ | \\ CH_3 \end{array} + NADPH + CO_2$$

7. What chemical features allow flavins (FAD, FMN) to mediate electron transfer from NAD(P)H to cytochromes or iron-sulfur proteins?

8. Write the mechanisms that show the involvement of biotin in the following reactions.

(a)

$$CH_3{-}\overset{\overset{O}{\|}}{C}{-}SCoA + HCO_3^- + ATP \rightarrow$$

$$^-O_2C{-}CH_2{-}\overset{\overset{O}{\|}}{C}{-}SCoA + ADP + HOPO_3^{2-}$$

(b)

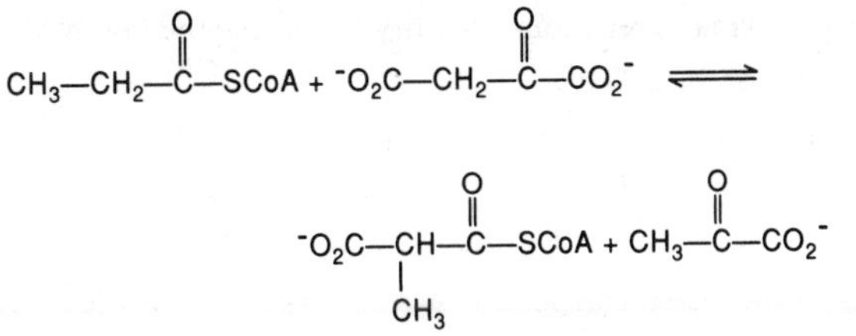

$$CH_3-CH_2-\overset{O}{\overset{\|}{C}}-SCoA + {}^-O_2C-CH_2-\overset{O}{\overset{\|}{C}}-CO_2^- \rightleftharpoons$$

$$ {}^-O_2C-\underset{CH_3}{\underset{|}{CH}}-\overset{O}{\overset{\|}{C}}-SCoA + CH_3-\overset{O}{\overset{\|}{C}}-CO_2^-$$

9. (a) What metabolic advantage is gained by having flavin cofactors covalently or tightly bound to the enzyme?

(b) Would covalently bound NAD^+ ($NADP^+$) be a metabolic advantage or disadvantage?

10. Given the amino acid of the general structure

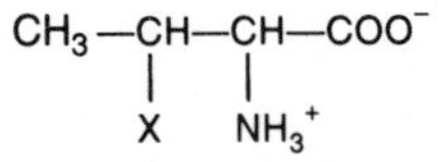

$$CH_3-\underset{X}{\underset{|}{CH}}-\underset{NH_3^+}{\underset{|}{CH}}-COO^-$$

we could use pyridoxal-5′-phosphate to eliminate X, decarboxylate the amino acid, or oxidize the α carbon to a carbonyl with formation of pyridoxamine-5′-phosphate. The metabolic diversity afforded by PLP, unchanneled, could wreak havoc in the cell. What other components are required to channel the PLP-dependent reaction along specific reaction pathways?

11. Amino acid oxidases catalyze the flavin-dependent reaction

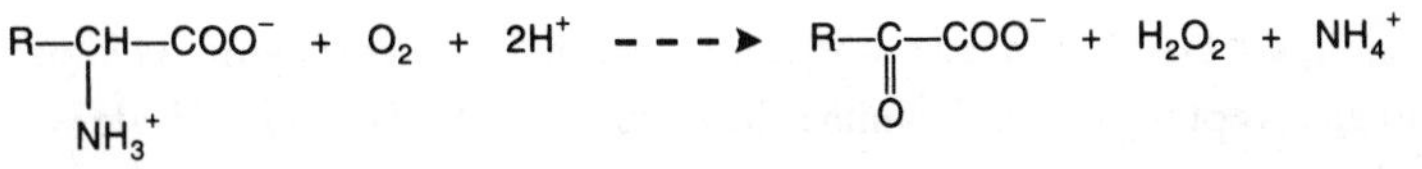

$$R-\underset{NH_3^+}{\underset{|}{CH}}-COO^- + O_2 + 2H^+ \dashrightarrow R-\underset{O}{\underset{\|}{C}}-COO^- + H_2O_2 + NH_4^+$$

What advantages are gained by the cell in using the PLP-dependent transamination reaction rather than the FAD-dependent deamination reaction to convert α-amino acids to the corresponding α-keto acids?

12. How is flavin adenine dinucleotide involved in the following reaction? Show possible mechanisms.

$$\alpha\text{-D-Glucose} + O_2 \rightarrow \text{D-gluconolactone} + H_2O_2$$

13. For each of the following enzymatic reactions, identify the coenzyme involved.

(a)

$$R\text{—}\underset{\displaystyle NH_3^+}{\underset{|}{CH}}\text{—}COO^- + 2H^+ + O_2 \dashrightarrow R\text{—}\overset{\displaystyle O}{\overset{\|}{C}}\text{—}COO^- + NH_4^+ + H_2O_2$$

(b)

$$HO\text{—}CH_2\text{—}\underset{\displaystyle NH_3^+}{\underset{|}{CH}}\text{—}CO_2^- \rightarrow CH_3\text{—}\overset{\displaystyle O}{\overset{\|}{C}}\text{—}CO_2^- + NH_4^+$$

(c)

$$CH_3\text{—}CH_2\text{—}\overset{\displaystyle O}{\overset{\|}{C}}\text{—}SCoA + HCO_3^- + ATP \rightarrow$$

$$^-O_2C\text{—}\underset{\displaystyle CH_3}{\underset{|}{CH}}\text{—}\overset{\displaystyle O}{\overset{\|}{C}}\text{—}SCoA + ADP + P_i$$

(d)

$$\underset{\displaystyle O(P)}{\underset{|}{CH_2}}\text{—}\underset{\displaystyle OH}{\underset{|}{CH}}\text{—}\underset{\displaystyle OH}{\underset{|}{CH}}\text{—}\overset{\displaystyle OH}{\overset{|}{CH}}\text{—}\overset{\displaystyle O}{\overset{\|}{C}}\text{—}\overset{\displaystyle OH}{\overset{|}{CH_2}} + \underset{\displaystyle O(P)}{\underset{|}{CH_2}}\text{—}\underset{\displaystyle OH}{\underset{|}{CH}}\text{—}CHO \rightleftharpoons$$

$$\underset{\displaystyle O(P)}{\underset{|}{CH_2}}\text{—}\underset{\displaystyle OH}{\underset{|}{CH}}\text{—}\underset{\displaystyle OH}{\underset{|}{CH}}\text{—}CHO + \underset{\displaystyle O(P)}{\underset{|}{CH_2}}\text{—}\underset{\displaystyle OH}{\underset{|}{CH}}\text{—}\overset{\displaystyle OH}{\overset{|}{CH}}\text{—}\overset{\displaystyle O}{\overset{\|}{C}}\text{—}\underset{\displaystyle OH}{\underset{|}{CH_2}}$$

14. Lipid-soluble polycyclic aromatic hydrocarbons are in part excreted from the body conjugated to one of several hydrophilic substituents, principally glucuronic acid. What role does the liver microsomal cytochrome P450 (polysubstrate monoxygenase) system play in converting a polycyclic aromatic hydrocarbon to forms that could be conjugated to glucuronic acid?

15. Some bacterial toxins use NAD^+ as a true substrate rather than as a coenzyme. The toxins catalyze the transfer of ADP-ribose to an acceptor protein. Examine the structure of NAD^+ and indicate which portion of the molecule is transferred to the protein. What is the other product of the reaction?

Solutions

1. Lipoic acid and biotin are each covalently bound to their respective enzymes through amide bonds involving the carboxyl group of the cofactor side chain and the ε-amino group of lysine in the protein. The combined length of the lysyl side chain and the acyl group on the cofactor allows free rotation of the dithiol group (lipoic acid) or the imidazolone group (biotin) at the end of a tether that may extend to approximately 14 to 16 Å long. The

chemically reactive portion of the cofactor may diffuse between several catalytic centers present either on different subunits of the enzyme or different enzymes of a multienzyme complex but the cofactor is not physically released from the complex. Moreover, the effective concentration of the cofactor at the active site is greater than if the equivalent amount of cofactor were free in solution, thus increasing the probability of a chemical reaction.

2. (Mechanism shown in figure 2a and 2b is based on structures shown in figure 11.5 in text.)

2a. Refer to figure 11.5b in the text for the structures of each intermediate step.

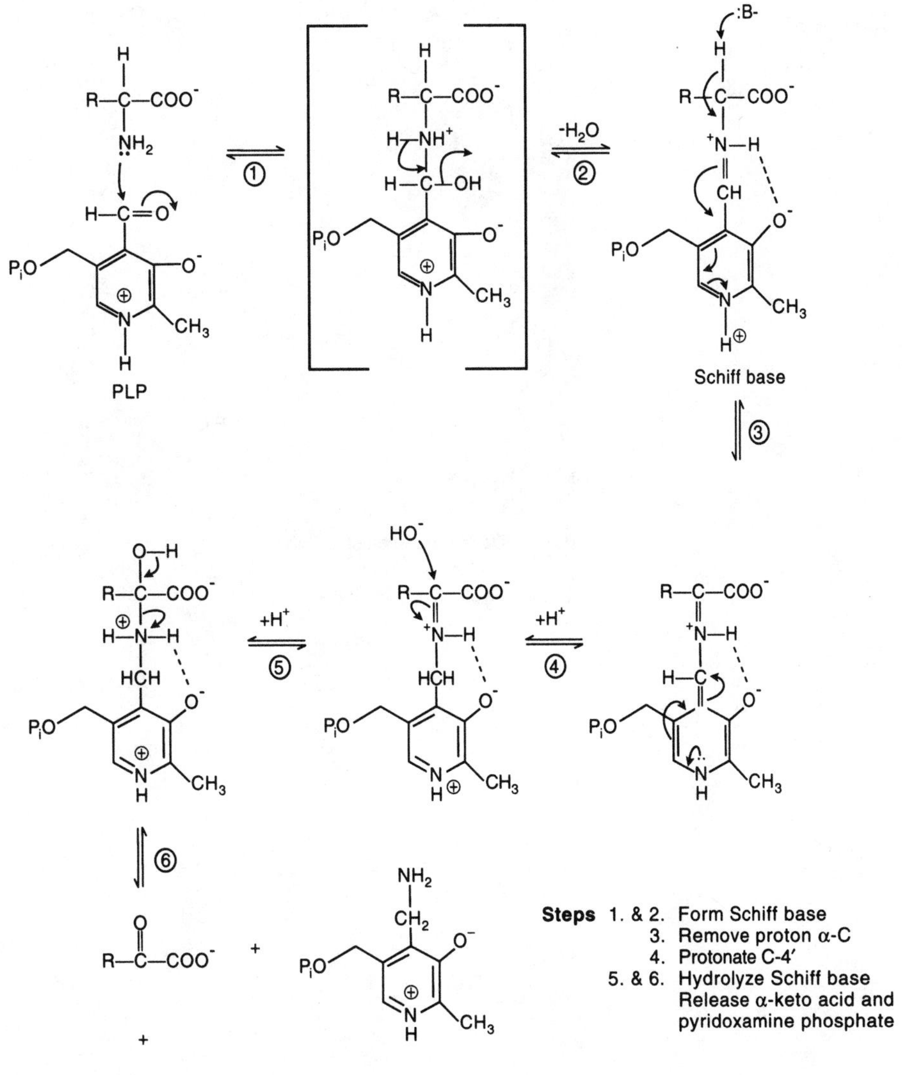

PLP

$-H_2O$

Schiff base

$+H^+$

$+H^+$

Steps 1. & 2. Form Schiff base
3. Remove proton α-C
4. Protonate C-4′
5. & 6. Hydrolyze Schiff base
Release α-keto acid and pyridoxamine phosphate

Transfer of amino group from pyridoxamine phosphate to pyruvate, forming alanine occurs by reversal of the steps shown above. Other transaminases use other α-amino acids and α-keto acids.

2b. Figure 11.5b in the text gives the structures of intermediates shown in the proposed mechanism. Alternate mechanisms are also described by Walsh and Boeker and Snell. (See References at end of chapter.)

α-Decarboxylation occurs from Schiff base form Reactions 1–2, part 2a

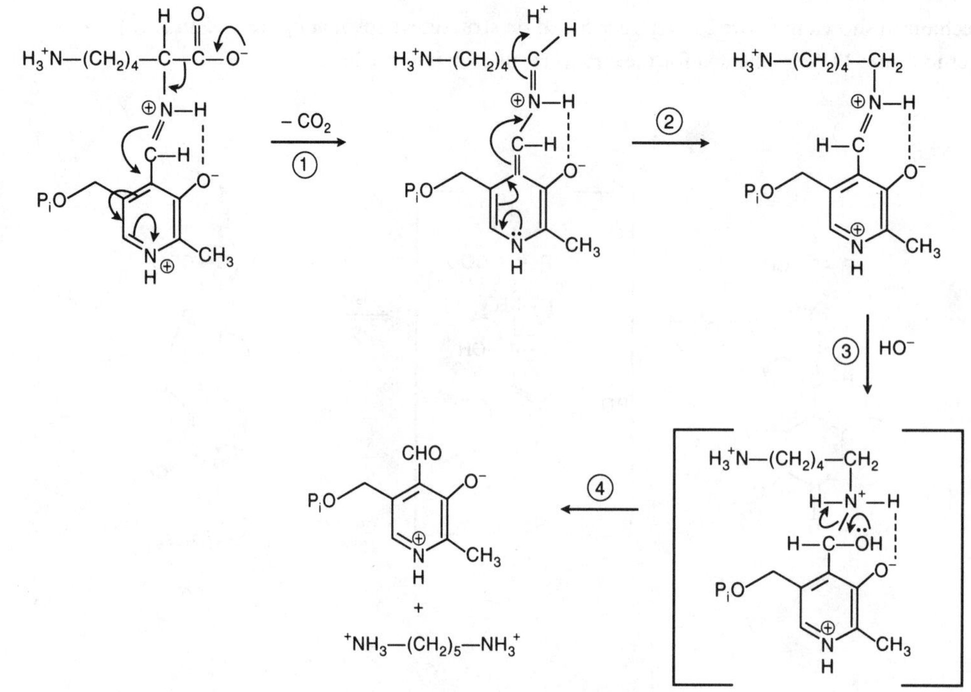

$H_3^+N{-}(CH_2)_4{-}CH(COO^-){-}N^+H{=}CH{-}$(pyridoxal phosphate, P_iO, O^-, CH_3, N^+H) $\xrightarrow[①]{-CO_2}$ $H_3^+N{-}(CH_2)_4{-}CH{=}N^+H{-}CH{=}$(quinonoid intermediate) $\xrightarrow{② \; H^+}$ $H_3^+N{-}(CH_2)_4{-}CH_2{-}N^+H{=}CH{-}$(PLP) $\xrightarrow{③ \; HO^-}$ [$H_3^+N{-}(CH_2)_4{-}CH_2{-}N^+H_2{-}CH(OH){-}$(PLP)] $\xrightarrow{④}$ PLP (CHO) + $^+NH_3{-}(CH_2)_5{-}NH_3^+$

Steps 1. Decarboxylation of α-COO−
2. Protonate α-Carbon
3. & 4. Hydrolyze Schiff base

2c. **β-Decarboxylation. Form Schiff base steps 1–3, solution 2a**

① $^-O{-}C(=O){-}CH_2{-}C(H)(COO^-){-}N^{\oplus}H{=}CH{-}$(pyridoxal: P_iO, O^-, $N^{\oplus}H$, CH_3)

$\xrightarrow{-\alpha H^+}$

② $^-O{-}C(=O){-}CH_2{-}C(COO^-){=}N^{\oplus}H{-}C(H){=}$(ring: P_iO, O^-, $\ddot{N}H$, CH_3), H^+

$\xrightarrow[\text{C-4}']{\text{Protonate}}$

③ $^-O{-}C(=O){-}CH_2{-}C(COO^-){=}N^{\oplus}H{-}CH_2{-}$(ring: P_iO, O^-, $N H^{\oplus}$, CH_3)

$\downarrow -\beta\ COO^-$

④ $[\,^-CH_2{-}C(COO^-){=}N^{\oplus}H{-}CH_2{-}$(ring: P_iO, O^-, $N^{\oplus}H$, CH_3)$\,]$ ⇌ $[\,CH_2{=}C(COO^-){-}\ddot{N}H{-}CH_2{-}$(ring: P_iO, O^-, $N^{\oplus}H$, CH_3)$\,]$

⇌ Protonate β-Carbon

⑤ $CH_3{-}C(COO^-){=}N^{\oplus}H{-}CH(H){-}$(ring: P_iO, O^-, $N H^+$, CH_3)

$\xleftarrow[\text{C-4}']{-H^+}$

⑥ $CH_3{-}C(H^+)(COO^-){=}N^{\oplus}H{-}CH{=}$(ring: P_iO, O^-, $\ddot{N}H$, CH_3)

$\xrightarrow[\alpha C]{+H^+}$

⑦ $CH_3{-}C(H)(COO^-){-}N^{\oplus}H{=}C(H){-}$(ring: P_iO, O^-, $N H^+$, CH_3)

$\xrightarrow[\text{Schiff base}]{\text{Hydrolyze}}$ $CH_3{-}CH(NH_3^{\oplus}){-}COO^-$ + PLP

3. Pyridine nucleotide-dependent dehydrogenases typically have almost absolute specificity for either NAD^+ or $NADP^+$ as a cofactor. The specificity of the dehydrogenases provides a way to differentiate anabolic and catabolic processes. For example, the mitochondrial β-oxidation of fatty acids requires NAD^+ as an oxidant of the β-hydroxyacyl group to the β-ketoacyl group. In fatty acid biosynthesis, NADPH is used to reduce the β-ketoacyl group to a β-hydroxyacyl group and to reduce the β-enoyl group.

The ratios of $NADH/NAD^+$ and $NADPH/NADP^+$ may contribute to differential regulation of metabolic pathways.

4a.

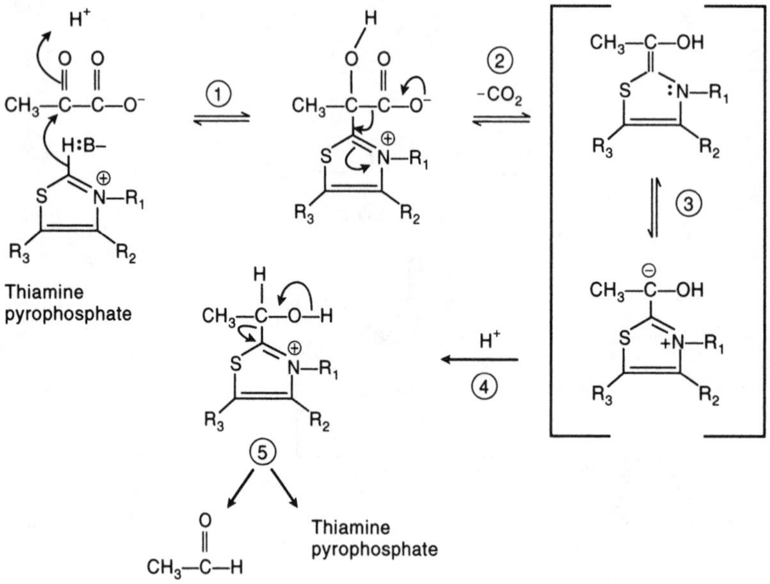

H^+
O O
CH_3—C—C—O^-
H:B–
⊕
S N—R_1
R_3 R_2
Thiamine pyrophosphate

①

H
O O
CH_3—C—C—O^-
⊕
S N—R_1
R_3 R_2

②
$-CO_2$

CH_3—C—OH
S :N—R_1
R_3 R_2

③

⊖
CH_3—C—OH
S +N—R_1
R_3 R_2

H^+
④

H
CH_3—C—O—H
⊕
S N—R_1
R_3 R_2

⑤

O
CH_3—C—H

Thiamine pyrophosphate

Steps
1. Deprotonation of TPP to ylid form, nucleophilic addition of ylid to α-ketogroup
2. Decarboxylation
3. Resonance stabilization
4. Protonation, elimination of TPP

4b.

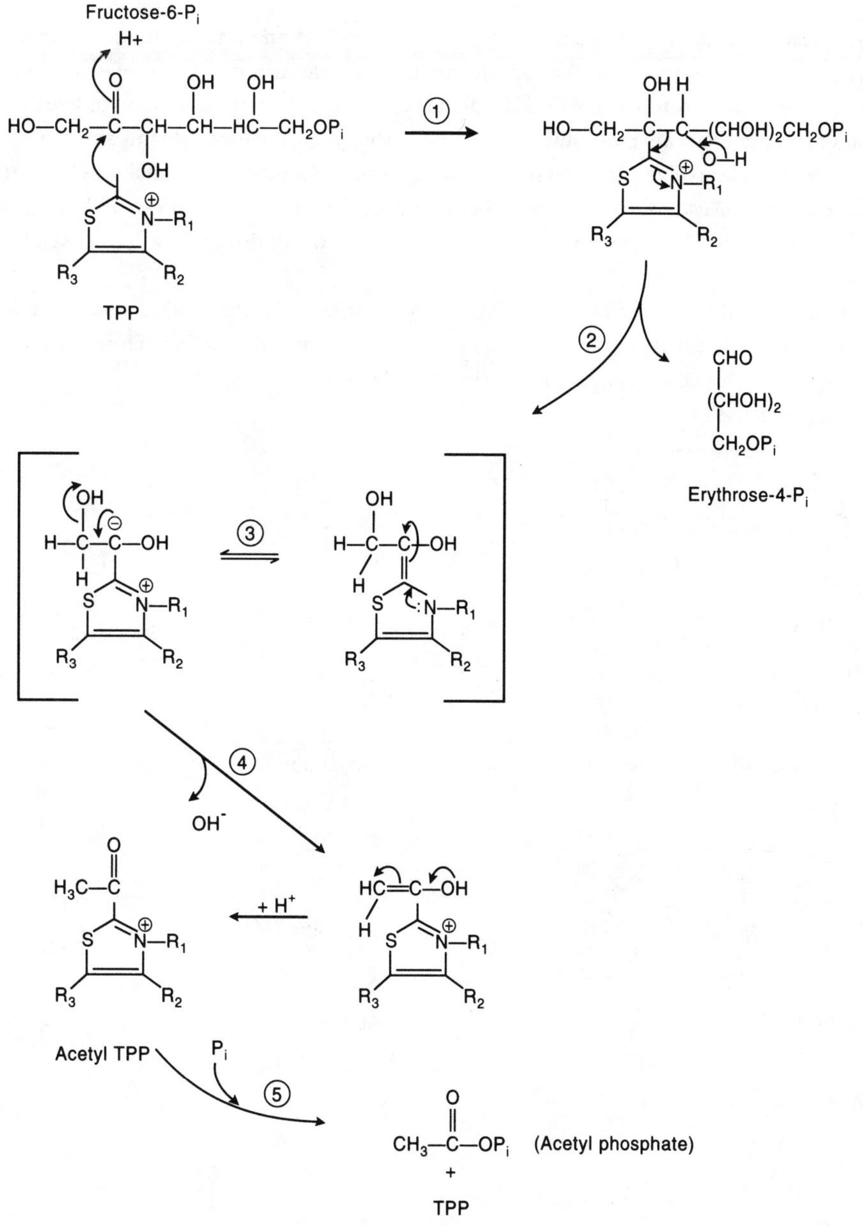

Fructose-6-P_i

H+

$HO{-}CH_2{-}C(=O){-}CH(OH){-}CH(OH){-}HC(OH){-}CH_2OP_i$

TPP

① →

$HO{-}CH_2{-}C{-}C(OH)(H){-}(CHOH)_2CH_2OP_i$

② → Erythrose-4-P_i (CHO, $(CHOH)_2$, CH_2OP_i)

③ ⇌

④ OH^-

$+ H^+$

Acetyl TPP

P_i ⑤ →

$CH_3{-}C(=O){-}OP_i$ (Acetyl phosphate)

+

TPP

Steps
1. Deprotonation to ylid form, nucleophilic addition of ylid to fructose-6-phosphate
2. Oxidation of C-3, release of erythrose-4-phosphate
3. Resonance stabilization of intermediate
4. Elimination of -OH from C-2
5. Transfer of acyl group to phosphate

5a.

$$H{-}CH_2{-}\overset{O}{\overset{\|}{C}}{-}SCoA$$

$$+$$

$$CHO{-}COO^- \dashrightarrow ({}^-OOC{-}CHOH{-}CH_2{-}\overset{O}{\overset{\|}{C}}{-}SCoA) \dashrightarrow CoASH + \text{Malate}$$

5b. Malate synthase catalyzes the Claisen condensation of acetyl-CoA and glyoxalate. In that reaction, an α-proton is abstracted and the carbanion thus formed adds to the aldehyde carbonyl group of glyoxalate. The pK of the α proton is decreased (the proton is more acidic) in the acetyl-CoA thioester compared to free acetate. α- carbanions of thioesters are more stable than if the carboxyl group were either in an oxygen ester or were unesterified. Abstraction of an α proton from acetate, forming the dianion, is much less likely to occur. (See Walsh, C. *Enzymatic Reaction Mechanisms.* San Francisco: W. H. Freeman, 1977, 759–62, *and* Higgins, M. J. P., J. A. Kornblatt, and H. Rudney in *The Enzymes,* Vol. VII, ed. P. Boyer. New York: Academic Press, 1972, 412–22.)

5c. Coenzyme A is a good leaving group and hydrolysis of the thioester is thermodynamically favorable. Hydrolysis of the thioester contributes approximately –10 kcal/mole to the overall reaction. Thus, the products malate and CoASH are more stable when compared with the malylthioester.

6a. Malate dehydrogenase

(NAD$^+$) + (L-malate) ⇌ (NADH) + (Oxaloacetate) + H^+

ADPR — C—NH$_2$ — B: — HO—C—O—H — CH_2 — COO^- — $C{=}O$

Hydride transfer from C-2 of malate to nicotinamide ring, loss of H+ from C-2—OH

6b. Malic enzyme

(NADP$^+$) + Malate —①→ (NADPH) + COO^-—C=O—CH_2—C=O—O^- + H^+ —②→ Pyruvate (COO^-—C=O—CH_3) + CO_2

ADPRP — C—NH$_2$ — B: — H—O—C—OH — —AH

Steps 1. Transfer of hydride from C-2 of malate to NADP$^+$.
2. β-decarboxylation of oxaloacetate.

7. Flavins (FMN or FAD) are fully reduced in biological systems upon accepting two electrons. The reduced flavins (FMNH or FADH) may be reoxidized by the loss of one or two electrons. The flavin free radical remaining after transfer of a single electron from the reduced flavin is resonance stabilized by the isoalloxazine ring system. Thus, flavins are frequently used in biological systems to oxidize molecules that obligatorily transfer two electrons (NADH, NADPH) and subsequently reduce molecules such as cytochromes and iron-sulfur

proteins that accept only one electron. In addition, the redox potential of the flavin bound to a flavoprotein positions it between redox potential of pyridine nucleotides and many iron-sulfur proteins or cytochromes.

8a. Acetyl-CoA carboxylase

(1) Carboxylation of biotin

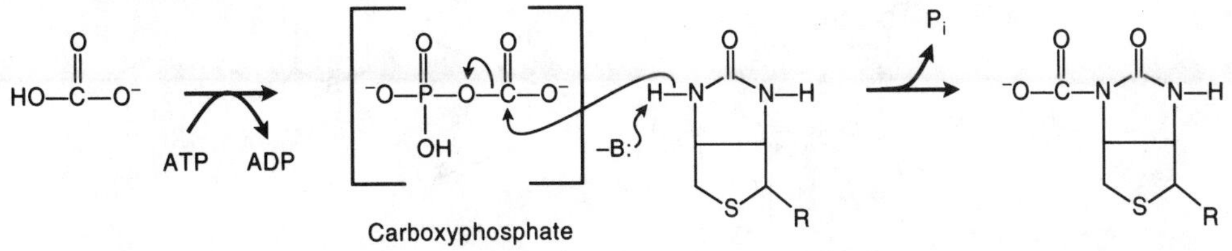

$HO-C(=O)-O^-$ + ATP → ADP; [$^-O-P(=O)(OH)-O-C(=O)-O^-$] + biotin (H–N, –B:) → P_i + $^-O-C(=O)-N$ (carboxybiotin, N–H, S, R)

Carboxyphosphate

(2) Proton abstraction from methyl of acetyl-CoA, addition of carbanion to carboxyl group of carboxybiotinyl enzyme.

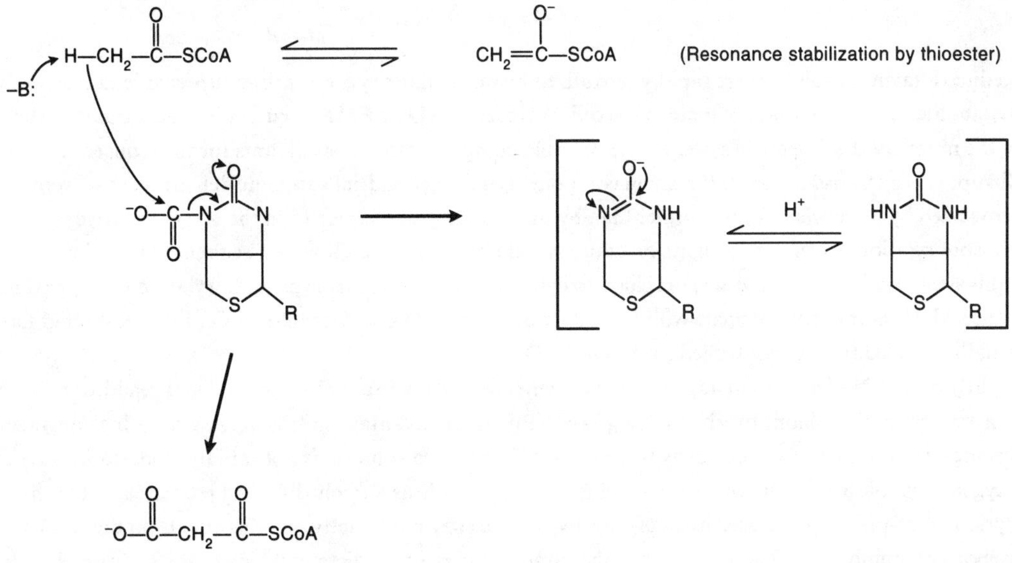

–B: $H-CH_2-C(=O)-SCoA$ ⇌ $CH_2=C(O^-)-SCoA$ (Resonance stabilization by thioester)

$^-O-C(=O)-N$ (carboxybiotin) → [O^-, N=, NH ⇌ H^+, HN, NH (biotin, S, R)]

$^-O-C(=O)-CH_2-C(=O)-SCoA$

Malonyl-CoA

8b. Transcarboxylase

(1) Transfer of β-carboxyl group from oxaloacetate to biotin

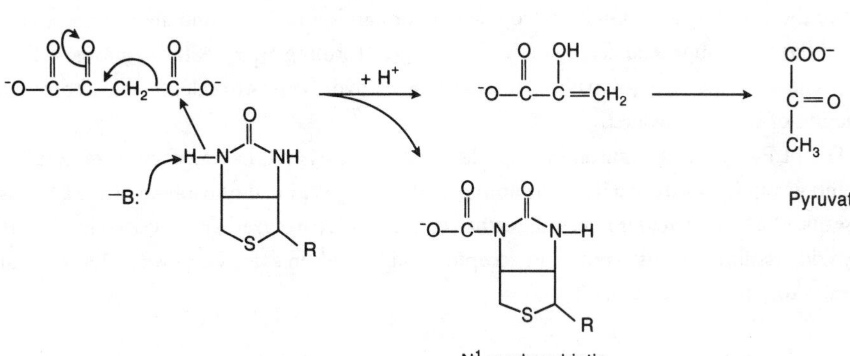

$^-O-C(=O)-C(=O)-CH_2-C(=O)-O^-$ + biotin (H–N, NH, –B:, S, R) → + H^+ → $^-O-C(=O)-C(OH)=CH_2$ → $COO^--C(=O)-CH_3$

Pyruvate

$^-O-C(=O)-N$ (N–H, S, R)

N^1- carboxybiotin

(2) Transfer of carboxyl group from carboxy biotin to propionyl-CoA

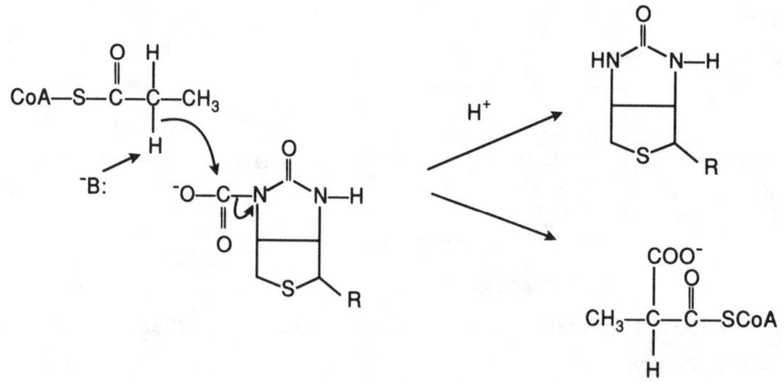

Methyl malonyl-CoA

9a. Reduced flavins in solution are rapidly reoxidized by molecular oxygen forming superoxide radical and H_2O_2, metabolites of oxygen that are toxic to the cell. Were free FAD or FMN used in biological transfer, for example in the mitochondrial electron system, oxygen could compete with other oxidants for the reduced flavin and disrupt energy transduction. If the flavin component of mitochondrial succinate dehydrogenase were directly reoxidized by molecular oxygen, rather than by ubiquinunt (see chapter 15 in the text), oxidative phosphorylation driven by succinate oxidation would be abolished. However, sequential transfer of the reducing equivalents via the electron transport chain to cytochrome oxidase drives phosphorylation of approximately 1.5 moles ADP per mole of succinate with the reduction of O_2 to H_2O rather than H_2O_2. Enzyme-bound flavins are usually shielded from uncontrolled oxidation by O_2.

9b. Tightly bound $NAD(P)^+$ is an advantage to enzymes catalyzing rapid H:– removal and readdition to a substrate in a stereospecific fashion. Freely diffusing NAD(P)H is an advantage in transferring reducing equivalents among catalytic sites of various dehydrogenases. NAD(P)H in solution is not rapidly oxidized by molecular oxygen as is soluble FADH or FMNH and is suited in its role as a freely diffusing redox reagent in the cell.

10. Pyridoxal-5′-phosphate is used as a cofactor by specific enzymes to activate substituents on the α- or β-carbons of amino acids. The specificity of the catalyst, the enzyme, dictates which of the activated bonds will specifically be altered. Thus, amino transferases utilize PLP to remove the α-amino group without α-decarboxylation as an interfering reaction.

11. The flavoprotein (amino acid oxidase) oxidizes the α-amino acid to an α-imino acid with reduction of the enzyme-bound flavin. In aqueous solution, the imino acid hydrolyzes with the release of the ammonium ion and formation of the α-keto acid. If not excreted, the ammonium ion may accumulate to toxic levels within the cell. The reduced flavoprotein is oxidized by molecular oxygen, forming H_2O_2. Although most cells have antioxidant enzymes (catalase, glutathione peroxidase) to scavenge the hydrogen peroxide, these defenses may be breached if the production of H_2O_2 is elevated.

The PLP-dependent transamination, the amino group is retained in organic molecules of low toxicity. The α-amino group is transferred from the amino acid to the pyridoxal phosphate at the active site of the amino transferase and the α-keto acid is released. Rather than being hydrolyzed and released into solution, the amino group of pyridoxamine is transferred to an acceptor α-keto acid, in some cases with ultimate transfer to a nontoxic excretory product (urea).

12a. **Oxidation of glucose-6-P_i by direct H: transfer to flavin**

D-glucose

FAD

FADH

D-gluconolactone

12b. **Formation of adduct of glucose and the flavin with electron rearrangement**

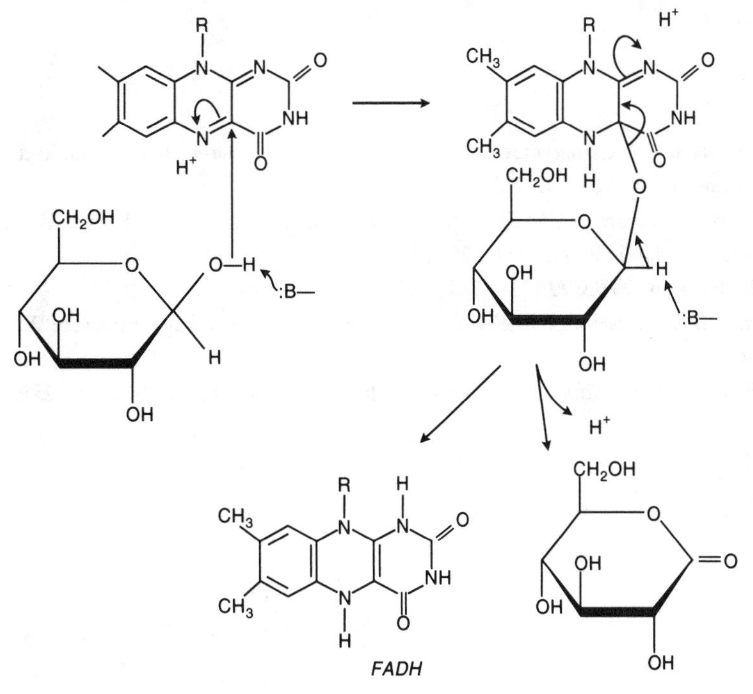

D-gluconolactone

13a. **Flavin. The enzyme catalyzes the oxidation of the α carbon, forming an α-imino acid with transfer of two electrons to the flavin ring. The reduced flavin on the flavoprotein subsequently is reoxidized by transfer of two electrons to molecular oxygen, forming H_2O_2.**

13b. **PLP. The cofactor is used to catalyze the β-elimination of the –OH group from serine, forming an imino acid. The imino acid hydrolyzes in solution to yield the α-keto acid plus ammonium ion.**

13c. **Biotin. ATP and bicarbonate are used in the carboxylation of enzyme-bound biotin. The carboxyl group of carboxybiotin is subsequently transferred to propionyl-CoA, forming methylmalonyl-CoA.**

13d. **Thiamine pyrophosphate. The enzyme transketolase catalyzes the transfer of C1-C2 as a unit from fructose-6-phosphate to thiamine pyrophosphate, with the release of the 4-carbon sugar, erythrose-4-phosphate. The 2-carbon unit bound to thiamine pyrophosphate is subsequently transferred to an acceptor, glyceraldehyde-3-phosphate. The product is the 5-carbon sugar, xylulose-5-phosphate.**

14. Polycyclic aromatic hydrocarbons, derived primarily from the pyrolysis of organic materials, are hydrophobic and dissolve in the lipid fraction of the cell. Their hydrophobicity in general makes their excretion from the body difficult. In addition, their lack of reactive functional groups precludes attachment of hydrophilic substituents to form water soluble adducts. The liver microsomal cytochrome P-450 (polysubstrate monoxygenase) system oxidizes the polycyclic aromatic hydrocarbon (PAH) by addition of an oxygen atom, derived from O_2, to a double bond, to form an epoxide. Hydrolysis of the epoxide, catalyzed by epoxide hydrolase, yields hydroxyl groups (dihydrodiol), thus generating functional groups that accept glucuronic acid from UDP-glucuronic acid. The glucuronide adduct is significantly more hydrophilic than either the unmetabolized PAH or the hydroxylated PAH and is subsequently excreted. (See Grover, P. L., *Xenobiotica.* 16:915–931 (1986)).

15. NAD^+ is cleaved at the N-glycosyl bond between nicotinamide and the ribose ring. The cleavage of this bond releases approximately 8 kcal/mole and is thermodynamically favorable. Adenosine diphosphate ribose is transferred to the acceptor, and nicotinamide and a proton are released. ADP-ribosylation or polyADP-ribosylation is a posttranslational modification of the acceptor protein in which NAD^+ is consumed as a substrate in the reaction, rather than functioning as a cofactor (1). ADP- or polyADP-ribosylation of several proteins has been documented, including polyADP-ribosylation of nuclear proteins and ADP-ribosylation of elongation factor-2, catalyzed by the toxin of *Corynebacterium diphtheriae* (2,3).

References

1. Ueda, K., in *Pyridine Nucleotide Coenzymes* Vol. II, Part B. eds. D. Dolphin, R. Poulson, and O. Avramovic'. New York: J. Wiley and Sons, 1987; 549–97.
2. Pekala, P. H., and B. M. Anderson, in *The Pyridine Nucleotide Coenzymes,* eds. J. Everse, B. Anderson, and K–S You. New York: Academic Press, 1982; 325–377.
3. Gaal, J. C., and C. K. Pearson, *TIBS. 11:*171–175 (1986).
4. Walsh, C., in *Aspartate β-Decarboxylase in Enzyme Reaction Mechanisms.* San Francisco: W. H. Freeman and Co., 1970; 811-814.
5. Boeker, E. A., and E. E. Snell, in "Amino Acid Decarboxylase" Academic Press, 1972; 217-253.

12 Metabolic Strategies

Problems

1. Explain why thermodynamics and kinetics are each important in life processes.

2. Explain why each of the following statements is false in terms of efficient metabolic regulation.
   (a) Most enzymes operate *in vivo* near V_{max}.

   (b) End-product inhibition usually occurs at the last or next-to-last enzyme in a metabolic pathway.

   (c) Catabolic pathways tend to diverge from a single metabolite.

   (d) The enzymes regulated in a metabolic pathway usually exhibit simple Michaelis-Menten kinetics.

   (e) Energy charge is unimportant in regulation of anabolic sequences but is of primary importance in the regulation of catabolic sequences.

   (f) Enzymes that catalyze a sequence of reactions are rarely grouped in multienzyme complexes.

   (g) Compartmentalization of metabolic pathways is seldom a regulatory strategy the cell uses.

3. Catabolic and anabolic pathways often differ in the specific reductant used (NADH or NADPH) and in the use of an energy source. How is each of these factors a metabolic advantage?

4. How is it possible that both the glycolytic degradation of glucose to lactate and the reverse process, formation of glucose from lactate (gluconeogenesis), are energetically favorable?

5. What is the metabolic advantage of having the "committed step" of a pathway under strict regulation?

6. Theoretically, the reactions shown below constitute a futile cycle. Explain.

$$\text{Glucose} + \text{ATP} \xrightarrow{\text{Glucokinase}} \text{Glc-6-P} + \text{ADP}$$

$$\text{Glc-6-P} + \text{HOH} \xrightarrow{\text{glucose-6-phosphatase}} \text{Glucose} + \text{phosphate}$$

In the liver cell, the enzymes are spatially separated, glucokinase in the cytosol and glucose-6-phosphatase in the endoplasmic reticulum. Does this separation influence the futile cycle?

Solutions

1. Thermodynamics dictates the feasibility of a reaction and the direction a reaction proceeds to reach equilibrium but does not specify the rate of the reaction or the reaction pathway. Kinetics describes the rate at which a reaction approaches equilibrium. Although catalysts alter the reaction kinetics, they do not alter the equilibrium position of the reaction. The most thermodynamically favorable reaction is of no biological value if it is not adequately catalyzed. Therefore, both kinetics and thermodynamics must be considered in evaluating metabolic processes.

2a. The concentration of most metabolites measured under steady-state conditions in the cell usually does not exceed the K_m value. For an enzyme whose reaction can be described by simple Michaelis-Menten kinetics, the observed velocity, v, is 0.5 V_{max} if substrate concentration equals the K_m value. The velocity of most enzymes is likely significantly less than V_{max} *in vivo.*

2b. End-product inhibition usually occurs at the committed step in a metabolic pathway or at a branch point in the pathway. Regulation of the enzyme(s) at these steps in the metabolic pathway prevents the accumulation of intermediates in the pathway when the cell has limited demand for the end product. Regulation by individual metabolites at a branch point of the pathway inhibits their own production without simultaneously inhibiting the production of product from the other branch. Each product may cumulatively inhibit an enzyme catalyzing a precursor common to each end product.

2c. Catabolic pathways tend to be convergent rather than divergent. Metabolic convergence of precursors into common intermediates of a metabolic pathway provides an efficient route for the metabolism of a variety of metabolites by a limited number of enzymes. For example, the catabolism of glucose, fructose, and glycerol to the common intermediate glyceraldehyde-3-phosphate in the liver illustrates the convergence of the glycolytic pathway. Anabolic pathways tend to be divergent, with the synthesis of several end products from a common precursor.

2d. Enzymes that are regulated in metabolic pathways most frequently exhibit cooperative kinetic responses rather than hyperbolic response with respect to substrate concentration and are frequently responsive to allosteric regulation by products, energy charge, or concentration ratio of $NAD(P)H/NAD(P)^+$. The rate of enzymatic activity over a narrow range of substrate concentration can be changed dramatically by allosteric activators or inhibitors. Such dramatic changes in velocity in response to small changes in substrate concentration is not observed with enzymes exhibiting Michaelis-Menten kinetics.

2e. Energy charge is a means of expressing the fraction of adenylate nucleotides that are high free energy compounds: ATP and ADP. The expression

$$E.C. = \frac{[ATP] + [ADP]/2}{[ATP] + [ADP] + [AMP]}$$

where E.C. represents the energy charge varying theoretically between 1 (all ATP) to 0 (all AMP). The energy charge is a metabolic signal for both anabolic (biosynthetic) and catabolic (degradative) metabolic pathways. Low energy charge signals the cell that a need for ATP formation exists, and pathways (glycolysis and Krebs cycle) leading to ATP formation are activated. Anabolic pathways that demand high ATP concentrations are inhibited at low energy charge. In the latter case, the ATP required to drive biosynthesis is in low supply. Conversely, increased energy charge inhibits pathways leading to ATP formation and activates anabolic pathways.

2f. Formation of multienzyme complexes is a strategy frequently used for efficient catalysis and control of metabolic pathways. Substrates diffuse shorter distances between active sites in multienzyme complexes than if the enzymes were not organized. Frequently, intermediates covalently bound to cofactors (e.g., lipoamide, biocytin) are moved among active sites within the complex, effectively trapping intermediates in the complex and increasing the concentration of substrates at the enzyme active site.

2g. Separation of catabolic and anabolic pathways diminishes the likelihood of futile cycling of metabolites. Enzymes catalyzing the β-oxidation of fatty acids are located in the mitochondrial matrix, whereas enzymes catalyzing the synthesis of palmitate are located in the cytosol.

3. Using different high-energy intermediates, or different portions of the same high-energy intermediate, and different forms of reductant allows the cell ample sites to control catabolism and anabolism differentially. Catabolic (degradative) pathways provide phosphorylated intermediates that are used to phosphorylate ADP, or provide reducing equivalents for oxidative phosphorylation of ADP. The ATP in turn may be used in biosynthesis. ATP may be cleaved at the phosphoric anhydride bond with the release of pyrophosphate and transfer of AMP (adenylate) to activate an intermediate in the biosynthetic pathway.

NAD(H) is found as the electron transfer cofactor central to catabolism of carbohydrates and pyruvate (lactate) via Krebs cycle. NADPH is frequently the reductant used in biosynthetic pathways. The two processes may be responsive to the relative amounts of each reductant, the amounts of which are under separate control.

4. Glycolytic catabolism of glucose is a thermodynamically favorable process that results in the formation of lactate and net phosphorylation of ADP to form ATP. If gluconeogenesis were simply the reversal of glycolysis,

the process would be thermodynamically unfavorable. Although glycolysis and gluconeogenesis have some reactions in common, there are key steps in gluconeogenesis that allow the input of metabolic energy and in this way overcome the thermodynamic barriers. ATP (4 moles) and GTP (2 moles) are required per mole of glucose synthesized from lactate in the liver. The gluconeogenic pathway becomes thermodynamically favorable because of the ATP and GTP energy input.

5. The "committed step" in a reaction sequence steers the metabolite to a sequence of reactions whose intermediates usually have no other function in the cell. Control of the committed step prevents wasteful accumulation of these single-purpose intermediates and obviates the necessity of controlling each enzyme in a pathway. Enzymes that catalyze reactions just past a branch point in a pathway are likely inhibited by their respective products or the end product of the branch.

6. Futile cycles are series of metabolic reactions or pathways whose combined reaction is that of hydrolyzing high energy metabolic intermediates for no apparent metabolic advantage. The reactions

(1) Glucose + ATP $\rightarrow$ Glc-6-P_i + ADP (glucokinase)

(2) Glc-6-P_i + HOH $\rightarrow$ glucose + P_i (glucose-6-phosphatase)

in sum catalyze the hydrolysis of ATP.

(3) ATP + HOH $\rightarrow$ ADP + P_i (ersatz ATPase)

The net standard-state free energy change ($\Delta G^{o\prime}$) for reaction (3) is about −7.5 kcal/mole, and the equilibrium will strongly favor product formation.

If the glucokinase and glucose-6-phosphatase reactions were unregulated, rapid hydrolysis of ATP with release of heat energy would ensue. However, glucokinase, a liver cytosolic enzyme, and glucose-6-phosphatase, an enzyme located in the liver endoplasmic reticulum, are separated by a membrane that imposes a permeability barrier to Glc-6-P, a common metabolite. Hence, regulating the transport of Glc-6-P from the cytosol to the endoplasmic reticulum diminishes futile cycling.

13 Glycolysis, Gluconeogenesis, and the Pentose Phosphate Pathway

Problems

1. Suppose that you have isolated a facultative microorganism that you are growing anaerobically in a medium containing a carbohydrate. Explain, on the basis of your knowledge of metabolism, why each of the following statements about the fermentation is false.

   (a) The culture must be growing on glucose because bacteria ferment few other compounds.

   (b) The products of the fermentation must be more highly oxidized than the substrates, otherwise no energy is conserved.

   (c) The culture cannot be producing any CO_2.

   (d) Addition of fluoride to the anaerobic culture will not alter the ratio of 2-phosphoglycerate to phosphoenolpyruvate.

2. A yeast culture is fermenting glucose to ethanol. In order to ensure that the CO_2 released during fermentation is radio-labeled, what carbon(s) of glucose must be labeled with ^{14}C?

3. Assume that a mutant form of glyceraldehyde-3-phosphate dehydrogenase was found to hydrolyze the oxidized enzyme-bound intermediate with water rather than phosphate.

   (a) Write a chemical reaction that describes the hydrolysis, showing the structures of the products.

(b) What would be the effect, if any, on the ATP yield from glycolysis of glucose to lactate?

(c) What would be the effect of the mutation on an obligately aerobic microorganism?

4. Suppose that you are seeking bacterial mutants with altered triose phosphate isomerase (TPI). The organism of interest is known to use the glycolytic pathway with the production of lactate.

(a) Explain why the absence of TPI would be lethal to an organism fermenting glucose exclusively through the glycolytic pathway.

(b) Suppose that you have an organism that uses glycolysis and an oxidative pathway as energy sources. Mutants of that organism having only 10% of the TPI activity present in the wild-type cells grew slowly on glucose under anaerobiosis but grew faster aerobically. Explain the metabolic basis of the observation.

(c) You have constructed a plasmid that would direct the synthesis of dihydroxyacetone phosphate phosphatase and have introduced the plasmid into the mutant organism described in part (b). Predict whether the plasmid-bearing organism with an active DHAP phosphatase could grow on glucose or glycerol either anaerobically or aerobically. What are the metabolic considerations you used to make your predictions?

5. There are two sites of ADP phosphorylation in glycolysis. These processes are called substrate level phosphorylations. Arsenate (AsO_4^{3-}), an analog of phosphate, uncouples ATP formation resulting from glyceraldehyde-3-phosphate oxidation but not that resulting from dehydration of glycerate-2-phosphate. Explain.

6. The disaccharide sucrose can be cleaved by either of two methods:

$$\text{Sucrose} + H_2O \xrightarrow{\text{Invertase}} \text{glucose} + \text{fructose}$$

$$\text{Sucrose} + P_i \xrightarrow{\text{Sucrose phosphorylase}} \text{glucose-1-phosphate} + \text{fructose}$$

(a) Given that the $\Delta G^{\circ\prime}$ value for the invertase reaction is –7.0 kcal/mole, calculate the $\Delta G^{\circ\prime}$ value for sucrose phosphorylase. Assume that the $\Delta G^{\circ\prime}$ for hydrolysis of glucose-1-phosphate is –5 kcal/mole. Based on the calculated value of $\Delta G^{\circ\prime}$, calculate the equilibrium constant for sucrose phosphorylase at 25°C.

(b) Explain the metabolic advantage to the cell of cleaving sucrose with phosphorylase rather than with invertase.

7. What concentration of glucose would be in equilibrium with 1.0-mM glucose-6-phosphate, assuming that hexokinase is present and the concentration ratio of ATP to ADP is 5? Is it reasonable to expect an actively metabolizing cell to maintain the concentration of glucose necessary to sustain the concentration of glucose-6-phosphate at 1mM?

8. 2-Phosphoglycerate and phosphoenolpyruvate differ only by dehydration between C-2 and C-3, yet the difference in $\Delta G^{\circ\prime}$ of hydrolysis is about –12 kcal/mole. How does dehydration "trap" so much chemical energy?

9. Explain the metabolic role of alcohol dehydrogenase in yeasts that are (a) fermenting glucose to ethanol and (b) aerobically oxidizing ethanol.

10. Elevated pyruvate concentration inhibits the heart muscle lactate dehydrogenase (LDH) isoenzyme but not the skeletal muscle LDH isoenzyme.

(a) What would be the consequence of having only the heart isoenzyme form in skeletal muscle?

(b) Would there necessarily be a negative consequence if the skeletal muscle isoenzyme were the only LDH in heart muscle?

11. Beginning with pyruvate, show which reactions of gluconeogenesis introduce the four chiral centers into glucose.

12. We have stated that the equilibrium constant for the pyruvate kinase reaction is 10^6. Assume that the steady-state concentration of ATP is 2mM and of ADP is 0.2 mM. Calculate concentration ratio of pyruvate to PEP under these conditions. Does your calculation support or refute the assertion that pyruvate kinase reaction is metabolically irreversible?

13. In gluconeogenesis, the thermodynamic barrier imposed by pyruvate kinase is overcome by coupling two separate reactions for the synthesis of PEP from pyruvate.

(a) Write the two chemical reactions used to bypass the pyruvate kinase reaction.

(b) Calculate the overall $\Delta G°'$ of the two reactions you wrote in part (a). (Assume that GTP is the thermodynamic equivalent of ATP). What can you now surmise about the feasibility of PEP formation from pyruvate by this route?

14. Write a chemical reaction for the $NADP^+$-dependent oxidation of 6-phosphogluconate to ribulose-5-phosphate.

15. Fructose-2,6-bisphosphate is a potent activator of the liver phosphofructokinase (PFK-1) and a potent inhibitor of liver fructose-1,6-bisphosphate phosphatase (FBPase-1). Fructose-2,6-bisphosphate is the product of a second phosphofructokinase (PFK-2) and is hydrolyzed to fructose-6-phosphate by FBPase-2. The activities of PKF-2 and FBPase-2 reside on a single, bifunctional protein in liver. The bifunctional protein is under glucagon control imposed via cAMP. (H. -G. Hers, *Arch. Biol Med. Exp.* 18:243–251(1985).)

(a) Under what metabolic conditions would PKF-2 be active? FBPase-2?

(b) Gluconeogenesis in liver is stimulated by the hormone glucagon. The activity of PFK-2/FBPase-2 bifunctional enzyme under glucagon regulation shifts from an active PFK-2 to an inactive PFK-2. Inactivation of the PFK-2 alone would still not be adequate to stimulate gluconeogenesis sufficiently for the organism. Explain.

(c) cAMP-dependent phosphorylation of PFK-2/FBPase-2 not only inhibits PFK-2 but stimulates FBPase-2. Under these conditions, gluconeogenesis is sufficiently rapid to meet cellular demand. Explain.

(d) What would you predict as the relative activities of the following enzymes in the liver of a rat made diabetic through chemical means (administration of alloxan or streptozotocin)? PFK-2, FBPase-2, PFK-1, FBPase-1, pyruvate carboxylase, PEP carboxykinase.

(e) If the diabetic animal were treated with insulin, what changes would you predict in the activities of the liver enzymes cited in part (d)? In the concentration of cAMP?

16. Muscle pyruvate kinase (PK) responds hyperbolically to its substrate, PEP, but the liver form of the enzyme responds sigmoidally. Fructose-1,6-bisphosphate is an allosteric activator of liver pyruvate kinase, but it apparently has no effect on the muscle enzyme.

(a) If liver PK responded hyperbolically to PEP and were otherwise unregulated, how might gluconeogenesis be affected?

(b) What is the metabolic advantage of having the liver PFK activated by fructose-1,6-bisphosphate?

Solutions

1a. Bacteria are adept at fermenting numerous sugars in addition to glucose. The sugars are used as a carbon source for metabolic energy transduction and for biosynthesis.

1b. Glycolytic conversion of glucose to glyceraldehyde-3-phosphate ($Ga3P_i$) occurs without net oxidation or reduction. $Ga3P_i$ is oxidized to a thioester that remains covalently bound to the enzyme (Ga3P dehydrogenase) with the concomitant reduction of the electron transfer coenzyme, NAD^+, to NADH. The 3-phosphoglyceryl group is released from the enzyme by phosphorolysis and is subsequently metabolized to pyruvate. Pyruvate, the α-keto acid, is reduced by the NADH-dependent lactate dehydrogenase in order to resupply NAD^+ to Ga3PDH. Hence, there is no net oxidation, but the product (lactate) is of lower free energy than the substrate (glucose). The energy difference between product and substrate is used to drive the phosphorylation of ADP.

1c. No CO_2 is released upon conversion of glucose to lactate. However, the culture could also be decarboxylating pyruvate to acetaldehyde and CO_2. The acetaldehyde may subsequently be reduced by the NADH-dependent alcohol dehydrogenase to ethanol using the NADH from the glyceraldehyde-3-phosphate dehydrogenase reaction. As noted in part (b), pyruvate is the terminal electron acceptor in glycolysis. During fermentation, acetaldehyde is the terminal electron acceptor. In either case, NAD^+ is resupplied to the Ga3PDH to prevent inhibition of glycolysis at the $Ga3P_i$ dehydrogenase step.

1d. The reaction of fluoride with phosphate forms an inhibitor of enolase, the enzyme that catalyzes the dehydration of 2-phosphoglycerate to form phosphoenolpyruvate. Inhibition of enolase would increase the amount of 2-phosphoglycerate, but the amount of PEP would decline because of the pyruvate kinase activity. The phenomenon of increased substrate and diminished product concentrations at a specific site is characteristic of specific metabolic inhibitors.

2. The carboxyl group of pyruvate is lost as CO_2 during the pyruvate decarboxylase-catalyzed formation of acetaldehyde. The pyruvate carboxyl group is formed by the oxidation of glyceraldehyde-3-phosphate. The aldehyde (C-1) carbon is derived directly from the aldolase-dependent cleavage of the fructose-1,6-bisphosphate. In this cleavage, C-4 of glucose becomes C-1 of glyceraldehyde-3-P_i while C-3 of glucose becomes C-3 of dihydroxyacetone phosphate. DHAP is isomerized to $Ga3P_i$. In this isomerization, C-3 of DHAP (originally C-3 of glucose) becomes C-1 of Ga3P. Thus, labeling either C-3 or C-4 of glucose will ensure that label is released as CO_2 upon fermentation to ethanol. The labeling scheme shown here indicates the carbons of fructose-1,6-bisphosphate, DHAP, $Ga3P_i$ and ethanol numbered to correspond to the corresponding carbon of glucose.

Glc	FBP	DHAP	$Ga3P_i$	pyruvate	ethanol
C_1HO	$C_1H_2OP_i$	$C_1H_2OP_i$	C_4HO	$C_{3,4}OOH$	$HC_{2,5}OH$
HC_2OH	$C_2{=}O$	$C_2{=}O$	C_5HOH	$HC_{2,5}{=}O$	$C_{1,6}H_3$
HOC_3H	HOC_3H	C_3H_2OH	$C_6H_2OP_i$	$C_{1,6}H_3$	
HC_4OH	HC_4OH				
HC_5OH	HC_5OH				
C_6H_2OH	$C_6H_2OP_i$				

3a. Hydrolysis rather than phosphorolysis of the Ga3PDH-bound thioester releases 3-phosphoglycerate rather than 1,3-bisphosphoglycerate.

E—SH (NAD$^+$) + H—C(=O)—CHOH—CH_2OP_i —Thiohemiacetal formation→ E—S—C(OH)(H)—R (NAD$^+$)

NAD$^+$ → NADH (Oxidation)

H^+ + E—S—C(=O)—R (NAD$^+$) + OH^- —Hydrolysis→ E—SH (NAD$^+$) + COO^-—CHOH—CH_2OP_i

3-phosphoglycerate

E—SH represents glyceraldehyde—3—phosphate dehydrogenase

3b. There would be no net yield of ATP if glycolysis began with glucose. Conversion of glucose to fructose-1,6-bisphosphate (FBP) requires 2 moles of ATP per mole of FBP formed. Cleavage of FBP by aldolase yields the equivalent of 2 moles of glyceraldehyde-3-phosphate ($Ga3P_i$). The oxidation and phosphorolysis of $Ga3P_i$ by Ga3PDH is one of the two phosphate activation steps in glycolysis. The dehydration of 2-phosphoglycerate by enolase to form PEP is the second phosphate-activation step. Metabolism of the two moles of $Ga3P_i$ to pyruvate yields 4 moles of ATP, a net 2 moles per glucose. Hydrolysis rather than phosphorolysis of the enzyme-bound thioester at the Ga3PDH step therefore abolishes net ATP yield from glycolysis.

3c. Considerably more free energy, and greater potential for ATP formation, is available from the aerobic metabolism of glucose compared with the anaerobic glycolysis of the sugar. Oxidation of the pyruvate formed in glycolysis should yield sufficient ATP for growth of the aerobe, even if the net ATP yield from glycolysis was zero.

4a. Triose phosphate isomerase deficiency would inhibit conversion of DHAP to Ga3P and would cause accumulation of DHAP, preventing half of the glucose molecule (C1-C3) from being metabolized through the remainder of the glycolytic pathway. There would be a recovery of only 2 of the possible 4 moles of ATP from glucose, resulting in no net formation of ATP. In addition, DHAP, a product of the aldolase reaction, would likely reverse the aldolase (reaction) and eventually inhibit glycolysis. Either result would be lethal to a cell whose only energy source was glycolysis.

4b. The small amount of TPI activity would likely allow glycolysis to proceed slowly, but low energy (ATP) level will limit the growth rate under anaerobiosis. However, the yield of ATP is significantly greater when the pyruvate, formed during glycolysis, is oxidized to CO_2 and H_2O. Hence, the growth rate of the mutant should be correspondingly greater under aerobic growth conditions but not as great as the wild type.

4c. Cells expressing DHAP phosphatase would likely not grow anaerobically if glycolysis of glucose to lactate were the only pathway for ATP formation. The combined activities of TPI and DHAP phosphatase would be predicted to deplete the pool of triosephosphate and the yield of ATP per glucose would likely be less than 1.

The cells might grow aerobically, depending on the competition among Ga3PDH, TPI, and DHAP phosphatase for the triosephosphate pool.

Glycerol can be phosphorylated to α-glycerolphosphate and oxidized to DHAP. The organism expressing DHAP phosphatase would likely not grow anaerobically on glycerol, but might grow aerobically for the reasons described above.

5. Arsenate is structurally similar to inorganic phosphate and substitutes for phosphate in the Ga3PDH reaction. The enzyme-bound thioester is cleaved by arsenolysis rather than phosphorolysis to form a mixed anhydride of the phosphoglycerate and arsenate. The mixed anhydride spontaneously hydrolyzes, regenerating the arsenate and forming 3-phosphoglycerate. Arsenate acts catalytically rather than stoichiometrically in dispelling the energy of the thioester bond, and the substrate-level formation of ATP is bypassed because the phosphoanhydride (1,3-bisphosphoglycerate) is not formed. Arsenate does not uncouple phosphorylation at the pyruvate kinase step because there is no labile organic intermediate with which arsenate can react to displace or replace the phosphate. The activated phosphate at C-2 of 2-phosphoglycerate was derived initially from ATP, not orthophosphate.

6a. Consider the reactions:

$$\text{H}_2\text{O} + \text{Sucrose} \rightarrow \text{fructose} + \text{glucose} \quad \Delta G^{\circ\prime} = -7 \text{ kcal/mole}$$
$$\text{Glucose} + \text{P}_i \rightarrow \text{Glucose-1-P}_i + \text{H}_2\text{O} \quad \Delta G^{\circ\prime} = +5 \text{ kcal/mole}$$
$$\text{Sucrose} + \text{P}_i \rightarrow \text{Glucose-1-P}_i + \text{fructose} \quad \Delta G^{\circ\prime} = -2 \text{ kcal/mole}$$

$\Delta G^{\circ\prime} = -2.3RT \log K'_{eq}$ where 2.3RT is 1.36 kcal/mole at 25° C.

$$-2 \text{ kcal/mole} = -1.36 \text{ kcal/mole} \log K'_{eq}$$
$$\text{Log } K'_{eq} = 1.47$$
$$K'_{eq} = 30$$

6b. Hexoses brought into the glycolytic pathway must be phosphorylated to provide the appropriate substrate for the glycolytic enzymes and to trap the sugar within the cell. Phosphorylation at the expense of ATP or group translocation at the expense of PEP are common methods to activate the sugar molecules. Sucrose phosphorylase uses the exergonic lysis of the glycosidic bond between the hemiacetal OH group of glucose and the hemiketal OH group of fructose to drive the endergonic phosphorylation of the hemiacetal C-1 OH group of glucose. Transfer of the phosphate to C-6, catalyzed by phosphoglucomutase, provides substrate for entry into the glycolytic pathway without addition of ATP. The net ATP yield will therefore be 3 rather than 2 moles of ATP per mole of glucose derived from sucrose. The fructose can be phosphorylated by ATP and used in the glycolytic pathway.

7. If the $\Delta G^{\circ\prime}$ for the hexokinase reaction is –5 kcal/mole, the K'_{eq} is 4,800 at 25° C. The glucose concentration in equilibrium with 1 mM glc-6-P_i must be 40 nM, a concentration easily attainable in the cell. However, the K_m of hexokinase for glucose is in the range 10–100 μM. The glucose concentration in actively metabolizing cells must be several orders of magnitude greater than 40 nM for the cell to have reasonable hexokinase activity. At concentrations of substrate less than saturating, the enzymatic activity is dependent on both the concentration or relative activity of the enzyme and the substrate concentration.

8. The free energy available from a reaction depends on the energies of the products compared with the substrates. Dehydration of 2-phosphoglycerate "traps" phospho-(enol)pyruvate in the enolate form. The hydrolysis products of PEP are phosphate and the enol form of pyruvate, but (enol) pyruvate is significantly less stable that (keto) pyruvate and rapidly tautomerizes to the more stable keto form. The tautomerization drives the reaction strongly toward products, resulting in a larger free energy difference between substrate and product.

9a. Yeast cells anaerobically metabolize glucose to pyruvate with the formation of a net 2 moles of ATP per mole of glucose. Pyruvate is decarboxylated to acetaldehyde by the pyruvate decarboxylase, and the aldehyde group of acetaldehyde is subsequently reduced to ethanol. The alcohol dehydrogenase-catalyzed reduction of acetaldehyde regenerates NAD^+ for Ga3PDH and thus prevents the supply of the oxidized cofactor (NAD^+) from becoming rate limiting for the glycolytic pathway. Recall that in many organisms, pyruvate is reduced to lactate to regenerate NAD^+.

9b. Yeasts oxidizing ethanol must convert the 2-carbon metabolite to a substrate for oxidative metabolism. In addition, the concentrations of dicarboxylic acids in the Krebs cycle must be sufficient to ensure efficient oxidation of substrates. Ethanol is oxidized to acetaldehyde by the alcohol dehydrogenase, and the acetaldehyde is further oxidized and activated to acetyl-CoA. The acetyl-CoA is subsequently used both as a substrate for the Krebs cycle and as a carbon source in anaplerotic reactions to replenish Krebs cycle intermediates. Yeast cells use the glyoxalate bypass (see Chapter 14 in the text) to form malate, a Krebs cycle intermediate. Acetyl-CoA is oxidized in the Krebs cycle to provide metabolic energy.

10a. During anaerobic (oxygen-debt) exercise, when pyruvate is being formed in large quantity in the muscle, the heart type LDH would be inhibited by the substrate, pyruvate. Inhibition of Ga3PDH and subsequently of glycolysis would occur because of an inadequate supply of NAD^+ generated from the NADH-dependent reduction of pyruvate to lactate. The muscle isozyme of LDH is not inhibited by pyruvate and has a larger turnover number than does the heart LDH isozyme.

10b. There would not necessarily be negative consequences to the muscle type LDH in the heart. Heart muscle is an aerobic tissue and oxidizes pyruvate supplied in part from the oxidation of lactate taken from the blood stream. Were the muscle isozyme the only LDH in the heart, lactate from the bloodstream would be oxidized to pyruvate, and there could possibly be a larger steady-state level of pyruvate in the heart.

11. PEP formed by the PEP carboxykinase reaction is hydrated by enolase to form 2-phosphoglycerate (2-PGA) whose C-2 is a chiral center. Conversion of 2-PGA to 3-phosphoglycerate by phosphoglyceromutase, phosphorylation by phosphoglycerokinase, and reduction to $Ga3P_i$ by Ga3PDH retains the chirality at C-2 but introduces no additional chiral centers. Aldolase-catalyzed condensation of DHAP, formed from the $Ga3P_i$, with $Ga3P_i$ introduces chiral centers at C-3 and C-4 of the fructose-1,6-bisphosphate. The chiral center at C-5 arises from C-2 of $Ga3P_i$. The remaining chiral center at C-2 is introduced upon conversion of the Fru-6-P_i to Glc-6-P_i catalyzed by phosphoglucoisomerase.

12. Given the ratio of ATP/ADP and the K'_{eq} of 10^6, the equilibrium ratio of [Pyr]/[PEP] would be about 10^5. This calculation supports the metabolic irreversibility of the pyruvate kinase reaction.

13a. (1) Pyruvate carboxylase (Review biotin-dependent carboxylation in chapter 11 of text.)

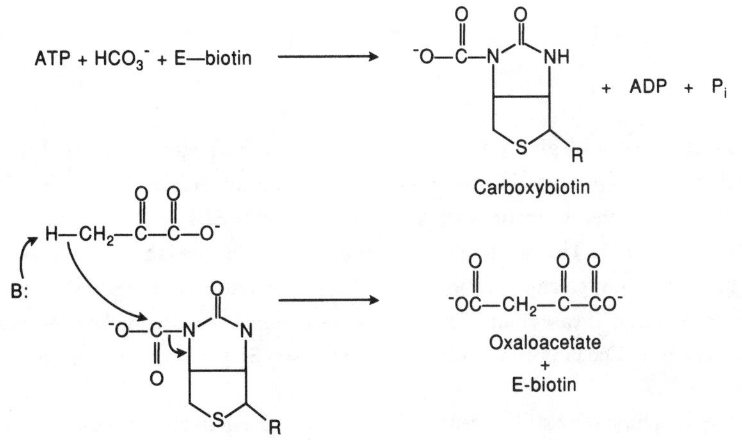

$ATP + HCO_3^- + E{-}biotin \longrightarrow$ Carboxybiotin $+ ADP + P_i$

Pyruvate + Carboxybiotin $\longrightarrow$ Oxaloacetate + E-biotin

(2) GTP-dependent PEPcarboxykinase (See review by Utter and Kollenbrander in *The Enzymes*, vol. 6 (ed. P. Boger), Academic Press, NY, pp. 136–54, 1972.)

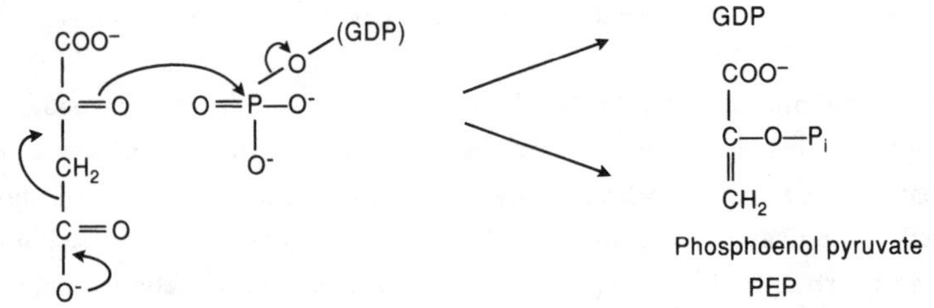

(GDP) → GDP

Phosphoenol pyruvate
PEP

13b. $\Delta G^{o\prime}$ overall is approximately –1.0 kcal/mole if the hydrolysis of PEP releases 14 kcal/mole and if the hydrolysis of ATP (GTP) releases 7.5 kcal/mole. If the overall free energy change for the coupled reactions is as calculated, the formation of PEP from pyruvate is metabolically feasible. Consider the following reactions and remember that the overall $\Delta G^{o\prime}$ is calculated from the sum of the component reactions.

$$\text{Pyr} + \text{ATP} + CO_2 \rightarrow \text{OAA} + \text{ADP} + P_i$$
$$\text{OAA} + \text{GTP} \rightarrow \text{PEP} + \text{GDP} + CO_2$$
$$\text{Pyr} + \text{ATP} + \text{GTP} \rightarrow \text{PEP} + \text{ADP} + \text{GDP} + P_i$$

but

$$\textbf{ATP} \rightarrow \textbf{ADP} + P_i \quad \Delta G^{o\prime} = -7.5 \text{ kcal/mole}$$
$$\textbf{GTP} \rightarrow \textbf{GDP} + P_i \quad \Delta G^{o\prime} = -7.5 \text{ kcal/mole}$$
$$\textbf{Pyr} + P_i \rightarrow \textbf{PEP} \quad \Delta G^{o\prime} = +14 \text{ kcal/mole}$$

$$\text{Pyr} + \text{ATP} + \text{GTP} \rightarrow \text{PEP} + \text{ADP} + \text{GDP} + P_i \; \Delta G^{o\prime} = -1 \text{ kcal/mole}$$

14.

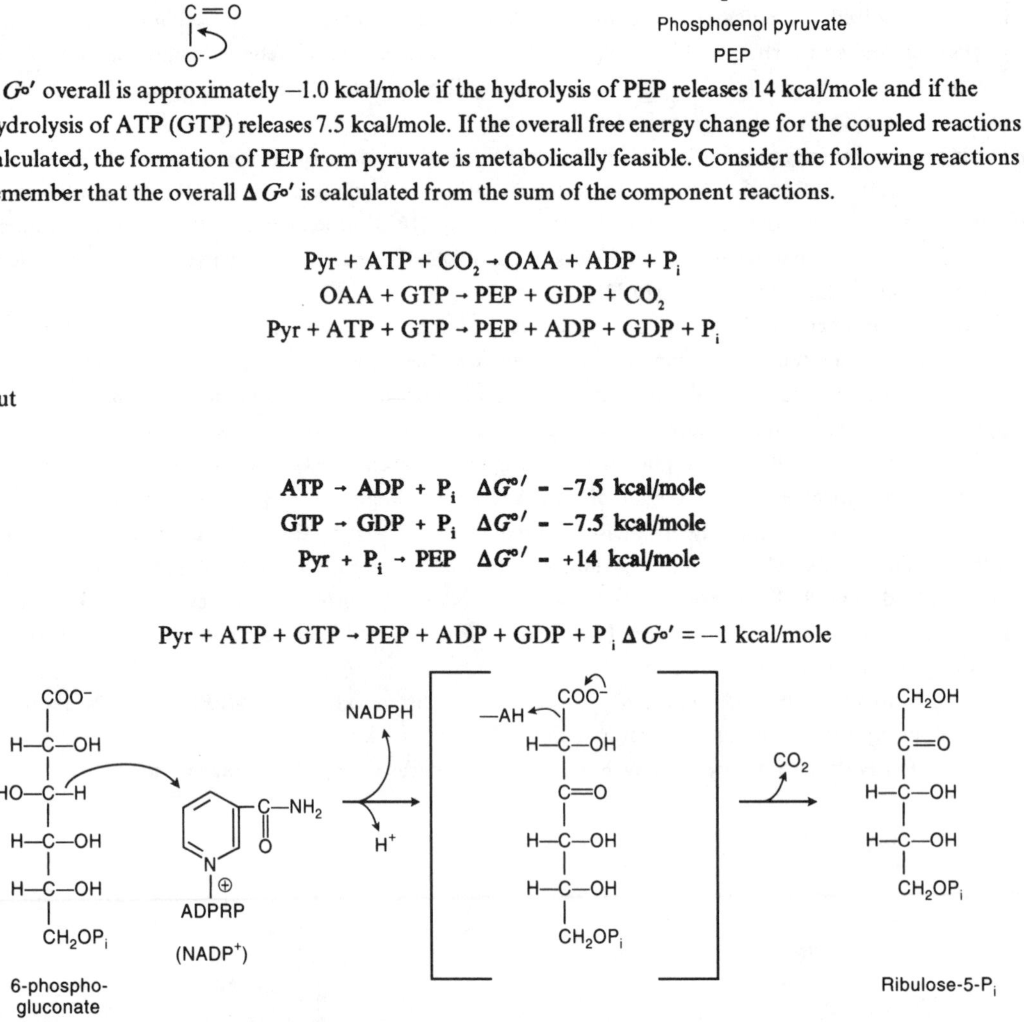

NADPH, ADPRP, $(NADP^+)$, H^+, CO_2

6-phospho-gluconate → Ribulose-5-P_i

15a. PFK-2 is active when the insulin/glucagon ratio is high (for example, when the animal is fed a carbohydrate-rich meal). FBPase-2 will be active when the insulin/glucagon ratio is low (blood glucose is low). Thus, the level of fructose-2,6-bisphosphate, the activator of PFK-1 and inhibitor of FBPase-1, is increased when the insulin/glucagon ratio is high. The rate of glycolysis is increased, whereas the rate of gluconeogenesis is diminished. When the insulin/glucagon ratio is low, PFK-2 is inactive but FBPase-2 is active. The level of fructose-2,6-bisphosphate decreases and PFK-1 is much less active because the level of its activator (fructose-2,6,-bisphosphate) is low. The inhibition of FBPase-1, imposed by fructose-2,6-bisphosphate, is relieved and gluconeogenesis occurs.

15b. If fructose-2,6-bisphosphate were still present, although not being actively synthesized, it would continue to stimulate PFK-1 and inhibit FBPase-1. Thus, simply halting the synthesis of the PFK-1 activator and FBPase-1 inhibitor may not be sufficient to inhibit glycolysis and activate gluconeogenesis.

15c. Activation of FBPase-2 causes the hydrolysis of fructose-2,6-bisphosphate and alleviates inhibition of FBPase-1 and prevents activation of PFK-1. Glycolysis is effectively inhibited due to inhibition of the rate controlling enzyme, PFK-1, whereas gluconeogenesis is activated by alleviation of the inhibition of FBPase-2.

15d. The insulin level will markedly decline so that the insulin/glucagon ratio will decline. As indicated in part (a), FBPase-2, FBPase-1, pyruvate carboxylase, and PEP-carboxykinase activities should increase. PFK-2 and PFK-1 activities will decline.

15e. The administered insulin will restore the insulin/glucagon ratio. The cAMP level in the liver cell is predicted to decrease, as are the activities of FBPase-2, FBPase-1, pyruvate carboxylase, and PEP-carboxykinase as discussed in part (a).

16a. Unregulated hepatic PK theoretically could become part of a futile cyle.

$$\mathrm{PEP + ADP \rightarrow Pyr + ATP}$$
$$\mathrm{Pyr + CO_2 + ATP \rightarrow oxaloacetate_{mito} + ADP + P_i}$$
$$\mathrm{OAA_{mito} + NADH \rightarrow L\text{-}malate_{mito} + NAD^+}$$
$$\mathrm{L\text{-}malate_{mito} \rightarrow L\text{-}malate_{cyto}}$$
$$\mathrm{L\text{-}malate_{cyto} + NAD^+ \rightarrow OAA_{cyto} + NADH}$$
$$\mathrm{OAA_{cyto} + GTP \rightarrow PEP + GDP + CO_2}$$

Net: GTP → GDP + P_i plus formation of cytosolic NADH at the expense of mitochondrial NADH.

16b. Activation by fructose-1,6-bisphosphate decreases $S_{0.5}$ for PEP, and increases PK activity at a given PEP concentration. During gluconeogenesis, the fructose-1,6-bisphosphate concentration should diminish due to the hydrolytic activity of the FBPase-1. The low FBP concentration, coupled with the elevated ATP levels, could inhibit the hepatic pyruvate kinase. (See Cardenas, J. M. in *Methods Enzymol.* 90:140–49, 1982.)

14 The Tricarboxylic Acid Cycle

Problems

1. Using your knowledge of metabolism, determine whether the following statements are true or false and explain the reasoning behind your decision.
   (a) Dihydrolipoamide dehydrogenase catalyzes the only oxidation-reduction reaction in the pyruvate dehydrogenase complex.

   (b) Hydrolysis of the thioester bond of acetyl-CoA yields insufficient energy to drive phosphorylation of ADP.

   (c) The methyl group of each acetyl-CoA molecule entering the TCA cycle is derived from the methyl group of pyruvate.

   (d) Even if aconitase were unable to discriminate between the two ends of the citrate molecule, the CO_2 released would still come from the oxaloacetate rather than the acetyl-CoA substrate of the citrate synthase reaction.

   (e) Malate cannot be converted to fumarate because the TCA cycle is unidirectional.

2. Assume that you have a buffered solution containing pyruvate dehydrogenase and all the enzymes of the TCA cycle but none of the cycle intermediates.
   (a) If you add 3 µmoles each of pyruvate, CoASH, NAD^+, GDP, and P_i, how much CO_2 will be evolved? What other products are formed?

(b) In addition to the reagents in (a), you add 3 μmoles each of the TCA cycle intermediates. How much CO_2 is evolved? Explain.

(c) If you were to add to the system described in (a) an electron acceptor that reoxidized NADH, would there be increased CO_2 evolution? Why or why not?

(d) Explain the effect on CO_2 evolution of adding the NADH-reoxidizing system to the system described in (b), assuming that you also added excess GDP and P_i?

3. Suppose that you supply [1-^{14}C]-labeled glucose to an aerobic bacterial culture and rapidly isolate the intermediates of the glycolytic and TCA pathways. Indicate which of the carbons will be initially labeled in the following intermediates, and explain the reasoning behind your answer. Assume that the activity of pyruvate carboxylase can be ignored in formulating your answer.

(a) Glyceraldehyde-3-phosphate

(b) Acetyl-CoA

(c) Citrate

(d) α-Ketoglutarate

(e) Succinate

(f) Malate

4. What would you expect to be the metabolic consequences of the following mutations in yeast?

(a) Inability to synthesize malate synthase.

(b) Pyruvate carboxylase that is not activated by acetyl-CoA.

(c) Pyruvate dehydrogenase that is inhibited by acetyl-CoA more strongly than is the wild-type enzyme.

5. The substrate hydroxypyruvate

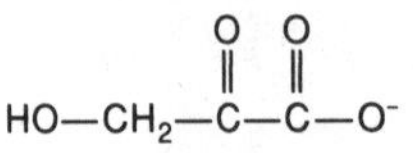

$$HO{-}CH_2{-}\overset{O}{\overset{\|}{C}}{-}\overset{O}{\overset{\|}{C}}{-}O^-$$

Hydroxypyruvate

is metabolized to pyruvate in a five-step process requiring the four intermediates whose structures are shown below.

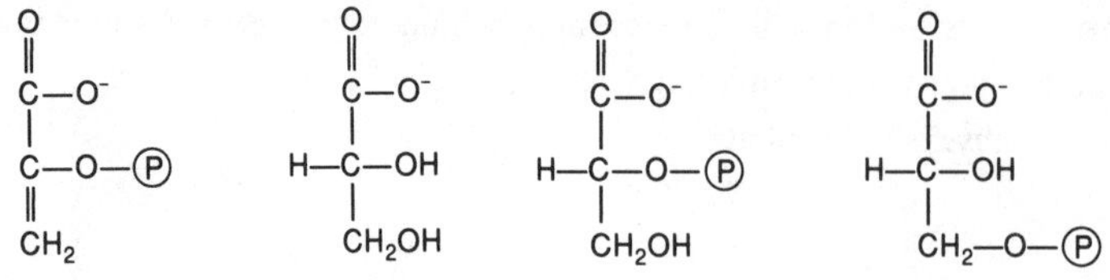

A: $^-O{-}\overset{O}{\overset{\|}{C}}{-}C(O{-}Ⓟ){=}CH_2$

B: $^-O{-}\overset{O}{\overset{\|}{C}}{-}CH(OH){-}CH_2OH$

C: $^-O{-}\overset{O}{\overset{\|}{C}}{-}CH(O{-}Ⓟ){-}CH_2OH$

D: $^-O{-}\overset{O}{\overset{\|}{C}}{-}CH(OH){-}CH_2{-}O{-}Ⓟ$

The letter designating the intermediate does not necessarily reflect the order in which it is used. The metabolic conversion requires NADH and catalytic quantities of both ATP and ADP. Assume that the pathway begins with an NADH-mediated reaction.

(a) Designate the order in which the intermediates are used in the metabolism of hydroxypyruvate to pyruvate.

(b) Write an overall equation for the metabolism of hydroxypyruvate to pyruvate.

(c) Name each of the intermediates (A–D) and indicate which are intermediates in the glycolytic pathway.

(d) Explain why only catalytic rather than stoichiometric amounts of ADP and ATP are required in the pathway.

6. Under anaerobic conditions, *E. coli* synthesizes an NADH-dependent fumarate reductase rather than succinate dehydrogenase, the flavoprotein that oxidizes succinate to fumarate.

   (a) Write an equation for the reaction catalyzed by fumarate reductase.

   (b) NADH produced by the glyceraldehyde-3-phosphate dehydrogenase reaction is reoxidized by reducing an organic intermediate. Rather than reduce pyruvate to lactate, anaerobic *E. coli* utilize the fumarate reductase. However, under anaerobiosis, the activity of α-ketoglutarate dehydrogenase is virtually nonexistent. Show how fumarate is formed, using reactions beginning with PEP and including the necessary TCA cycle enzymes. (Spiro, S., and J. R. Guest, *TIBS.* 16:310–314 (1991).)

   (c) What is the metabolic advantage to anaerobic *E. coli* in using the fumarate reductase pathway rather than lactate dehydrogenase to reoxidize NADH?

7. Cite some metabolic advantages in having lipoic acid covalently bound to the acyltransfer enzymes of pyruvate dehydrogenase and α-ketoglutarate dehydrogenase complexes.

8. Consider the glyceraldehyde-3-phosphate dehydrogenase-phosphoglycerokinase enzymes of glycolysis and the succinate thiokinase of the TCA cycle. Compare the mechanisms of incorporation of inorganic phosphate into the respective nucleoside diphosphates.

9. Explain why a genetic deficiency of dihydrolipoamide dehydrogenase decreases the oxidation rate of pyruvate and α-ketoglutarate in mitochondria. Assume that each of the dehydrogenase complexes were assayed with their respective substrates. What effect would the defect have in ATP production in the mitochondria?

10. The pyruvate dehydrogenase complex may have been regulated by phosphorylation of any one of the three different enzymes in the complex, yet regulation occurs on the first enzyme of the complex. How is regulation of the complex consistent with the regulation observed in metabolic pathways whose enzymes are not physically associated?

11. *E. coli* growing on acetate as the sole carbon source partition isocitrate between the TCA cycle and the glyoxalate bypass by reversible phosphorylation of isocitrate dehydrogenase (ICDH). Phosphorylated ICDH is catalytically inactive. Explain how decreasing the ICDH activity effectively shifts isocitrate to the glyoxalate bypass. (Nimmo, H. G., *TIBS.* 9:475–478 (1984).)

12. A bifunctional regulatory enzyme (ICDH kinase/phosphatase) catalyzes the phosphorylation/dephosphorylation of ICDH and is under metabolic control. Predict the relative activities of the kinase and phosphatase function of the bifunctional enzyme when *E. coli* is using acetate as the sole carbon source under each of the following conditions.

(a) Oxalacetate is being diverted to an anabolic pathway.

(b) AMP levels are high.

(c) The culture is shifted to a glucose-containing medium.

13. Plants effectively accomplish the glyoxalate bypass without resort to phosphorylation of the mitochondrial ICDH. How is partitioning of the glyoxalate bypass substrates and products accomplished in plants? (Tolbert, N. E., *Ann. Rev. Biochem.* 50:133–157 (1981).)

14. Net movement of acetyl-CoA from the mitochondria to the cytosol is an ATP-consuming process. When the acetyl-CoA transfer is accompanied by formation of NADPH at the expense of NADH, the reaction stoichiometry is as follows:

$$2\ \text{ATP} + \text{NADH} + \text{NADP}^+ + \text{acetyl-CoA}_{mit} \rightarrow \text{NAD}^+ + \text{NADPH} + \text{acetyl-CoA}_{cyt} + 2\ \text{ADP} + 2\ \text{P}_i$$

The stoichiometry belies the number of reactions involved in the separate metabolic compartments.

(a) Construct a metabolite flow diagram that illustrates the net movement of acetyl-CoA from the mitochondria to the cytosol and that accommodates all the intermediate steps.

(b) For each of the enzyme-catalyzed reactions, indicate which cofactors or coenzymes (if any) are required.

(c) Explain why there is no net consumption of TCA cycle intermediates in the reaction sequence you derived in part (a).

(d) What are some sources of cytoplasmic NADPH other than $NADP^+$-malic enzyme?

15. Although there is no net synthesis of glucose from acetyl-CoA in mammals, acetyl-CoA has two major functions in gluconeogenesis. Explain the functions of acetyl-CoA in the synthesis of glucose from lactate in mammalian liver.

Solutions

1a. False. Pyruvate dehydrogenase catalyzes reduction of the lipoamide disulfide concomitantly with oxidation and transfer of the hydroxyethyl group from thiamine pyrophosphate. The hydroxyethyl group is oxidized to form a thioester with one sulfhydryl group of lipoamide. The acetyl group is subsequently transferred to CoASH, forming the thioester adduct, acetyl-CoA. During the oxidation of the α-keto carbon, the disulfide (oxidized form) of lipoamide is reduced to the dithiol form. Dihydrolipoamide dehydrogenase, a flavoprotein, catalyzes oxidation of dihydrolipoamide and reduction of NAD^+ to NADH.

1b. False. Hydrolysis of acetyl-CoA thioester should yield as much free energy as succinyl-CoA hydrolysis. Succinate thiokinase catalyzes the substrate level phosphorylation of GDP at the expense of succinyl-CoA hydrolysis. However, in the TCA cycle, there is no enzymatic pathway to couple the hydrolysis of the acetyl-CoA to the activation of orthophosphate and subsequent transfer of the activated phosphate to ADP. Succinate thiokinase provides such an enzymatic pathway to couple the energy of succinyl-CoA hydrolysis to GDP.

1c. False. The TCA cycle is a versatile pathway for the oxidation of various substrates, including carbohydrates, fatty acids, and lipids. Thus, the methyl group of acetyl-CoA could be derived from pyruvate, from β-oxidation of long chain fatty acids, or from amino acid catabolism.

1d. False. The CO_2 molecule released by oxidative decarboxylation of isocitrate (ICDH) is derived from the carboxyl group of oxaloacetate with which the acetyl-CoA was condensed. However, the CO_2 released from α-ketoglutarate does depend on discrimination between the ends of the citrate molecule. If the aconitase reaction were random, half the CO_2 would arise from an oxaloacetate carboxyl group and half from the acetate carboxyl group. Such is not the case because aconitase discriminates between the two arms of citrate.

1e. False. Malate can easily be dehydrated to fumarate by reversal of the fumarase reaction.

2a. There will be stoichiometric conversion of 3 μmoles pyruvate to acetyl-CoA with release of 3 μmoles CO_2 and production of 3 μmoles of NADH. However, there will be no further metabolism of the acetyl-CoA because the TCA cycle intermediates are lacking. The amounts of GDP and P_i will remain unchanged.

2b. Upon addition of the TCA cycle intermediates, acetyl-CoA will condense with oxaloacetate to form citrate. Aconitase will catalyze the formation of isocitrate, in equilibrium with citrate. However, isocitrate dehydrogenase will be inactive because little, if any, NAD^+ will be available. The NAD^+ initially added to the mixture was reduced stoichiometrically by the pyruvate dehydrogenase activity. Hence, there should be no detectable increase in CO_2 released from the TCA cycle.

2c. Reoxidation of the NADH in the absence of TCA cycle intermediate(s) or in the absence of pyruvate carboxylase would have no affect on CO_2 evolution. The TCA cycle intermediates are lacking and there is no mechanism to replenish them in the system described in the problem.

2d. Reoxidation of the NADH will regenerate NAD^+ and alleviate the inhibition of the TCA cycle, and CO_2 will be evolved: 6 additional μmoles from the 3 μmoles of acetyl-CoA and some CO_2 from the citrate, isocitrate, and α-ketoglutarate that were added. These intermediates will be oxidized. GDP will be phosphorylated to GTP by the succinate thiokinase reaction.

3. a. Ga3P, C-3 formed from C-3 of dihydroxyacetone phosphate; b. Acetyl-CoA, methyl of acetate derived from C-3 of pyruvate; c. Citrate, C-2 derived from methyl of acetyl-CoA; d. α-Ketoglutarate, C-4; e. Succinate, C-2 (equivalent to C-3); f. Malate, both C-2 and C-3 because fumarase hydrates the double bond of fumarate by adding the –OH randomly to either carbon of the double bond.

4a. Without malate synthase, the yeast would be unable to grow on 2-carbon precursors as sole carbon source because TCA cycle intermediates could not be synthesized. The glyoxalate bypass provides a pathway for the net synthesis of malate from acetyl-CoA. Entry of malate, a dicarboxylate intermediate in the TCA cycle, would provide increased concentrations of each of the TCA cycle intermediates with subsequent increase in oxidation of acetyl groups. ATP formation by oxidative phosphorylation could then occur. In addition, the glyoxalate bypass provides a carbon source for synthesis of carbohydrate and precursors of other cellular constituents. The absence of malate synthase would adversely affect these reactions.

4b. Pyruvate carboxylase catalyzes the carboxylation of pyruvate to form oxaloacetate. Oxaloacetate thus formed enters the TCA cycle to replenish cycle intermediates. Pyruvate carboxylase is normally activated by acetyl-CoA; thus the need for oxaloacetate is tied to an increased supply of acetyl-CoA.Were the pyruvate carboxylase in the mutant less responsive to acetyl-CoA, it is possible that the TCA cycle activity would markedly diminish if the cycle intermediates were being used in biosynthetic pathways. In addition, the biosynthetic pathways (lipids, amino acids, and carbohydrates) dependent on TCA cycle intermediates would also be inhibited due to lack of metabolites.

4c. If the PDH were inhibited more strongly than usual by acetyl-CoA, one might suspect that acetyl-CoA concentration in the mitochondrial matrix would markedly decrease, in turn limiting activity of citrate synthase and diminishing TCA cycle activity. Hence the organism may become growth-limited because of lowered energy production and because of diminished concentrations of biosynthetic precursors supplied by the TCA cycle.

5a. Hydroxypyruvate → B → D → C → A → pyruvate

5b. Hydroxypyruvate + NADH + H^+ → pyruvate + NAD^+ + HOH

5c. (A) Phosphoenolpyruvate; (B) glycerate; (C) glycerate-2-phosphate; (D) glycerate-3-phosphate. A, C, and D are glycolytic intermediates.

5d. ATP is used by the glycerate kinase to phosphorylate glycerate (intermediate B); that step provides the phosphorylated substrate required by the phosphoglyceromutase. In the phosphoglyceromutase reaction there is a net transfer of phosphate from C-3 to C-2 of the glycerate moiety. The 2-phosphoglycerate is dehydrated by the catalytic action of enolase, and the product, PEP, is used by pyruvate kinase to phosphorylate ADP. Hence, the ATP used to phosphorylate glycerate is regenerated when pyruvate is formed from PEP.

6a. Fumarate + NADH + H^+ → succinate + NAD^+.

6b. Fumarate is formed from the oxidation of succinate, a TCA cycle intermediate formed from the decarboxylation of α-ketoglutarate. α-Ketoglutarate dehydrogenase is an integral enzyme in the TCA cycle and is active aerobic metabolism of acetyl-CoA. However, under anaerobic conditions in *E. coli*, there is little, if any, α-ketoglutarate dehydrogenase activity. Succinate, and therefore fumarate, are not produced by this route. Some reactions of the TCA cycle are reversible, however. For example, the reduction of oxaloacetate, formed by carboxylation of PEP pyruvate, yields L-malate, which may be dehydrated to fumarate. These reversible reactions form fumarate even if succinate dehydrogenase or α-ketoglutarate dehydrogenase reactions are blocked. (See Spiro, S., and J. R. Guest, *TIBS.* 16:310-314 (1991).) Consider the reactions

Phosphoenolpyruvate + CO_2 → oxaloacetate + P_i
(PEP carboxylase)

Oxaloacetate + NADH + H^+ → L-malate + NAD^+
(malate dehydrogenase)

L-Malate → fumarate + HOH
(fumarase)

Fumarate + NADH + H^+ → succinate + NAD^+
(fumarate reductase)

6c. In the reactions shown in part (6b), four reducing equivalents (two hydride groups) are transferred to carbon acceptors. Reduction of oxaloacetate to malate by malate dehydrogenase (MDH) and reduction of fumarate to succinate by fumarate reductase each requires hydride (or equivalent) transfer from NADH to the organic substrate. In the reduction of pyruvate to lactate via LDH only one hydride is used. Thus, two equivalents of NAD^+ are resupplied to glycolysis by the activities of MDH and fumarate reductase, whereas only one equivalent of NAD^+ is regenerated by LDH. Fumarate is one of the terminal electron acceptors used during anaerobic respiration in *E. coli.* However, each mole of PEP carboxylated is at the expense of 1 mole equivalent of ATP that could have been formed as a product of pyruvate kinase.

7. Lipoic acid covalently bound to the acyltransferase in the multienzyme complexes (a) provides a greater effective coenzyme concentration at the active site than if the cofactor must diffuse into and away from the complex and (b) decreases the diffusion path required for a product of one active site to move to the next catalytic center. At less than saturating concentrations of substrate, enzyme activity is dependent on both substrate and enzyme

concentrations. Hence, the effective increase in substrate concentration near the enzymatic active sites increases the rate of catalysis.

8. Glyceraldehyde-3-phosphate is bound to glyceraldehyde-3-phosphate dehydrogenase in a thiohemiacetal bond that is subsequently oxidized to an enzyme-bound thioester. Phosphorolytic cleavage of the thioester forms free enzyme, and the product, 1,3-bisphosphoglycerate (a mixed anhydride), diffuses from the enzyme. 1,3-Bisphosphoglycerate is substrate for the ADP-dependent phosphoglycerokinase whose products are 3-phosphoglycerate and ATP. In the reaction sequence described, the inorganic phosphate is activated by addition to an organic substrate.

   The formation and transfer of "activated phosphate" by the succinate thiokinase follows a route similar to that described except that an enzyme-bound phosphoryl group is involved. Succinyl-CoA is formed by α-ketoglutarate dehydrogenase from the oxidative decarboxylation of α-ketoglutarate. The succinate thiokinase (a) catalyzes phosphorolysis of the succinyl-CoA to succinylphosphate, a mixed anhydride. (b) The phosphate from the mixed anhydride is transferred to an imidazole side chain of an active site histidine, forming a phosphoramide. The phosphate is subsequently transferred from the phosphoramide on the enzyme to GDP. Each of the enzyme pairs (glyceraldehyde-3-phosphate dehydrogenase, phosphoglycerokinase and α-ketoglutarate dehydrogenase, succinate thiokinase) catalyze substrate-level phosphorylation.

9. The dihydrolipoamide dehydrogenase that is associated with the pyruvate dehydrogenase and with the α-ketoglutarate dehydrogenase is apparently produced by a single gene. Hence, a deficiency of dihydrolipoamide dehydrogenase affects the activity of each dehydrogenase complex. If the complexes were deficient in dihydrolipoamide dehydrogenase, dihydrolipoamide formed during the catalytic cycle of each enzyme would be reoxidized to lipoamide more slowly than necessary to maintain optimal activity of the complexes. Oxidation of pyruvate and subsequent ATP production would be lowered as a consequence of diminished TCA cycle activity.

10. The committed step in a metabolic pathway is usually under metabolic control. Inhibition of the committed step in a metabolic sequence or pathway prevents the accumulation of unneeded intermediates and effectively precludes activity of the enzymes using those intermediates as substrates. The decarboxylation of pyruvate and the oxidative transfer of the hydroxyethyl group by pyruvate dehydrogenase constitutes the committed step in the pyruvate dehydrogenase catalytic sequence and is a logical control point.

11. In bacteria, acetate is oxidized via the TCA cycle and electron transport system and is also used as a carbon source for synthesis of intermediates for cell growth and division. Competition between ICDH and isocitrate lyase for isocitrate determines whether acetate is oxidized via the TCA cycle or is used for biosynthesis. Metabolism of isocitrate through the glyoxalate bypass is increased by phosphorylation of ICDH, and the flux through TCA cycle is increased upon dephosphorylating (reactivating) ICDH. The fraction of ICDH in the active form dictates the level of isocitrate available to isocitrate lyase. The isocitrate lyase is apparently not regulated at the metabolite level. (Nimmo, H. G., *TIBS.* 9:475–478 (1984).)

12a. Anaplerotic reactions, e.g., PEP carboxylase and formation of malate via the glyoxalate bypass, replenish intermediates of the TCA cycle that are diverted to biosynthetic pathways. Without the anaplerotic reactions, TCA cycle activity rapidly becomes limited by the low concentrations of intermediates. Such would be the case if oxaloacetate were being diverted to an anabolic pathway. In order to replenish oxaloacetate, the glyoxalate bypass activity must increase. The ICDH kinase will be more active, the ICDH phosphatase will be less active, and a larger fraction of ICDH will be inhibited. Thus, more isocitrate will be available for the isocitrate lyase.

12b. High levels of AMP are consistent with high ATP energy demand by the cell. The TCA cycle activity, coupled to oxidative phosphorylation, is needed to provide an adequate supply of ATP. Hence, ICDH must be active, and the glyoxalate bypass activity must be decreased in order to provide metabolite for the TCA cycle. To accomplish this, the ICDH kinase is less active and the ICDH phosphatase is more active, increasing the fraction of ICDH that is in the active, dephosphorylated form.

12c. If the culture is growing on glucose, the glucose carbon pool will be used for anabolism and for oxidative metabolism. The TCA cycle activity will increase, but the glyoxalate bypass activity should diminish markedly. Thus, ICDH kinase is less active, but the ICDH phosphatase must be more active.

13. In plants, the TCA cycle and glyoxalate bypass are located in different subcellular organelles. The TCA cycle is located in mitochondria and the glyoxalate bypass is located in glyoxosomes. The glyoxosomes also contain the enzymes for β oxidation of fatty acids, yielding acetyl-CoA. Hence, a supply of fatty acids to the glyoxosomes results in the net synthesis of malate. Communication between the compartments includes flow of succinate to the mitochondria to replenish the TCA cycle intermediates.

14a.

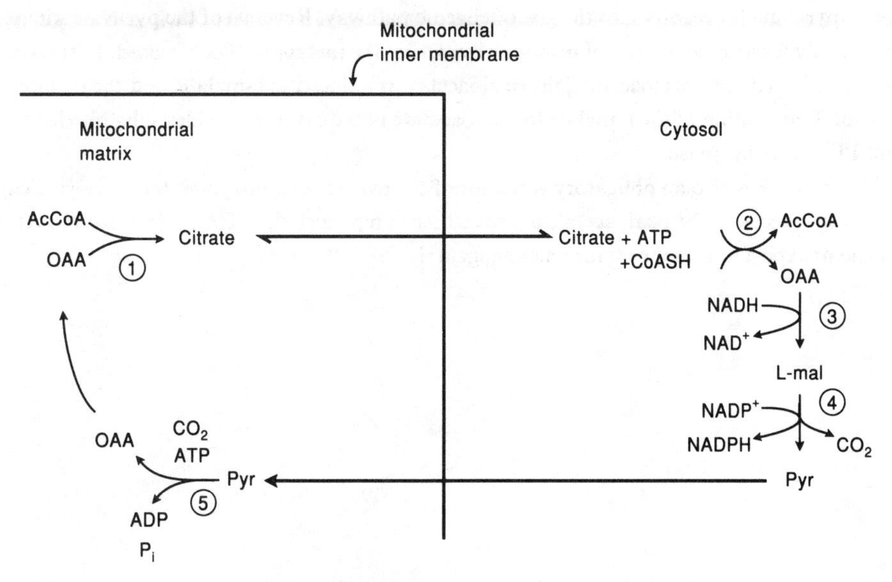

Rxn. 1. Citrate synthase
Rxn. 2. ATP-dependent citrate lyase (Coenzyme A)
Rxn. 3. Cytosolic malate dehydrogenase (NAD(H))
Rxn. 4. $NADP^+$-dependent malic enzyme (NADP(H))
Rxn. 5. Pyruvate carboxylase (biotin)

14b. The enzyme-catalyzed reactions and the coenzymes are listed below and are numbered to correspond to the reactions shown in the scheme in part (a).

Rxn. 2 ATP-dependent citrate lyase (coenzyme A)
Rxn. 3 Cytosolic malate dehydrogenase (NAD(H))
Rxn. 4 $NADP^+$-dependent malic enzyme (NADP(H))
Rxn. 5 Pyruvate carboxylase (biotin)

14c. Oxaloacetate in the mitochondria is condensed with acetyl-CoA to form citrate (Rxn 1 in the scheme). Citrate is transported to the cytosol and subsequently cleaved to acetyl-CoA and oxaloacetate (Rxn 2). Were there no mechanism to replenish the TCA cycle, a net six carbons would be lost from the mitochondrial pathway. However, oxaloacetate is replenished in the TCA cycle in the following manner: Oxaloacetate is reduced by malate dehydrogenase to L-malate. Malic enzyme catalyzes the oxidative decarboxylation of malate, forming pyruvate and the reduced cofactor NADPH. The malate dehydrogenase and malic enzyme activities catalyze a net transfer of reducing equivalents from NADH to $NADP^+$. NADPH is subsequently used in fatty acid biosynthesis. Pyruvate, released from the malic enzyme, is transported into the mitochondrial matrix for carboxylation to oxaloacetate. This oxaloacetate replenishes the four-carbon dicarboxylic acid pool initially used in citrate formation in the mitochondrial matrix.

14d. Oxidation of glucose-6-phosphate, catalyzed by the $NADP^+$-dependent glucose-6-phosphate dehydrogenase, and oxidation of 6-phosphogluconate by the $NADP^+$-dependent phosphogluconate dehydrogenase provide NADPH. The oxidation of isocitrate by the cytosolic TCDH also provides NADPH.

15. In mammalian liver, acetyl-CoA provides reducing equivalents by its oxidation in the TCA cycle and subsequent oxidative phosphorylation to provide ATP. The complete oxidation of acetyl-CoA provides approximately 10 moles of ATP per mole of acetyl-CoA oxidized by the mitochondrial TCA and electron transport system. Hence, the ATP required to resynthesize glucose from lactate, pyruvate, or their equivalent is provided by the net oxidation of acetyl-CoA. The carbon source for gluconeogenesis, pyruvate (lactate), must be converted to phosphoenolpyruvate for reentry into the gluconeogenic pathway. Reversal of the pyruvate kinase reaction is not metabolically feasible, so a series of reactions bypassing the metabolic block is used. In these reactions, pyruvate is carboxylated to oxaloacetate, the oxaloacetate is reduced to L-malate, and the malate is transported to the cytosol. Reoxidation of the L-malate to oxaloacetate in the cytosol provides substrate for the GTP-dependent PEP carboxykinase.

Acetyl-CoA is also an obligatory activator of the pyruvate carboxylase, that catalyzes carboxylation of pyruvate to oxaloacetate. The oxaloacetate is used either to replenish the TCA cycle or for reduction to malate and subsequent export to the cytosol for gluconeogenesis.

15 Electron Transport and Oxidative Phosphorylation

Problems

1. Compare iron-sulfur proteins, flavoproteins, and quinones with respect to the following.

   (a) Chemical nature of the functional group that undergoes oxidation-reduction.

   (b) Number of reducing equivalents per redox center involved in electron donor/acceptor reactions of physiological importance. If semiquinones are formed, so indicate and include them in the reduction scheme.

   (c) Stoichiometry of protons taken up per electron.

2. (a) Describe how heme is bound to the protein portion of the a/a_3-, *b*-, and *c*-type cytochromes.

   (b) Although there are three types of cytochromes in rat liver mitochondria, CO and CN^- inhibit electron transfer only at the cytochrome a/a_3 complex. Why do these inhibitors interact with cytochrome a/a_3 but not with cytochrome *b* or cytochrome *c*?

3. Calculate the standard redox potential change ($\Delta E°'$) and the standard free energy change ($\Delta G°'$) for the following reactions at pH 7.0. Write a balanced equation for each reaction.

   (a) Cyt c + (Fe^{2+}) cyt a_3(Fe^{3+}) $\rightarrow$ cyt c(Fe^{3+}) + cyt a_3(Fe^{2+})

   (b) 4 cyt c(Fe^{2+}) + O_2 + $4H^+$ $\rightarrow$ 4 cyt c(Fe^{3+}) + 2 HO

(c) Oxidation of succinate by succinate: cytochrome *c* reductase.

(d) Reduction of extramitochondrial NAD^+ by dihydroubiquinone, via the α-glycerolphosphate shuttle.

4. (a) In biological oxidation-reduction reactions, does the stoichiometry of electron transfer (reducing equivalents/mole) differ among the 1Fe, 2Fe-2S, and 4Fe-4S centers?

   (b) 4Fe-4S centers function in electron transport over a wide range of reduction potentials. There is nothing inherent in the iron-sulfur cluster to suggest this range of reduction potentials. Therefore, what other component(s) must dictate reduction potential?

5. What percent of cytochrome *c* will be in the oxidized form in a solution held at +0.30 V and pH 7.0?

6. Given the standard reduction potentials for cytochrome *c* and ubiquinone at pH 7.0 (see text), calculate the corresponding values at pH 6.0 and 8.0.

7. (a) Acetylation of one or more lysines near the edge of the heme in cytochrome *c* decreases both the rate of electron transfer to the cytochrome *c* from complex III and transfer of electrons from the reduced cytochrome *c* to cytochrome oxidase. What does this suggest concerning operation of the respiratory chain?

   (b) What types of amino acid residues on complex III and complex IV would you expect to interact with cytochrome *c?*

8. Rotenone, which blocks the transfer of electrons from $FMNH_2$ of the NADH dehydrogenase to ubiquinone, is a potent insecticide and fish poison.

(a) Explain why rotenone is lethal to insects and fish.

(b) Would you expect the use of rotenone as an insecticide to be potentially hazardous to other animals (e.g., humans)? Why or why not?

(c) If isolated mitochondria were respiring with succinate as substrate, would you expect a change in O_2 consumption upon addition of rotenone? If β-hydroxybutyrate were the respiratory substrate?

9. We can estimate the overall ATP yield for the oxidation of specific metabolic intermediates by considering both oxidative and substrate level phosphorylation. In principle, total molar yields of "high-energy phosphate" (ATP or equivalent) from cytosolic and mitochondrial processes divided by the molar consumption of O in the mitochondria yield theoretical P/O (or ADP/O) ratios. Using the value of P/O = 2.5 for NADH oxidation and P/O = 1.5 for succinate oxidation in the mitochondria, calculate theoretical P/O ratios for the oxidations given below. Assume that all required enzymes and cofactors are present and that extramitochondrial NADH is oxidized via the α-glycerolphosphate shuttle.

(a) Oxidation of lactate to CO_2 and HO.

(b) Oxidation as in part (a), with 2,4-dinitrophenol present.

(c) Oxidation of dihydroxyacetone phosphate to CO_2 and HO.

(d) Oxidation as in part (c), with 2,4-dinitrophenol present.

10. (a) Explain the necessity of having ubiquinone concentration in excess of other mitochondrial electron-transfer components.

(b) Suppose that you are examining muscle tissue mitochondria in which the UQ content is well below normal. You find that concentrations of all other electron-transfer components are within the normal range. Predict the effect of UQ deficiency on oxidation of: (i) NADH-producing substrates, (ii) succinate, (iii) ascorbate plus a redox mediator.

(c) Explain why UQ-deficient muscle tissue has greater than normal concentrations of lactate. (Ogasahara, Engel, Frens, and Mack, *PNAS.* 86:2379–2382 (1989).)

11. (a) Explain what is meant by "tightly coupled" mitochondria. How can we determine whether mitochondria are tightly coupled?

(b) What is the importance of "respiratory control" in oxidation of metabolites?

(c) In what metabolic circumstance is it advantageous to the organism to have mitochondria uncoupled?

12. The uncoupling reagent, 2,4-dinitrophenol (2,4-DNP), is highly toxic to humans, causing marked increase in metabolism, body temperature, profuse sweating, and in many instances, collapse and death. For a brief period in the 1940s, however, doses of 2,4-DNP presumed to be sublethal were prescribed as a means of weight reduction in humans.

(a) Explain why administration of 2,4-DNP results in increased metabolic rate as evidenced by increased O_2 consumption.

(b) How are the metabolic events that occur upon administration of 2,4-DNP pertinent to regulation of glycolysis and the TCA cycle?

(c) Why would consumption of 2,4-DNP lead to hyperthermia and profuse sweating?

(d) Explain how 2,4-DNP uncouples oxidative phosphorylation.

13. A suspension of mitochondria is incubated with pyruvate, malate, and ^{14}C-labeled triphenylmethylphosphonium [TPP] chloride under aerobic conditions. The mitochondria are rapidly collected by centrifugation and the amount of ^{14}C that they contain is measured. In a separate experiment, the volume of the mitochondrial matrix space has been determined so that the concentration of TPP cation in the matrix can be calculated. The internal concentration is found to be 1,000 times greater than that in the external solution.

(a) What is the apparent electrical potential difference ($\Delta\Psi$) across the inner membrane? Express your answer in the appropriate units and indicate which side of the membrane is positive.

(b) Qualitatively, how might $\Delta\Psi$ be affected by the addition of an uncoupler?

14. Differentiate between electrogenic and neutral transport systems in mitochondria. How is electrogenic transport influenced by the membrane potential? What is the effect of neutral transport on the pH gradient?

15. (a) Suppose you prepare submitochondrial particles (SMP) that can oxidize succinate and catalyze oxidative phosphorylation. Protons are transported to the interior of the formed particles. How does the orientation of the vectorial H^+ transport in SMP compare with the orientation in mitochondria?

(b) Uncouplers do not prevent succinate oxidation, but the particle interior is no longer acidified. Explain.

(c) Atractyloside, an inhibitor of adenine nucleotide translocator, blocks oxidative phosphorylation in intact mitochondria. Would atractyloside inhibit oxidative phosphorylation in the SMPs? Would you expect the SMPs to oxidize added NADH? Explain.

Solutions

1a. The chemical nature of the functional group that is oxidized and reduced is: (1) Iron-sulfur protein: iron-sulfur clusters, whose molecular composition may be 1 iron, no inorganic sulfide; 2 irons, 2 inorganic sulfides; 4 irons, 4 inorganic sulfides. (2) Flavoproteins use either FAD or FMN, whose redox center is an isoalloxazine ring. (3) Quinone: The benzoquinone ring (quinone ring) is fully reduced to the hydroquinone upon addition of 2 electrons.

1b. The number of electrons required to reduce the oxidized form are: (1) Iron-sulfur proteins: 1 electron. (2) Flavoproteins: 2 electrons. A stable semiquinone (1-electron reduction product) may be produced. (3) Quinone: 2 electrons. Stable semiquinone may be formed.

1c. Stoichiometry of protons taken up per electron: (1) Iron-sulfur proteins: no proton taken up. (2) Flavoprotein: 1 proton per electron. (3) Quinone: 1 proton per electron.
The flavin and the quinone redox centers may accept or donate either 1 or 2 electrons and therefore are positioned as electron-transfer agents between obligatory 2 electron donors (NADH) and obligatory 1 electron acceptors (cytochromes or iron-sulfur centers).

2a. Heme of the mitochondrial *b*-type cytochrome interacts hydrophobically with adjacent hydrophobic residues from the membrane-spanning α helices. The heme iron is fully coordinated through two imidazole groups from histidines in the protein. Heme in the *c*-type cytochromes is covalently bound to the protein through thioethers formed by the addition of cysteinyl sulfhydryl groups to the vinyl substituents on the heme ring. The iron is also fully liganded to a nitrogen from the imidazole group of histidine and a sulfur from the thioether linkage of methionine providing the fifth and sixth ligands. Heme *a* differs from the protoheme (heme) of the *b*- and *c*-type cytochromes by substitution of a formyl group at ring position 8 and a 17-carbon isoprenoid chain at position 2. The hydrophobic isoprenoid chain provides added hydrophobic interaction between the heme *a* and the protein. Iron of heme *a* is fully coordinated, but heme a_3 has an open ligand position available for binding of oxygen.

2b. Carbon monoxide binds to the reduced heme a/a_3 and cyanide binds to the oxidized heme a/a_3 presumably at the oxygen-binding site; they inhibit transfer of electrons from reduced cytochrome *c* to O_2. Electron transport is effectively blocked and ADP phosphorylation ceases. Neither CO nor CN^- at low concentration interact with the cytochrome *b* or *c* heme iron because there is no open ligand position to the iron available in these heme proteins.

3a. The oxidation-reduction reactions being considered are

$$\text{Cyt } a_3 (\text{Fe}^{3+}) + 1\ e^- \rightarrow \text{cyt } a_3 (\text{Fe}^{2+}) \qquad E°' = +350 \text{ mV}$$
$$\text{Cyt } c(\text{Fe}^{2+}) \rightarrow \text{cyt } c(\text{Fe}^{3+}) + 1\ e^- \qquad E°' = -240 \text{ mV}$$

$$\text{Cyt } a_3 (\text{Fe}^{3+}) + \text{cyt } c(\text{Fe}^{2+}) \rightarrow \text{cyt } a_3 (\text{Fe}^{2+}) + \text{cyt } c(\text{Fe}^{3+}) \ \Delta E°' = +110 \text{ mV}$$

$\Delta G°' = -nF\Delta E°'$ where n is electron equivalents per mole and F is the Faraday constant (23.06 kcal eq^{-1} V^{-1}). In this reaction, n = 1 eq per mole.

$$\Delta G°' = -(1 \text{ eq/mole})(23.06 \text{ kcal V}^{-1} \text{ eq}^{-1})(+0.11\text{V})$$
$$\Delta G°' = -2.5 \text{ kcal/mole}$$

3b.

$$O_2 + 4\ e^- + 4\ H^+ \rightarrow 2\ H_2O \qquad E°' = +820 \text{ mV}$$
$$4 \text{ cyt } c(\text{Fe}^{2+}) \rightarrow 4 \text{ cyt } c(\text{Fe}^{3+}) + 4\ e^- \qquad E°' = -240 \text{ mV}$$

$$O_2 + 4 \text{ cyt } c(\text{Fe}^{2+}) + 4\ H^+ \rightarrow 2\ H_2O + 4 \text{ cyt } c(\text{Fe}^{3+})$$

Therefore, using the approach shown in part (a), $\Delta E°'$ = +580 mV, $\Delta G°'$ = –53 kcal/mole O_2 reduced (n = 4 equivalents).

3c.

$$\text{Succinate} \rightarrow \text{fumarate} + 2\ e + 2\ H^+ \qquad E°' = +30 \text{ mV}$$
$$2 \text{ cyt } c(\text{Fe}^{+3}) + 2\ e^- \rightarrow 2 \text{ cyt } c(\text{Fe}^{2+}) \qquad E°' = +240 \text{ mV}$$

$$\text{Succinate} + 2 \text{ cyt } c(\text{Fe}^{+3}) \rightarrow \text{fumarate} + 2 \text{ cyt } c(\text{Fe}^{2+}) + 2\ H^+$$

Therefore, $\Delta E°'$ = +270 mV, $\Delta G°'$ = –12.4 kcal/mole (n = 2 equivalents).

3d. Consider the reactions

$$NAD^+ + 2\ e^- + H^+ \rightarrow NADH \qquad E°' = -320 \text{ mV}$$
$$UQH_2 \rightarrow UQ + 2\ e^- + 2\ H^+ \qquad E°' = -110 \text{ mV}$$

$$UQH_2 + NAD^+ \rightarrow UQ + NADH + H^+$$

Therefore, $\Delta E°'$ = – 430 mV, $\Delta G°'$ = +19.8 kcal/mole (n = 2 equivalents). Note that $\Delta E°'$ values are positive for exergonic reactions and are negative for endergonic reactions.

4a. In biological systems, the iron-sulfur centers are obligatorily single electron donors/acceptors, regardless of the number of Fe atoms in the center or their initial oxidation state. For example, the 4 (Fe-S) cluster iron-sulfur proteins accept electrons from flavoproteins or quinones in only 1-electron transfer cycles.

4b. The 4Fe-4S cluster is found in iron-sulfur proteins that transfer electrons at low and at high potential. The reduction potential is a measure of the ease of addition of an electron to the couple, compared to the standard hydrogen electrode. Thus the protein component of the iron-sulfur protein affects the reduction potential. For example, an electron-withdrawing environment at the Fe-S cluster in principle should yield a more positive reduction potential than if the Fe-S cluster were in an electron-donating environment. Spatial constraints and physical interaction between the protein and the iron-sulfur cluster may also affect the redox potential. Moreover, the initial redox state of the cluster affects reduction potential, for example, the oxidized form of high potential iron-sulfur protein cluster is (3 Fe^{3+}-1 Fe^{2+}) whereas the oxidized form of the low potential iron-sulfur cluster is (2 Fe^{2+}-2 Fe^{3+}).

5. The ratio of reduced to oxidized cytochrome may be calculated using the Nernst expression.

$$E = E^{\circ\prime} \frac{-2.3RT}{nF} \log (\text{Red})/(\text{Ox})$$

where $2.3RT/F$ is appropriately 60 mV/n. (Red) and (Ox) represent the concentration (or fraction) of the reduced component and the oxidized component of the couple, respectively.

$$E = E^{\circ\prime} - 60 \text{ mV}/1 \log [(\text{Red})/(\text{Ox})]$$

where

$$\text{E} = +300 \text{ mV}, E^{\circ\prime} = +240 \text{ mV}$$
$$+300 \text{ mV} = 240 \text{ mV} - 60 \text{ mV} \log [(\text{Red})/(\text{Ox})]$$
$$60 \text{ mV} = -60 \text{ mV} \log [(\text{Red})/(\text{Ox})]$$

Therefore, (Red) /(Ox) = 10^{-1} but (Red) + (Ox) = 1

$$(\text{Red}) = 0.1 (\text{Ox})$$
$$0.1 (\text{Ox}) + (\text{Ox}) = 1$$
$$(\text{Ox}) = .9$$

Thus, at +300 mV, the cytochrome *c* half-cell will be 90% oxidized and 10% reduced.

6. The standard reduction potential $E^{\circ\prime}$ changes by the ratio $\Delta E^{\circ\prime}/\Delta\text{pH} = -(60 \text{ mV})(n_{H^+}/n)$ where n_{H^+} is the proton equivalents taken up per equivalent (n) of electrons transferred in the reaction.

$$E^{\circ} = E^{\circ\prime} (\text{pH } 7) - E^{\circ} (\text{pH})$$

The standard reduction potential of cytochrome *c* is unaltered at either pH 6 or pH 8 because no protons are taken up in the reduction reaction. The iron in the heme is reduced from ferric to ferrous state. Reduction of ubiquinone results in the addition of two equivalents of protons per 2 electron equivalents in reducing the quinone to the dihydroquinone. Thus at pH 6, ΔpH is (7–6) = +1, so

$$E^{\circ\prime} - E^{\circ} (\text{pH } 6) = -60 \text{ mV}/1$$
$$E^{\circ\prime} (\text{pH } 7) - E^{\circ} (\text{pH } 6) = -60 \text{ mV}$$

$$E^{\circ} (\text{pH } 6) = E^{\circ\prime} + 60 \text{ mV} = (110 \text{ mV} + 60 \text{ mV}) = +170 \text{ mV}$$
$$E^{\circ} (\text{pH } 6) \text{ is } +170 \text{ mV}$$

At pH 8,

$$[E^{\circ\prime} - E^{\circ} (\text{pH } 8)] = -60 \text{ mV}/-1$$
$$E^{\circ\prime} - E^{\circ} (\text{pH } 8) = +60 \text{ mV}$$
$$E^{\circ} (\text{pH } 8) = E^{\circ\prime} - 60 \text{ mV}$$
$$= (110 \text{ mV} - 60 \text{ mV})$$
$$E^{\circ} (\text{pH } 8) = 50 \text{ mV}$$

Consider the reaction $UQ + 2 e^- + 2 H^+ \rightarrow UQH_2$. Decreasing pH (increasing H^+ concentration) favors the formation of UQH_2 by mass action. Hence the quinone should be more easily reduced to UQH_2 at pH 6 and the value of E° calculated is consistent with the prediction. Decreasing the proton concentration favors the oxidized form of the quinone again by mass action and the quinone should be more difficult to reduce.

7a. Cytochrome *c* accepts an electron from complex III and in turn transfers the electron to cytochrome oxidase for the reduction of O_2 to water. The cytochrome *c* may be tightly associated with the complexes (the linear chain model for electron-transfer system) or may diffuse between complex III and cytochrome oxidase during electron transfer. If the cytochrome *c* were tightly associated in the chain, we would predict that the electron acceptor site, associated with complex III, and the electron donor site, associated with cytochrome oxidase, would be at different sites on the cytochrome *c* protein. Acetylation of specific residues on the cytochrome *c* would be predicted to alter interaction with either but not likely both complexes. In fact, acetylation of lysine residues around the crevice of the heme ring alters electron transfer from complex III as well as transfer to cytochrome oxidase. Thus, the data are consistent with diffusion of cytochrome *c* between complex III and the cytochrome oxidase. Electrons are likely donated and accepted from the same area on the cytochrome *c* structure.

7b. The lysines surrounding the heme crevice on cytochrome *c* should interact electrostatically with anionic amino acid side chains (aspartic acid and glutamic acid) spatially located on complex III and on cytochrome oxidase to facilitate interaction with the cytochrome *c*. If the residues on the complexes at the site of interaction with cytochrome *c* were positively charged (lysine or arginine residues), the electrostatic repulsion would hamper the approach of cytochrome *c* and reduce the probability of electron transfer.

8a. Rotenone blocks transfer of electrons from $FMNH_2$ to ubiquinone in NADH dehydrogenase and prevents reoxidation of reduced NADH. An increase in NADH concentration will inhibit the TCA cycle and glycolysis. ATP production dependent on NADH oxidation ceases. Succinate could supply electrons to the ETS, but the TCA cycle is blocked. ATP stores dwindle and the organism dies (arrives at equilibrium with the local environment).

8b. Humans share the same requirement to oxidize NADH for ATP production as do fish and insects. Inhibition of the NADH dehydrogenase by application of rotenone should have the same deleterious effect in humans that it does in other life forms.

8c. O_2 consumption due to succinate oxidation in isolated mitochondria should not be inhibited by rotenone. Succinate is oxidized by succinate dehydrogenase with entry of electrons into the mitochondrial electron-transfer system on the O_2 side of the rotenone inhibition site. However, β-hydroxybutyrate is oxidized to acetoacetate by an NAD^+-dependent dehydrogenase. The NADH thus produced could not be oxidized in rotenone-blocked mitochondria, and O_2 consumption would be inhibited.

9. Examine each step in the metabolism described in the problem and determine whether reduced cofactor (NADH or FADH) or a high-energy compound (ATP or equivalent) is produced. Using the (P/O) ratios given in the problem, sum the high energy phosphates formed divided by the gram-atoms of oxygen reduced. The P/O ratio is sometimes expressed as P/2 e^- or moles of high-energy phosphate formed per two electron equivalents transferred to O_2.

9a. *Oxidation of lactate*

- Lactate dehydrogenase: Lactate + NAD^+ → pyruvate + NADH (NADH is extramitochondrial and is oxidized by the α-glycerolphosphate shuttle with P/O = 1.5/1).
- Pyruvate dehydrogenase: Pyruvate + CoASH + NAD^+ → acetyl-CoA + NADH + CO_2 (P/O = 2.5/1).

Oxidation of acetyl-CoA via TCA cycle

- Isocitrate dehydrogenase (NADH, P/O = 2.5/1)
- α-Ketoglutarate dehydrogenase (NADH, P/O = 2.5/1)
- Succinate thiokinase (GTP but no oxygen reduced, P/O = 1/0)
- Succinate dehydrogenase (FADH, P/O = 1.5/1)
- Malate dehydrogenase (NADH, P/O = 2.5/1)

In the complete oxidation of 1 mole of lactate, there are 14 moles of ATP (or equivalent) produced and 6 gram-atoms of oxygen reduced. The theoretical P/O ratio is 2.33.

9b. 2,4-Dinitrophenol is an uncoupler of oxidative phosphorylation. The weakly acidic uncoupler dissipates the mitochondrial H^+ gradient and thus prevents ADP phosphorylation via oxidative phosphorylation. The metabolic energy is released as heat. Electron transfer to O_2 will proceed more rapidly in the uncoupled mitochondria than in tightly coupled mitochondria but without ATP formation. In uncoupled mitochondria, there will be no change in the amount of oxygen reduced during lactate oxidation, and the GTP formed by the substrate level phosphorylation will be the only high-energy phosphate compound generated. The P/O ratio is 0.2.

9c. The complete oxidation of dihydroxyacetone phosphate (DHAP) is described in the following sequence.

- Triose phosphate isomerase: DHAP → glyceraldehyde-3-phosphate
- Ga3P dehydrogenase: Ga3P + NAD^+ + P_i → 1,3-bisphosphoglycerate + NADH (NADH, extramitochondrial, P/O = 1.5/1)
- Phosphoglycerokinase: 1,3-bisphosphoglycerate + ADP → 3-phosphoglycerate + ATP (P/O = 1/0)
- Phosphoglyceromutase: 3-phosphoglycerate → 2-phosphoglycerate
- Enolase: 2-phosphoglycerate → PEP
- Pyruvate kinase + ADP → pyruvate + ATP (P/O = 1/0)

Oxidation of pyruvate by PDH and TCA cycle: (12.5 ATP/5 oxygen)
Sum is 16 moles of ATP per 6 gram-atoms oxygen reduced. The P/O = 2.7.

9d. In the presence of 2,4-dinitrophenol, there will be 3 moles of ATP formed by substrate-level phosphorylation catalyzed by phosphoglycerokinase, pyruvate kinase, and succinate thiokinase. No ATP is formed by oxidative phosphorylation, but 6 gram-atoms of oxygen are reduced. The P/O ratio will be 0.5.

10a. Ubiquinone (coenzyme Q) is the electron acceptor for a number of dehydrogenases in the mitochondrial electron-transport system, including succinate dehydrogenase, NADH dehydrogenase, the flavoprotein α-glycerol phosphate dehydrogenase and the electron transfer flavoprotein dehydrogenase. Ubiquinol is oxidized by complex III. The large amount of UQ is necessary to ensure efficient transfer of electrons from the mitochondrial dehydrogenases to complex III. Although the ratio of UQ to Complex III is approximately 20:1, diffusion of UQ among the various dehydrogenases may be the rate-limiting step in electron transfer. (See Hackenbrock, C. R. *TIBS*, 6:151–54, 1981.)

10b. (i) If UQ (ubiquinone) were limiting, the rate of oxidation of NADH by the mitochondrial electron-transfer system would also be limited. NADH concentration would increase, the NAD^+ supply would decrease, and the NAD^+-dependent dehydrogenases would be inhibited. Subsequently, the rate of oxidation of NADH-producing substrates would decrease.

(ii) Electrons from succinate oxidation are also transferred to ubiquinone from the succinate dehydrogenase, so a deficiency of the quinone would limit succinate oxidation.

(iii) Ascorbate plus a redox mediator reduces cytochrome *c* but does not transfer electrons to ubiquinone. Limiting amounts of the quinone should not affect ascorbate-dependent reduction of cytochrome *c*.

10c. Lactate is formed in skeletal muscle by the NADH-dependent reduction of pyruvate. In resting muscle, the NADH formed by the glycolytic enzyme glyceraldehyde-3-phosphate dehydrogenase should be oxidized in the mitochondria via one of the reducing equivalent shuttles (α-glycerophosphate or malate). The deficiency of UQ will decrease the rate of extramitochondrial NADH oxidation resulting in an increase in the amount of pyruvate reduced to lactate. The lactate content would be expected to increase rapidly upon mild exercise.

11a. Tightly coupled mitochondria exhibit an obligatory codependence of electron transfer, as measured by O_2 uptake and phosphorylation of ADP. Mitochondria are said to be in state 3 when respiration is measured with substrate and ADP, phosphate, and O_2 present. Under these conditions, the substrate is oxidized, O_2 is reduced, and ADP is phosphorylated to ATP. In state 4 the conditions are identical to state 3 except that ADP is omitted. In perfectly coupled mitochondria, there should be no reduction of O_2 in state 4. However, perfect coupling has not been observed experimentally and the degree of coupling is estimated by the Respiratory Control Ratio, the ratio of rate of O_2 reduction by mitochondria in state 3 divided by the rate of O_2 reduction in state 4. Larger values of the ratio are consistent with a greater codependence on ADP phosphorylation and electron transfer.

11b. Substrates are oxidized in the mitochondria to provide ATP for cellular processes. Unregulated oxidation of substrate wastes energy resources and eliminates the response of the mitochondrial production of ATP to cellular need. Respiratory control reflects the inhibition of mitochondrial respiration when cellular ATP levels are high and the stimulation of respiration when ATP levels are low (ADP levels are high). Respiratory control is a function of coupled mitochondria. Oxidation of NADH and succinate is blocked when ATP levels are high. Glycolysis is slowed by the lack of ADP (or AMP) stimulation of phosphofructokinase and the TCA cycle is slowed by the increased NADH to NAD^+ ratio. When ATP demand is high, ADP levels will stimulate the oxidation of succinate and NADH. Elevated ADP concentration and high NAD^+/NADH ratio will stimulate glycolysis and the TCA cycle.

11c. Uncoupled mitochondria oxidize substrate rapidly but fail to phosphorylate ADP. The free energy from the oxidation of NADH (approximately 52 kcal/mole) and succinate (approximately 39 kcal/mole) is released as heat. For animals that hibernate, a mechanism to elevate body temperature upon emerging from hibernation is important. This mechanism, nonshivering thermogenesis, involves brown fat mitochondria that contain the uncoupler thermogenin. The electron-transport system is partially uncoupled and heat released by the oxidation of fat provides the increase in body temperature.

12a. The uncoupler 2,4-dinitrophenol circumvents respiratory control in the mitochondria by short-circuiting the proton gradient. The lipophilic weak acid transports H^+ across the membrane, bypassing the F_1-F_0 complex. Substrates will be oxidized independently of ADP or ATP concentrations, and O_2 reduction will be more rapid than in state 3 respiration. Although respiration is rapid, no ADP is phosphorylated.

12b. Uncoupled mitochondria oxidize NADH and succinate but fail to phosphorylate ADP. Cellular processes will continue to utilize ATP, causing an accumulation of ADP and AMP. Increased levels of NAD^+ and ADP or AMP activate both the TCA cycle and glycolysis. The rate of oxidation of carbohydrates and fatty acids would markedly increase.

12c. Oxidation of substrates with the subsequent reduction of oxygen releases 52 kcal per mole of NADH oxidized (or about 39 kcal per mole of succinate oxidized). The energy normally would be used in part to drive the vectorial accumulation of H^+ and subsequent phosphorylation of ADP. The mitochondria are uncoupled and use little, if any, energy to phosphorylate ADP. The energy is dissipated as heat, leading to elevated body temperature (hyperthermia) and profuse perspiration in an effort to decrease body temperature. The metabolic scenario is reminiscent of the brown fat mitochondria and nonshivering thermogenesis discussed in solution 11c.

12d. 2,4-Dinitrophenol is a lipid soluble weakly acidic compound thought to allow equilibration of protons across the inner mitochondrial membrane. Although protons are translocated from the matrix across the inner membrane during electron transfer, the proton gradient would immediately be depleted without passing through the F_0-F_1–dependent ADP phosphorylation system. Respiratory (ADP) control would be lost.

13a. The cationic lipid soluble TPP will be distributed across the membrane so that the net electrical potential is zero. The accumulation of the cation inside the inner mitochondrial membrane indicates that the inside of the membrane is negatively charged. The potential is calculated from the expression

$$\Delta G_j = F\,\Delta\Psi + RT \ln [C_j\text{(inside)}C_j\text{(outside)}]$$

where $\Delta\Psi$ is membrane electrical potential. At equilibrium, ΔG_j is zero, so

$$F\Delta\psi = -2.3\ RT \log [C_j\text{ (inside)}/C_j\text{ (outside)}]$$

$$\Delta\psi = \frac{-2.3RT}{F} \log [Cj\text{ (inside)}/C_j\text{ (outside)}]$$

$$\Delta\psi = -60\text{ mV} \log (10^3)$$

$$\Delta\psi = -180\text{ mV, negative on the inside of the inner membrane.}$$

13b. An uncoupler of oxidative phosphorylation allows H^+ equilibration across the membrane. The protons will equilibrate across the membrane to negate any charge differential and, qualitatively, will dissipate both the membrane potential and the H^+ gradient.

14. The inner mitochondrial membrane contains specific transmembrane proteins that transport charged molecules across the inner mitochondrial membrane. Charged molecules diffuse with little restriction across the outer membrane. The transport process is electrogenic if the export of one molecule coupled with the import of another molecule yields a net charge difference across the membrane. In general terms, transfer of A^{3-} from the matrix and A^{3-} into the matrix yields a net negative charge on the cytoplasmic side of the membrane. Electrogenic processes are driven by the membrane potential ($\Delta\Psi$). Transport that diminishes the membrane potential are energetically favorable and are driven in that direction. H^+-transport driven by the electron-transport system is another example of electrogenic (active) transport. Proton translocation increases the net charge differential across the membrane and energy is required to drive the transport.

Neutral transport processes exchange molecules of net identical charge (sign and magnitude) in the opposite vectorial direction or oppositely charged molecules in the same vectorial direction. For example, exchange of a dianions (succinate for malate) is a neutral transport process. Neutral processes in the mitochondria may be driven by the H^+ gradient, for example, cotransport of H^+ and A^-. Processes coupled to export of OH^- are equivalent to H^+ import, that is, to an energetically favorable decrease in the proton gradient.

15a. The hypothetical submitochondrial particles formed by disrupting the inner mitochondrial membrane are inverted with respect to the topology of the inner membrane. The ATPase (F_1), succinate dehydrogenase, and the NADH dehydrogenase (complex I) will face the exterior of the particle. Transport of H^+ is in the same orientation in these SMPs as in the mitochondria, but the cytoplasmic aspect of the inner mitochondrial membrane is facing the interior of the vesicle. Electron transport would acidify the inside of the vesicle.

15b. Mechanical or chemical uncoupling allows free equilibration of H^+ across the SMP membrane. Mechanical uncoupling results from removal of F_1 ATPase from the membrane with subsequent opening of the H^+ channel, and chemical uncouplers equilibrate protons across the membrane. Succinate dehydrogenase activity is not altered by the uncoupling process, but succinate is oxidized without vectorial accumulation of H^+. The SMP interior will not be acidified.

15c. Atractyloside inhibits ADP transport into mitochondria. In intact mitochondria, a supply of ADP into the matrix and export of ATP to the cytoplasm is necessary to maintain respiration. However, neither ADP nor ATP diffuses freely across the inner mitochondrial membrane. The ADP-ATP translocator catalyzes the ADP-ATP exchange. In SMP, the F_0-F_1 complex faces the bulk solvent and has free access to ADP. The ADP-ATP transporter is not required, so atractyloside would have no effect in this system.

In intact mitochondria, NADH does not diffuse across the inner membrane. Shuttle systems are used to transfer reducing equivalents from NADH produced in the cytosol to the matrix. The α-glycerol phosphate shuttle is metabolically unidirectional (for oxidation of extramitochondrial NADH), whereas the malate shuttle provides a reversible transfer of reducing equivalents. In SMP, the NADH dehydrogenase active site faces the solvent and there is no membrane barrier to prevent NADH oxidation.

16 Photosynthesis and Other Processes Involving Light

Problems

1. The lowest-energy absorption band of P870 occurs at 870 nm.
   (a) Calculate the energy of an einstein of light at this wavelength.

   (b) Estimate the effective standard redox potential ($E^{\circ\prime}$) of P870 in its first excited singlet state, given that the $E^{\circ\prime}$ for oxidation in the ground state is +0.45 V.

2. The traces below show measurements of optical absorbance changes at 870 and 550 nm when a suspension of membrane vesicles from photosynthetic bacteria was excited with a short flash of light. Downward deflection of the traces represent absorbance decreases. Explain the observations. (Absorption spectra of a *c*-type cytochrome in its reduced and oxidized forms are described in the previous chapter.)

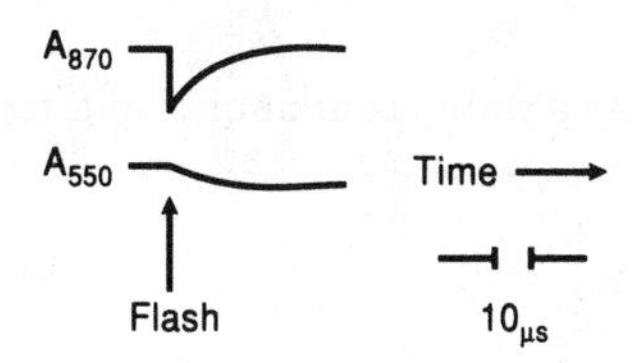

3. You add a nonphysiological electron donor to a suspension of chloroplasts. When you illuminate the chloroplasts, the donor becomes oxidized. How could you determine whether this process involves both photosystems I and II? (In principle, the donor could transfer electrons either to some component on the O_2 side of photosystem II or to a component between the two photosystems.)

4. Ubiquinone has an absorbance band at 275 nm. This band bleaches when the quinone is reduced to either the semiquinone or the dihydroquinone. The anionic semiquinone has an absorption band at 450 nm but neither the quinone nor the dihydroquinone absorbs at this wavelength. A suspension of purified bacterial reaction centers was supplemented with extra ubiquinone and reduced cytochrome *c* and was then illuminated with a series of short flashes of light. The absorbance at 275 nm decreased on the odd-numbered flashes as shown on the first curve below. The absorbance at 450 nm increased on the odd-numbered flashes but returned to the original level on the even-numbered flashes as shown in the second curve.

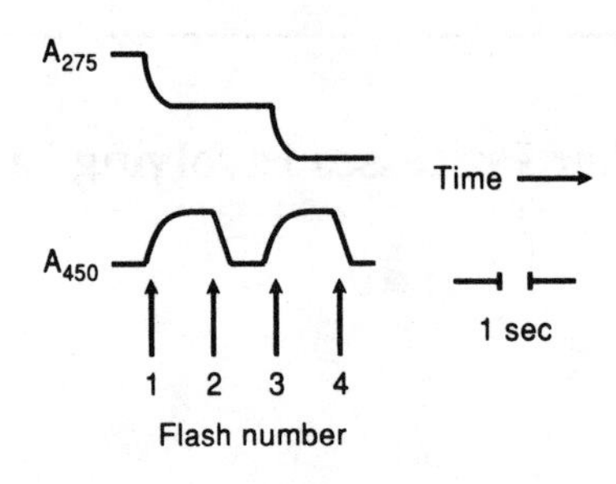

(a) Explain the patterns of absorbance changes at the two wavelengths.

(b) Why is it necessary to have reduced cytochrome *c* present in order to see these effects?

5. When chloroplasts are illuminated with a weak continuous white light, the intensity of fluorescence from the antenna chlorophyll is relatively low. (About 2% of the photons absorbed are remitted as fluorescence.) The intensity of the fluorescence increases severalfold if the redox potential is decreased to approximately 0 mV. (Assume that the plastoquinone pool is reduced.) Explain the increase in fluorescence.

6. When chloroplasts are illuminated with a strong continuous light, the intensity of fluorescence from the antenna chlorophyll increases with time as shown in curve A.

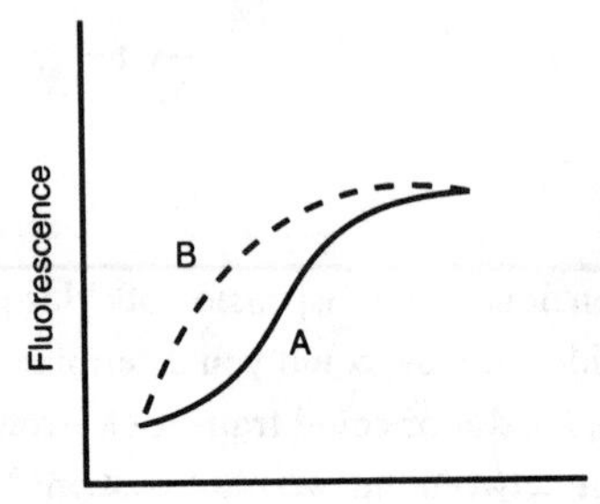

(a) Explain the increase in fluorescence.

(b) The herbicide DCMU (3-(3,4-dichlorophenyl)-1,1-dimethylurea) blocks electron transfer between the two plastoquinones in the reaction center of photosystem II. In the presence of DCMU, the fluorescence increases more abruptly (curve B). Explain.

(c) What effect would DCMU have on $NADP^+$ reduction? On O_2 evolution?

7. Explain why in green plants, the quantum efficiency of photosynthesis drops sharply at wavelengths longer than 680 nm.

8. Define photophosphorylation. Differentiate between cyclic and noncyclic photophosphorylation in green plants and describe how photosystems I and II are involved in each process.

9. Would you predict that plant cells lacking carotenoids would be more or less susceptible to damage by photooxidation? Explain.

10. O_2 is an alternate substrate for the ribulose bisphosphate carboxylase/oxygenase and is also a competitive inhibitor with respect to CO_2 fixation. Explain.

11. Compare and contrast the initial process of CO_2 fixation in C_3 and C_4 plants. In each case, indicate which stable organic molecule initially becomes labeled with ^{14}C-labeled CO_2.

12. C_4 plants ultimately use ribulose bisphosphate carboxylase/oxygenase, yet have lower rates of photorespiration than do C_3 plants. Explain.

13. Reduction of three moles of CO_2 to form one mole of triose phosphate requires nine moles ATP and six moles NADPH.

(a) What is the source of NADPH used in the reduction of CO_2?

(b) Account for the ATP consumed in the formation of triose phosphate.

(c) Assume that the CO_2 initially is added to PEP in the C_4 pathway. What is the additional cost in ATP per mole of CO_2 added?

(d) Is additional NADPH required for CO_2 fixation in the C_4 plant?

Solutions

1a. The relationship between the wavelength and the energy of light is

$$E = hc/\lambda$$

where

E is energy in electron volts
h is Plancks constant: 4.2×10^{-15} eV•s
c is speed of light: 3×10^{10} cm•$^{-1}$s
λ is wavelength: 8.7×10^{-5} cm (870 nm)

$$E - \frac{(4.2 \times 10^{-15}\ \text{eV}\cdot\text{s})\ (3 \times 10^{10}\ \text{cm}\cdot\text{s}^{-1})}{(8.7 \times 10^{-5}\ \text{cm})}$$

$$E - 1.4\ \text{eV}$$

and

$$E(\text{kcal/einstein}) = nF[E(\text{electron volts})]$$

where

n is 1 electron-equivalent per einstein and F(Faraday's Constant) is 23.06 kcal V^{-1} eq^{-1}.

$$E\ (\text{kcal/einstein}) - (1\ \text{electron-equivalent einstein})\ (23.06\ \text{kcal}\ V^{-1}\ eq^{-1})\ (1.45\ \text{eV})$$
$$E - 33\ \text{kcal/einstein}$$

1b. The total potential span between the first excited singlet state and the ground state is −1.45 electron volts as calculated in part (a). Thus absorption of a photon will cause a 1.45 volt separation of an electron for PS_{870} to a more negative (more strongly reducing) potential. Thus:

$$E^{\circ\prime}_{final} - E^{\circ\prime}_{initial} = -1.45 \text{ V}$$
$$E^{\circ\prime}_{final} - 0.45\text{V} = -1.45 \text{ V}$$
$$E^{\circ\prime}_{final} = -1.0 \text{ V}$$

2. Upon illumination, the chromatophore P_{870} is activated by absorption of a photon of light. The absorbance at 870 nm decreases because the pi-cation radical of the oxidized chromatophore has a lower absorbance at that wavelength. Thus, the trace monitoring 870 nm decreases upon illumination of the chromatophore. The *c*-type cytochrome is added initially in the reduced form (cyt c^{2+}) and absorbs at 550 nm. Oxidized cytochrome *c* (cyt c^{3+}) has only a small absorbance at 550 nm. Electron transfer from reduced cytochrome *c* to the pi-cation radical (P_{870+}) regenerates the ground state P_{870} and forms oxidized cytochrome *c*. The absorbance of the P_{870} increases to the initial level and the absorbance at 550 nm of the cytochrome *c* pool decreases. The explanation is consistent with the observed upward trace at 870 nm and the downward trace at 550 nm over 30 μs.

3. Oxidation of compound A by photosystem II (PS II) in chloroplasts may require the concerted operation of both photosystem I (PS I) and PS II. Inhibition of electron flow between the photosystems would effectively inhibit oxidation of compound A. DCMU is a herbicide that inhibits electron transfer between PS II and PS I. Therefore, if oxidation of compound A requires operation of both photosystem I and photosystem II, oxidation of compound A will be inhibited by DCMU. Moreover, if compound A is oxidized by PS II, the action spectrum for oxidation of the compound will parallel the absorbance spectrum of PS II. Oxidation will diminish if chloroplasts are illuminated with red light (>680 nm), a phenomenon sometimes called red drop. However, if compound A is oxidized by photosystem I, DCMU will not inhibit the oxidation nor will the action spectrum of compound A oxidation exhibit red drop.

4a. Illumination of the bacterial photocenter generates an active reductant that transfers an electron to ubiquinone, forming the semiquinone. Some, but not all, of the ubiquinone is reduced by a single flash. The absorbance at 275 nm decreases because the concentration of ubiquinone is diminished and because the semiquinone radical does not absorb 275 nm light. The absorbance increase at 450 nm is consistent with the formation of semiquinone radical that absorbs 450 nm light. The second flash activates transfer of a second electron from the photocenter to reduce the bound semiquinone to dihydroquinone. The dihydroquinone absorbs at neither 275nm nor 450 nm. Reduction of the semiquinone form abolishes the absorbance at 450 nm. The decrease in absorbance at 275 nm is consistent with the decrease in oxidized (ubiquinone) concentration. The sequence of events is repeated with the third and fourth flashes.

4b. Transfer of an electron from the activated photocenter (P_{870}*) to the ubiquinone leaves the oxidized pi-cation radical form of the reaction center (P_{870}+), which must be reduced to allow a second light-activation cycle. In this experiment, reduced cytochrome *c* is the electron donor to P_{870}+. Were the reductant omitted, the P_{870}+ might oxidize the reduced quinone, or the activated reaction center P_{870}* might return to ground state by emission of fluorescence.

5. Absorption of light energy activates chlorophyll to an excited singlet state whose energy is dissipated in one of several ways: (a) radiationless energy transfer to a neighboring chlorophyll molecule; (b) activation of a photocenter "trap" molecule; or (c) fluorescence. Little fluorescence is observed in undamaged chloroplasts because light energy absorbed by antenna chlorophyll molecules is eventually transferred to a photocenter trap (P_{680}) where an electron is transferred initially to pheophytin, then to bound plastoquinone. However, if the potential is decreased so that the quinone pool will be reduced, electron transfer from pheophytin to the quinone pool will diminish, and the steady-state level of activated antenna chlorophyll molecules will increase. The alternate energy dissipation pathway, fluorescence, therefore increases.

6a. Illumination of chloroplasts activates both photosystem I and photosystem II, with oxidation of H_2O to O_2, vectorial transport of H^+ to drive photophosphorylation, and reduction of $NADP^+$ to NADPH. If the activated photocenters transfer electrons to their respective acceptors at rates that maintain a low steady-state level of activated chlorophyll in the antenna system, the amount of fluorescence is small. The system transduces the energy efficiently. If the chloroplasts are illuminated with strong light, a larger fraction of the chlorophyll molecules will be activated. If the electron transfer between PS II and PS I now becomes rate-limiting, the steady-state level of chlorophyll in the excited state will increase and a greater fraction of the energy will be released as fluorescence. The lag observed in curve A reflects the gradual increase in fluorescence for excited chlorophyll as the electron-transfer process between the two photosystems becomes rate-limiting because all potential electron acceptors have become reduced.

6b. DCMU, a herbicide, blocks electron transfer within the pool of bound plastoquinone and immediately abolishes the transfer of electrons from photoactivated $P_{680}*$ to the quinone pool. Reduced pheophytin fails to become reoxidized and the steady-state level of activated antenna chlorophyll increases. Energy of the photoactivated chlorophyll is emitted as fluorescence. The steady-state levels of activated antenna chlorophyll increase rapidly upon inhibition of the electron transfer by DCMU, and an immediate fluorescence increase is observed. DCMU potentiates photooxidation in plants.

6c. The oxidized photosystem I (P_{700}+) is reduced by an electron via the electron-transfer system from activated photosystem II. DCMU inhibits electron transfer between the photosystems and inhibits reduction of $NADP^+$ to NADPH. Addition of an electron donor on the photosystem I side of the DCMU block should restore $NADP^+$ reduction.

O_2 is the product of H_2O oxidation by the photosystem II^+-Mn-protein complex. H_2O is the electron donor to the Mn-protein complex that reduces the P_{680}+ pi-cation radical. If photosystem II is activated but electron transfer to the bound quinone pool is blocked, the strong oxidant (P_{680}+) will not be formed and O_2 evolution will be inhibited.

7. Oxidation of H_2O to O_2 by plant chloroplasts requires the operation of both photosystem I and II, whose reaction center pigments absorb maximally at 700 and 680 nm, respectively. At wavelengths greater than about 700 nm the oxygenic photosystem (PS II) ceases to absorb light and fails to be activated, whereas the PS I absorbs light and is still activated at this wavelength. The quantum yield of O_2 evolution from plants as a function of light absorbed parallels the absorption spectrum until the wavelength of impinging light is greater than approximately 700 nm. Although the chloroplasts absorb wavelengths greater than 700 nm, O_2 evolution declines because the oxygenic photosystem II is no longer activated.

8. Electron transfer through the chloroplast electron-transfer system generates a proton gradient and a membrane potential gradient. The energy of the proton gradient is used to drive the phosphorylation of ADP. The chloroplast coupling factors CF_0 and CF_1 apparently provide a proton channel and an ATP, synthase, similar to the mitochondrial F_0 and F_1 components. The process of ADP phosphorylation by the light-driven proton gradient is called photophosphorylation. Noncyclic photophosphorylation requires net electron transfer from a donor (HO) to an acceptor ($NADP^+$) with vectorial H^+-translocation. In the noncyclic process, operation of both photosystems I and II is required. Cyclic photophosphorylation is driven by a proton gradient generated by electron transfer from photosystem I through the quinone pool and the electron-transfer system back to PSI. Photosystem II is not required for cyclic photophosphorylation.

9. Oxygen, a product of photosynthesis, may damage the chloroplast if the ground-state (triplet) oxygen is activated to the singlet state. Single oxygen may be formed by activated (singlet) chlorophyll that decays to the triplet state. Whereas energy transfer between singlet-state chlorophyll and triplet-state oxygen does not occur, triplet-state chlorophyll may activate triplet-state oxygen to the reactive singlet state. Singlet-state oxygen causes cumulative oxidative damage to the chloroplast. Carotenoids compete with the oxygen for the triplet-state chlorophyll and inhibit singlet oxygen production. Plants deficient in carotenoids risk photooxidative damage because of an increased flux of singlet oxygen. Although there are enzymes that scavenge the reduced, activated

oxygen intermediates (superoxide, hydrogen peroxide), there is no enzymatic activity thus far identified that catalytically scavenges singlet oxygen.

10. Ribulose bisphosphate carboxylase/oxygenase (RuBisCO) catalyzes the addition of CO_2 to ribulose-1,5-bisphosphate, yielding two moles of 3-phosphoglycerate per mole of CO_2 added. Alternatively, RuBisCO catalyzes the addition of O_2, to ribulose -1,5-bisphosphate, yielding one mole each of 3-phosphoglycerate and phosphoglycolate. Either CO_2 and O_2 binds to the catalytic site of the ribulose bisphosphate carboxylase/oxygenase (RuBisCO) and reacts with the ribulose-1,5-bisphosphate that is bound to the enzyme. Molecules that bind to the same form of the enzyme (active site) are competitive inhibitors. Thus O_2 competitively inhibits carboxylase activity and CO_2 competitively inhibits oxygenase activity.

11. The designation C_3 indicates plants whose product of CO_2 fixation is initially 3-phosphoglycerate. C_4 designates plants whose initial product of CO_2 fixation is a 4-carbon product (oxaloacetate). In C_4 plants, the CO_2 is released from the 4-carbon product at a concentration greater than atmospheric levels. Carboxylase activity of RuBisCO in C_4 plants is enhanced relative to carboxylase activity in C_3 plants because of the increased CO_2 concentration and consequent inhibition of the oxygenase activity.

12. The absorption of CO_2 from the air and addition of CO_2 to ribulose bisphosphate occur in separate compartments in the C_4 plant. In the mesophyll cell of the C_4 plant, CO_2 is added initially to PEP, yielding oxaloacetate and subsequently malate and/or aspartate. O_2 does not compete with CO_2 (HCO_3) for PEP carboxylation. The 4-carbon organic acids are translocated to the bundle sheath cells where aspartate is converted to oxaloacetate by transamination and the oxaloacetate is reduced to malate. CO_2 is released by oxidative decarboxylation of malate in the bundle sheath cells of C_4 plants, causing the CO_2/HCO_3^- concentration to exceed the CO_2/HCO_3_ concentration in the mesophyll cell. The larger concentration of CO_2 increases the fraction of ribulose-1,5-bisphosphate that is carboxylated rather than oxygenated. Net photorespiration is lower in C_4 plants than in C_3 plants.

13a. $NADP^+$ is reduced by the $NADP^+$-ferredoxin oxidoreductase whose reductant source is the noncyclic electron flow in the chloroplast. Electrons are transferred to PS II from the oxidation of water via the chloroplast electron-transfer system to PS I. Illumination of PS I generates the strong reductant required to reduce $NADP^+$ to NADPH.

13b. Addition of 3 moles of CO_2 to 3 moles of ribulose-1,5-bisphosphate yields 6 moles of 3-phosphoglycerate. Six moles of ATP are used by phosphoglycerokinase to form 6 moles 1,3-bisphosphoglycerate, which are subsequently reduced to glyceraldehyde-3-phosphate (requiring 6 moles NADPH). The 6 moles of glyceraldehyde-3-phosphate are used in the Calvin cycle to regenerate 3 moles of ribulose-5-phosphate with one mole of triose phosphate remaining. Phosphoribulokinase uses 3 moles of ATP to phosphorylate the 3 moles of ribulose-5-phosphate in order to sustain the cycle. (In addition to the 9 moles of ATP required, 6 moles of NADPH are used in the formation of 1 mole of triose phosphate.)

13c. The ATP consumption by the Calvin cycle in the bundle sheath cells of the C_4 plant does not differ from that of the C_3 plant (9 moles of ATP per mole of triose phosphate). However, the fixation and transfer of CO_2 consumes PEP in the mesophyll cells in the formation of oxaloacetate. Pyruvate is formed by the oxidative decarboxylation of malate in the bundle sheath cells. Resynthesis of PEP from pyruvate is catalyzed by the pyruvate, phosphate dikinase. The products of the reaction are PEP, AMP, and pyrophosphate. The pyrophosphate is hydrolyzed to 2 moles of orthophosphate by pyrophosphatase. The rephosphorylation of AMP to ATP requires two high-energy phosphate equivalents. Thus, two additional ATPs are required in the C_4 plant compared to the C_3 plant.

13d. There is no additional consumption of NADPH in the C_4 pathway of CO_2 fixation. NADPH is used in the mesophyll cell to reduce oxaloacetate to malate, but the NADPH is regenerated upon oxidative decarboxylation of malate to form CO_2 and pyruvate as catalyzed by the $NADP^+$-malic enzyme.

17 Metabolism of Fatty Acids

Problems

1. (a) Consider the complete oxidation of glucose (M_r = 180) via glycolysis and the TCA cycle and calculate the moles ATP generated during the oxidation of 1 mole of glucose to CO_2 and H_2O. Assume that the free energy of ATP hydrolysis under physiological conditions is –11 kcal/mole, and assume mitochondrial P/O ratios of 2.5 for NADH oxidation and 1.5 for succinate (or equivalent) oxidation. Estimate the free energy conserved as ATP during the oxidation. Calculate the free energy conserved per gram of glucose oxidized.

   (b) Repeat the calculations for ATP formation but consider the complete oxidation of 1 mole of palmitic acid (M_r = 256) via β-oxidation and the TCA cycle. Estimate the free energy conserved as ATP energy from palmitate oxidation. Calculate the free energy conserved per gram of palmitic acid oxidized.

   (c) Bearing in mind that respiratory metabolism is an oxidative process, how might you explain the differences in energy content of fat and carbohydrate on a weight basis?

   (d) Explain the rationale behind the use of fat rather than carbohydrate as energy reserve in plants and animals.

2. Explain the role of carnitine acyltransferases in fatty acid oxidation.

3. Carnitine deficiency in liver is correlated with hypoglycemia. Suggest a plausible explanation for hypoglycemia in the carnitine-deficient human.

4. Explain why α oxidation is an obligatory step in the oxidation of phytanic acid.

5. How might a deficiency of vitamin B_{12} affect oxidation of propionyl-CoA formed during β-oxidation of odd-chain-length fatty acids or of pristanic acid?

6. Oxidation of reduced fatty acyl-CoA dehydrogenase requires electron-transfer flavoprotein (ETF) and the enzyme ETF-ubiquinone oxidoreductase.

   (a) Write the reaction catalyzed by the ETF-ubiquinone oxidoreductase. Why may we consider the reaction essentially irreversible in the mitochondria under physiological conditions?

   (b) What effect would you expect a nonreducible structural analog of ubiquinone to have on ETF-ubiquinone oxidoreductase activity? On the fatty acyl-CoA dehydrogenase activity of mitochondria?

7. (a) Liver mitochondria convert long-chain fatty acids to ketone bodies (acetoacetate and β-hydroxybutyrate) that are subsequently transported in the plasma to nonhepatic tissues. Suggest some metabolic advantages of supplying ketone bodies to nonhepatic tissues.

   (b) In what way is β-hydroxybutyrate a better energy source than acetoacetate for nonhepatic tissues?

   (c) Outline the oxidation of β-hydroxybutyrate to acetyl-CoA in heart mitochondria.

8. Predict the effect on oxidation of ketone bodies and of glucose in nonhepatic tissue of individuals with markedly diminished β-oxyacid-CoA-transferase activity. Predict the effect if the activity were absent.

9. What is the role of the NADPH-dependent 2,4-dienoyl-CoA reductase in the oxidation of linolenoyl-CoA?

10. Except for malonyl-CoA formation, all the individual reactions for palmitate synthesis reside on a single multifunctional protein (fatty acid synthase) in animal cells. It has been shown that a dimer of the multifunctional protein is required to catalyze palmitate synthesis. Explain the molecular basis of this observation.

11. (a) For an *in vitro* synthesis of fatty acids with purified fatty acid synthase, the acetyl-CoA was supplied as the ^{14}C-labeled derivative

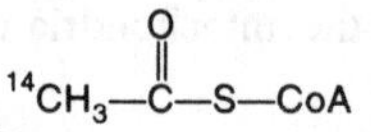

$$^{14}CH_3-\overset{O}{\overset{\|}{C}}-S-CoA$$

The other reactants, including the malonyl-CoA were not radioactive. Where would the ^{14}C-label be found in palmitic acid?

(b) If the malonyl-CoA were supplied as the only labeled compound deuterated as shown below, how many deuterium atoms would be incorporated in palmitate? On which carbon(s) would these deuterium atoms reside?

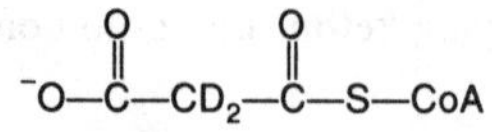

$$^{-}O-\overset{O}{\overset{\|}{C}}-CD_2-\overset{O}{\overset{\|}{C}}-S-CoA$$

(c) If [3-^{14}C]malonyl-CoA (shown below) were used in the reaction, which atoms in palmitate would be labeled? Why?

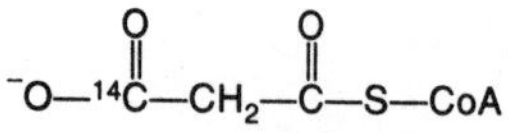

$$^{-}O-{}^{14}\overset{O}{\overset{\|}{C}}-CH_2-\overset{O}{\overset{\|}{C}}-S-CoA$$

12. Citrate is both a lipogenic substrate and a regulatory molecule in mammalian fatty acid synthesis.

(a) Explain each function of citrate in fatty acid synthesis.

(b) Write reactions (including structures) outlining the role of citrate as a lipogenic substrate.

13. What are the metabolic sources of NADPH used in fatty acid biosynthesis? How many moles of NADPH are required for the synthesis of 1 mole of palmitic acid from acetyl-CoA?

14. (a) Why is the location of biosynthesis and β oxidation of fatty acids in separate metabolic compartments essential to regulation of fatty acid metabolism in the hepatocyte?

(b) Would you expect an inhibitor of the extramitochondrial carnitine acyltransferase to mimic the effect of malonyl-CoA on β-oxidation? (Assume that the inhibitor can penetrate the cell membrane.) Explain the rationale for your answer.

15. Which catalytic activity of the mammalian fatty acid synthase determines the chain length of the fatty acid product?

Solutions

1a. Glycolytic oxidation of 1 mole glucose to pyruvate yields 2 moles pyruvate, 2 moles NADH, and 2 moles ATP. The extramitochondrial NADH is assumed to be reoxidized via the malate shuttle, yielding approximately 2.5 moles ATP. The mitochondrial oxidation of pyruvate to CO_2 plus H_2O yields 12.5 moles ATP each. The total ATP generation is 32 moles ATP per mole of glucose. Total energy conserved as ATP can be estimated: 32 moles ATP × 11 kcal/mole = 350 kcal/mole. Energy conserved per gram of glucose is 350 kcal/mole / 180 gm/mole = 1.9 kcal/gm^{-1}.

1b. Oxidation of palmitic acid requires the activation of the acid and formation of a thioester with coenzyme A. The synthase uses the equivalent of 2 moles ATP (hydrolysis of two phosphoanhydride bonds) in the reaction

$$RCOO^- + ATP + CoASH \rightarrow RCO\text{–}SCoA + AMP + pyrophosphate$$

Pyrophosphatase hydrolyzes the pyrophosphate. Thus, mole AMP plus 2 moles P_i may be considered end products.

β-Oxidation of 1 mole palmitoyl-CoA yields 8 moles acetyl-CoA in seven repetitions of the cycle. Seven moles FADH and 7 moles NADH are produced in the mitochondrial matrix. Transfer of these reducing equivalents to O_2 yields 28 moles ATP. Oxidation of the 8 moles acetyl-CoA in the TCA cycle yields 8 × 10 moles ATP. The total ATP yield from β oxidation plus the TCA cycle yields 108 moles ATP, but 2 moles ATP were used to activate the fatty acid. Net yield is 106 moles ATP per mole of palmitic acid oxidized. The energy stored as ATP is 106 moles ATP per mole of palmitic acid × 11 kcal per mole ATP = 1,170 kcal/mole conserved. Energy conserved per gram is 1,170 kcal/mole / 256 gm/mole = 4.6 kcal/gm.

1c. Lipids are more highly reduced than are carbohydrates (palmitic acid: $C_{16}H_{32}O_2$; glucose: $C_6H_{12}O_6$). Removal of reducing equivalents from substrate and subsequent reduction of O_2 releases energy that is conserved by oxidative phosphorylation. Lipids provide more reducing equivalents per gram than do carbohydrates, and they provide more energy for oxidative phosphorylation.

1d. Fat has approximately 2.4 times more energy stored per gram than does carbohydrate, and the energy is stored in compact hydrophobic droplets in specialized cells (adipocytes). On that basis, one would have to store 2.4 times the mass of carbohydrate to equal the energy stored by lipid. Carbohydrate mass likely would be even larger because of water of hydration present in the stored form.

2. The fatty acids are transported across the mitochondrial inner membrane as fatty acyl carnitine esters. The fatty acid is activated for metabolism as the fatty acyl-CoA derivative in the cytoplasm, a process requiring the equivalent of 2 moles ATP. Neither the free fatty acid nor the fatty acyl-CoA can freely traverse the mitochondrial inner membrane. Carnitine acyl transferase associated with the outside of the mitochondrial inner membrane (CAT I) catalyzes transfer of the fatty acyl group to carnitine. The fatty acyl carnitine ester is translocated across the inner membrane in exchange for carnitine. A carnitine acyl transferase on the matrix aspect of the inner membrane (CAT II) catalyzes the transfer of the acyl group to CoASH, reforming the fatty acyl-CoA. The result is translocation of the fatty acid from cytoplasm to mitochondrial matrix with preservation of the active thioester. The acyl-CoA is then available for β-oxidation in the mitochondrial matrix.

3. Carnitine is the cofactor required to transport long-chain fatty acids across the mitochondrial inner membrane to the matrix for β-oxidation. Carnitine deficiency would limit fatty acid transport to the mitochondrial matrix and limit energy conversion (as ATP) from the β-oxidation pathway. Glucose would thus be used as an energy source both in the fed and fasted state because of the limited fatty acyl oxidation.

Gluconeogenesis requires both a carbon source (e.g., lactate, pyruvate, or alanine but not acetyl-CoA) and a supply of ATP. Hepatocytes utilize fatty acid oxidation to supply ATP for gluconeogenesis. Without the energy supply, gluconeogenesis fails, and the blood glucose level fails (hypoglycemia). A plausible link between carnitine deficiency and hypoglycemia is suggested based on energy supply.

4. β-oxidation of fatty acyl-CoA requires the formation of a double bond between the α and β carbons of the acyl chain, hydration of the double bond, and subsequent oxidation of the β-hydroxyacyl-CoA to β-ketoacyl-CoA. The β carbon of phytanic acid has a methyl substient that prevents the β-oxidation of phytanic acid. The free phytanic acid would most likely not be substrate for fatty acyl-CoA dehydrogenase. In principle, the β-methyl substituent could prevent the formation of phytanoyl-CoA by the fatty acyl-CoA ligase, prevent oxidation of phytanoyl-CoA, if formed, or inhibit hydration of the double bond. These problems are circumvented by the oxidation of phytanic acid to an α-keto acid that is subsequently oxidatively decarboxylated. The product (pristanic acid) is a substrate for β-oxidation after activation to pristanoyl-CoA. In pristanoyl-CoA, the methyl group is an α substituent and does not appreciably interfere with β oxidation.

5. Oxidation of odd-chain-length fatty acids yields acetyl-CoA until the final 5-carbon product undergoes oxidation. The β-ketopentanoyl-CoA is cleaved by thiolase to acetyl-CoA plus propionyl-CoA. Three moles of propionyl-CoA are also produced during oxidation of pristanoyl-CoA. Propionyl-CoA is further metabolized by carboxylation to D-methylmalonyl-CoA and is enzymatically isomerized to the L-isomer. L-Methylmalonyl-CoA is substrate for the methylmalonyl-CoA mutase, a vitamin B_{12}-dependent enzyme. The product of the mutase

reaction, succinyl-CoA, is metabolized to succinate by mitochondrial succinate thiokinase, with GTP being formed from substrate level phosphorylation. A vitamin B_{12} deficiency would directly impede metabolism of the L-methylmalonyl-CoA and inhibit oxidation of propionyl-CoA.

6a. Consider the reactions

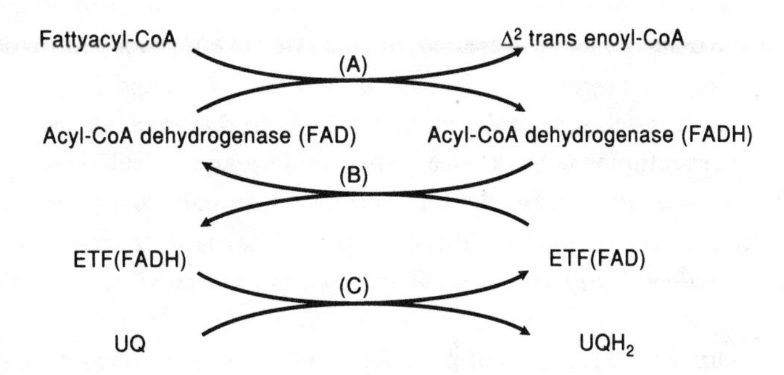

ETF is the electron-transfer flavoprotein, and UQ is the mitochondrial ubiquinone. Reaction (A) is catalyzed by fatty acyl-CoA dehydrogenase. Reaction (B) involves direct transfer of electrons between acyl-CoA dehydrogenase and ETF. Reaction (C) is catalyzed by the ETF: ubiquinone oxidoreductase. Oxidation of reduced ubiquinone (UQH_2) by the mitochondrial electron-transport system with reduction of O_2 represents an exergonic reaction with a large negative ΔG, rendering the reaction physiologically irreversible.

6b. A nonreducible analog of ubiquinone in principle would inhibit the ETF:UQ oxidoreductase competitively with respect to UQ. At sufficiently large concentrations, the oxidation of reduced ETF would halt. The reduced fatty acyl-CoA dehydrogenase has no known oxidant other than ETF. The dehydrogenase would soon accumulate in the reduced form and β-oxidation would cease. Moreover, the ubiquinone analog might inhibit oxidation of other metabolites whose reducing equivalents enter the electron-transport system at UQ.

7a. Nonhepatic tissues use ketone bodies as a rich energy source supplied by the liver. The ketone bodies are water soluble and easily transported in the blood. Oxidation of ketone bodies by heart muscle, for example, spares the oxidation of glucose, allowing the carbohydrate to be metabolized by erythrocytes and the brain. Oxidation of 1 mole acetoacetate yields a net 19 moles ATP.

7b. β-Hydroxybutyrate supplies an additional reducing equivalent per mole compared to acetoacetate. Oxidation of 1 mole β-hydroxybutyrate yields 1 mole acetoacetate plus 1 mole NADH. Subsequently, oxidation of the NADH provides 2.5 additional moles ATP compared to the oxidation of acetoacetate.

7c.

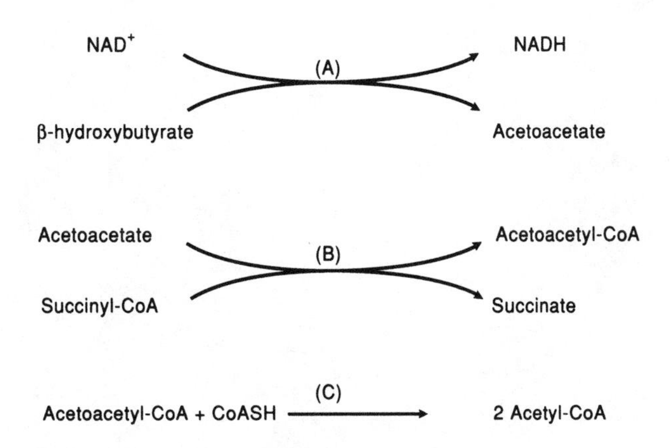

(A): NAD^+-dependent β-hydroxybutyrate dehydrogenase
(B): 3-ketoacyl-CoA transferase
(C): thiolase

8. Ketone bodies (β-hydroxybutyrate and acetoacetate) are synthesized by the liver-specific mitochondrial HMG-CoA synthase and HMG-CoA lyase and are transported in the blood to nonhepatic tissue. In nonhepatic tissues, β-hydroxybutyrate is oxidized to acetoacetate and subsequently activated to acetoacetyl-CoA. Acetoacetyl-CoA formation is catalyzed by β-oxyacid-CoA-transferase, an enzyme that transfers the coenzyme A group from succinyl-CoA to acetoacetate. If the transferase activity were markedly diminished, the rate of ketone body oxidation in nonhepatic tissue would be decreased, but the rate of ketone body formation in the liver likely would not diminish. Thus, there would be an increase in the acetoacetate and β-hydroxybutyrate concentration in the bloodstream. If the transferase were absent, ketone body oxidation in nonhepatic tissues would be abolished. The blood concentration of the ketone bodies would soar, particularly when the glucagon/insulin ratio increased, signaling increased lipid oxidation in the liver. Ketonuria and acidemia likely would result. In either case, the nonhepatic tissues could gain little or no metabolic energy from ketone body oxidation and would have greater dependency on glucose as a metabolic energy source. Glucose oxidation rates in those tissues would increase.
9. Linolenoyl-CoA is oxidized via three cycles of β-oxidation yielding the products $\Delta^{3,6,9}$-*cis*-dodecatrienoyl-CoA and 3 moles of acetyl-CoA. The Δ^{3}-*cis* double bond is isomerized to the Δ^{2}-*trans* isomer by enoyl-CoA isomerase. After one additional β-oxidation cycle, the product $\Delta^{4,7}$-*cis*-decadienoyl-CoA is oxidized to Δ^{2}-*trans*, $\Delta^{4,7}$-*cis*-decatrienoyl-CoA by the fatty acyl-CoA dehydrogenase. Reduction of this product by the NADPH-dependent 2,4-dienoyl-CoA reductase yields Δ^{3}-*trans*, Δ^{7}-*cis*-decadienoyl-CoA. Enoyl-CoA isomerase converts the Δ^{3}-*trans* to Δ^{2}-*trans* isomer, which is oxidized through two cycles of β-oxidation yielding Δ^{3}-*cis*-hexenoyl-CoA. Enoyl-CoA isomerase converts the isomer to the Δ^{2}-*trans* isomer and the product can now be oxidized completely to 3 acetyl-CoA's.

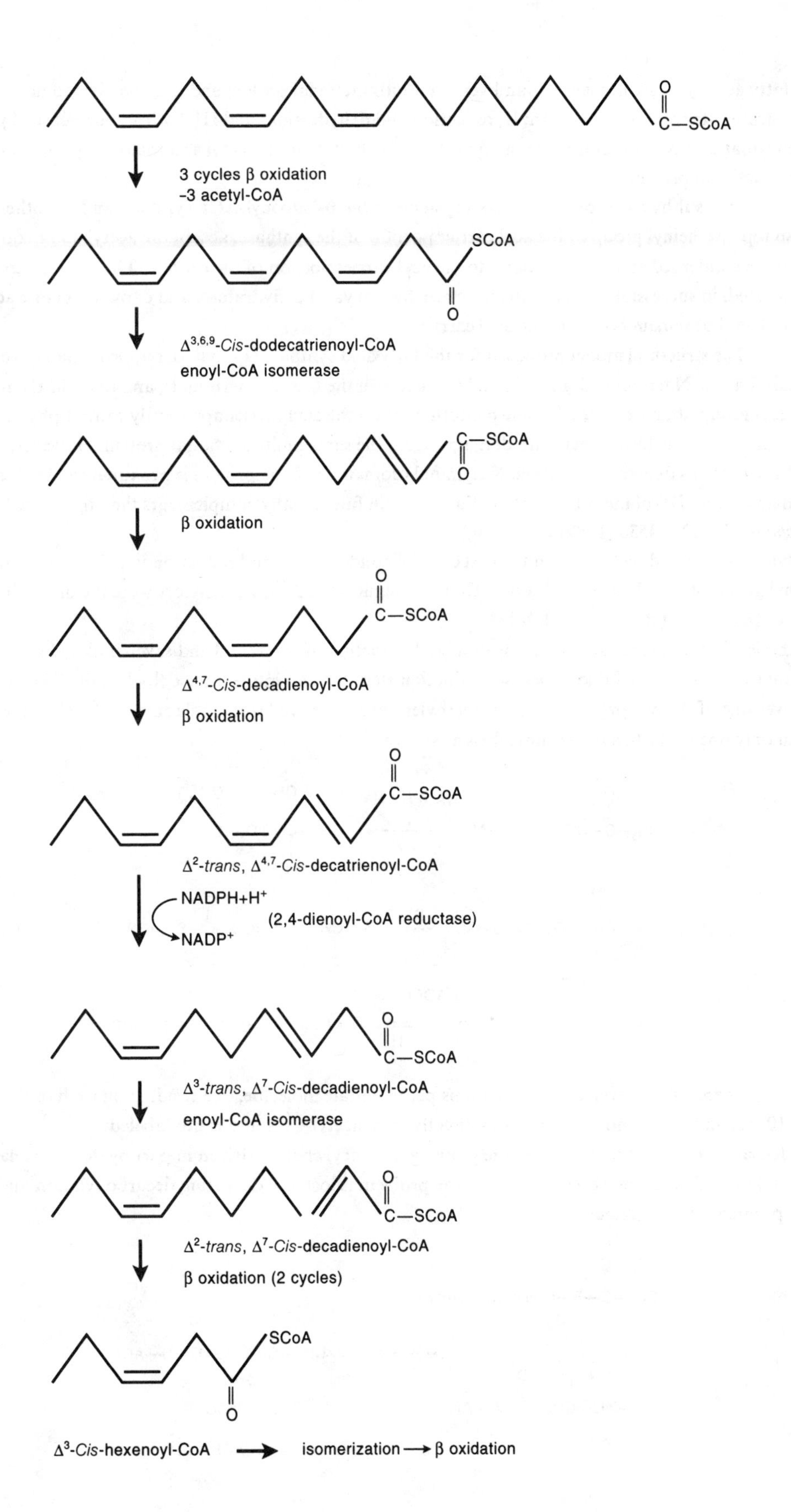
C—SCoA
3 cycles β oxidation
-3 acetyl-CoA
SCoA
Δ3,6,9-Cis-dodecatrienoyl-CoA
enoyl-CoA isomerase
C—SCoA
β oxidation
C—SCoA
Δ4,7-Cis-decadienoyl-CoA
β oxidation
C—SCoA
Δ2-trans, Δ4,7-Cis-decatrienoyl-CoA
NADPH+H+
NADP+
(2,4-dienoyl-CoA reductase)
C—SCoA
Δ3-trans, Δ7-Cis-decadienoyl-CoA
enoyl-CoA isomerase
C—SCoA
Δ2-trans, Δ7-Cis-decadienoyl-CoA
β oxidation (2 cycles)
SCoA
Δ3-Cis-hexenoyl-CoA
isomerization → β oxidation

10. The fatty acid synthase in mammalian liver is a multifunctional protein and may be divided into three domains: I, substrate entry and condensation; II, reduction and dehydration; and III, hydrolytic release of product. All of the enzymatic activities required for the synthesis of palmitate from acetyl and malonyl groups reside on the multifunctional protein.

Two sulfhydryl groups, one from cysteine in the β-ketoacyl-ACP synthase and the other from phosphopantetheinyl group of the acyl carrier portion of the synthase, accept the acetyl and malonyl groups, which are condensed and remain bound to the acyl carrier portion of the protein. The ketoacyl group must be transported, in succession, to the keto reductase, hydroxyacyl dehydratase, and enoyl reductase active sites on the protein, but remain bound to the acyl carrier.

The structural model proposed for the fatty acid synthase is a cylinder approximately 16 nm long, with domain I at the N terminal, domain II midway between the N and C terminals, and domain III at the C terminal. The acyl group attached to the flexible pantothenate on the acyl carrier apparently cannot physically reach, and thus cannot chemically interact, with domain I from the same multifunctional protein. However, the fatty acyl synthase proteins dimerize so that the N terminal (domain I) of one protein is proximal to the C terminal (domains II and III) of the other protein. Each protein functionally complements the other. (See Wakil, S., *Biochem.* 28: 4523–4530 (1989).)

11a. Acetyl-CoA is used directly only in the first cycle of condensation and reduction in palmitate synthesis. The methyl group of acetyl-CoA will become C-16 of palmitate. If 2-^{14}C acetyl-CoA were the only labeled metabolite, only C-16 of palmitate would be labeled.

11b. Synthesis of palmitate requires condensation and reduction of one acetyl and seven malonyl units. Each malonyl-CoA unit added undergoes two reduction steps with an intervening dehydration. The dehydration step removes one of the two protons from the methylene group derived from malaonyl-CoA. Thus the product will retain only one of the two deuterium labels as shown.

$$CH_3-\overset{O}{\overset{\|}{C}}-CD_2-\overset{O}{\overset{\|}{C}}-S-ACP \xrightarrow[+H^+]{NADPH \quad NADP^+} CH_3-\overset{OH}{\overset{|}{\underset{H}{\underset{|}{C}}}}-CD_2-\overset{O}{\overset{\|}{C}}-S-ACP$$

$$CH_3-\overset{OH}{\overset{|}{C}}H-CD_2-\overset{O}{\overset{\|}{C}}-S-ACP \xrightarrow{-D^+,\ OH^-} CH_3-CH{=}CD-\overset{O}{\overset{\|}{C}}-S-ACP$$

$$CH_3-CH{=}CD-\overset{O}{\overset{\|}{C}}-S-ACP \xrightarrow[+H^+]{NADPH \quad NADP^+} CH_3-CH_2-CHD-\overset{O}{\overset{\|}{C}}-S-ACP$$

There will be seven deuterium atoms per palmitate molecule, one residing on each of the carbons 2, 4, 6, 8, 10, 12, and 14. Carbon 16 originates directly from acetyl-CoA and is not labeled.

11c. Condensation of two-carbon units to the growing fatty acyl chain is driven in part by decarboxylation of the malonyl unit. The carbon shown labeled in the problem is lost as CO_2 during decarboxylation and is not incorporated into palmitate.

$$CH_3-\overset{O}{\overset{\|}{C}}-S\text{—acyl carrier protein}$$

$$^-O-{}^{14}C-CH_2-\overset{O}{\overset{\|}{C}}-S-ACP \dashrightarrow CH_3-\overset{O}{\overset{\|}{C}}-CH_2-\overset{O}{\overset{\|}{C}}-S-ACP \quad {}^{14}CO_2 \quad + \text{ ACP—SH}$$

12a. Citrate is formed from the condensation of acetyl-CoA and oxaloacetate in the mitochondrial matrix and is either oxidized by the TCA cycle enzymes or is exported to the cytosol to supply substrate for fatty acid biosynthesis. When metabolite supply to the mitochondria is elevated or in excess, and energy supply is being met, the excess metabolite, as citrate, is transported to the cytosol. The citrate is cleaved to acetyl-CoA and oxaloacetate by the ATP-dependent citrate lyase. Acetyl-CoA is the carbon source for palmitate synthesis.

Citrate plays a regulatory role in fatty acid synthesis as a potent activator of the acetyl-CoA carboxylase. Upon binding citrate, the liver acetyl-CoA carboxylase polymerizes to an aggregate (approximately 4–8 × 10^6 M_r) with a concomitant increase in catalytic activity. The rate of carboxylation of acetyl-CoA to form malonyl-CoA increases, as does palmitate synthesis. Thus, when energy and metabolite supply is increased, citrate is a source of substrate and an activator of fatty acid synthesis.

12b.

(A)

$$\text{Citrate}_{\text{mitochondrial}} \rightarrow \text{citrate}_{\text{cytoplasmic}}$$

(B)

$$\text{Citrate}_{\text{cytoplasmic}} + \text{ATP} + \text{CoASH} \rightarrow \text{acetyl-CoA} + \text{OAA}$$

(C)

$$\text{OAA} + \text{NADH} \rightarrow \text{L-malate} + \text{NAD}^+$$

(D)

$$\text{L-malate} + \text{NADP}^+ \rightarrow \text{pyruvate} + \text{CO}_2 + \text{NADPH}$$

(E)

$$\text{Pyruvate}_{\text{cytoplasmic}} \rightarrow \text{pyruvate}_{\text{mitochondrial}}$$

(A) Citrate transport system
(B) ATP-dependent citrate lyase
(C) Cytoplasmic malate dehydrogenase
(D) $NADP^+$-malic enzyme
(E) Pyruvate transporter

Citrate supplies acetyl-CoA and reducing equivalents (NADPH) for fatty acid synthesis. Alternatively, L-malate may be returned to the mitochondrial matrix via the malate shuttle, bypassing reactions (D) and (E).

13. NADPH for fatty acid biosynthesis is supplied by the glucose-6-phosphate and 6-phosphogluconate dehydrogenases in the pentose pathway and by the $NADP^+$-dependent-malic enzyme. Oxaloacetate released by the ATP-citrate lyase may be reduced to L-malate by the cytoplasmic NAD^+-dependent malate dehydrogenase. Malic enzyme oxidatively decarboxylates L-malate to pyruvate and reduces $NADP^+$ to NADPH. Pyruvate enters the mitochondrial matrix for resynthesis of oxaloacetate by carboxylation.

Palmitate formation requires seven cycles of malonyl-CoA condensation, each of which requires 1 mole NADPH for reduction of the ketoacyl group to β-hydroxyacyl-ACP and another to reduce the enoyl-ACP. Reduction of 1 mole of acetyl-CoA and 7 moles of malonyl-CoA to form 1 mole of palmitate requires 14 moles NADPH.

14a. Both the β-oxidation of fatty acids and palmitate synthesis occur in the liver cell. The oxidation and synthesis of fatty acids, although mechanistically different, may be considered metabolically opposing pathways and, if not regulated, could result in depletion of cellular ATP and reductant levels. Were the pathways in the same cellular compartment, the rate-limiting step of each pathway would likely be coordinately regulated. Recall the activation of phosphofructokinase-1 (glycolysis) and inhibition of fructose bisphosphatase-1 (gluconeogenesis) by the fructose-2,6-bisphosphate.

Another strategy used by the cell to regulate metabolism in opposing pathways is separation of the pathways into different cellular compartments. Thus, regulation of metabolite supply may then limit activity of the pathway. Enzymes catalyzing β-oxidation of fatty acids are mitochondrial; those synthesizing palmitate are cytoplasmic.

14b. The carnitine acyltransferase associated with the outside of the inner mitochondrial membrane (CAT I) is inhibited by malonyl-CoA with an inhibition constant (K_i) of approximately 1 to 2 micromolar. (See McGarry, J. D., and D. W. Foster, *Annu. Rev. Biochem.* 49:395–420 (1980).) Inhibition of the carnitine acyltransferase effectively prevents translocation of the long-chain fatty acid from the cytoplasm into the mitochondrial matrix, the site of β oxidation. An inhibitor of CAT I would be expected to exert the same inhibitory effect as malonyl-CoA on β oxidation.

15. Thiolesterase activity of the fatty acid synthase is more active with palmityl-ACP than with shorter fatty acyl ACPs. Palmitate, the 16-carbon fatty acid, is thus preferentially released.

18 Biosynthesis of Amino Acids

Problems

1. Molecules with structures as diverse as carbamoylphosphate, tryptophan, and cytidine triphosphate are feedback inhibitors of the *E. coli* glutamine synthase. The feedback inhibition is cumulative, with each metabolite exerting a partial inhibition of the enzyme. Why would complete inhibition of the glutamine synthase by a single metabolite be metabolically unsound?

2. Given the structural diversity of the compounds that feedback-inhibit glutamine synthase, would you predict that they interact at a common regulatory site? Why or why not?

3. A mutant of *E. coli* is discovered with a defect in serine hydroxymethyltransferase. What amino acid supplement(s) would you expect the mutant to require? Explain. Although the mutant does not require L-methionine, its growth is stimulated by addition of this amino acid to the medium. Suggest a reason for this observation.

4. Aspartate, asparagine, glutamate, glutamine, and proline are among the amino acids that are not essential for humans. Glucose may be considered the precursor for each of these amino acids. Explain.

5. How does increased synthesis of aspartate and glutamate affect the TCA cycle? How does the cell accommodate this effect?

6. Contrast the mechanisms of amidation of glutamate to glutamine and aspartate to asparagine.

7. For each carbon in serine (α carboxyl; α carbon; β carbon) indicate which carbons of glucose-6-phosphate are incorporated. Assume that glucose is supplied solely from the medium and is not being synthesized from precursors.

8. A genetic defect in cystathionine-β-synthase leads to homocystinuria in humans. Explain.

9. The biosynthesis of L-proline from glutamate involves an internal amination rather than a transamination. Explain what is meant by an "internal amination" in the context of proline biosynthesis.

10. Write a scheme illustrating the catalytic role of homocysteine in methyl transfer reactions involving *S*-adenosylmethionine and the pool of one-carbon metabolites.

11. In what sense may indole be viewed as an "intermediate" in L-tryptophan biosynthesis?

12. The accumulation of biosynthetic intermediates, or of metabolites derived from these intermediates, has proven to be valuable in the analysis of biosynthetic pathways in microorganisms. It was found that these accumulations occurred only after the required amino acid had been consumed and growth had stopped. How might you account for this observation?

13. When ^{14}C-labeled 4-hydroxyproline was administered to rats, the 4-hydroxyproline in newly synthesized collagen was not radiolabeled. Explain.

14. Although phenylalanine is an amino acid essential to humans, tyrosine is not. Explain why this is true.

Solutions

1. Glutamine synthase catalyzes the formation of glutamine, the organically bound nitrogen of which is used in the biosynthesis of a number of structurally unrelated nitrogen-containing compounds. Glutamine synthase activity is well regulated, as you would predict for an enzyme at the beginning of a multibranched biosynthetic system. The *E. coli* glutamine synthase is regulated by feedback inhibition exerted by products of pathways dependent on glutamine, and by covalent modification.

   The metabolites carbamoylphosphate, tryptophan, and cytidine triphosphate, as well as glucosamine-6-phosphate, histidine, and AMPs, are synthesized by pathways that incorporate nitrogen directly from glutamine. The supply of glutamine must respond to the demand generated by the several pathways. Hence, total inhibition of the glutamine synthase by a single product could inhibit biosynthesis of critically needed end products from other pathways. Cumulative inhibition of the glutamine synthase by end products of the various pathways modulates the supply of glutamine in direct response to the demand.

2. Glutamine synthase is allosterically inhibited upon binding each of several metabolites to specific sites on the enzyme. The regulation requires that the structure of an inhibitory metabolite bind to a specific portion of the protein in a process not unlike that of substrate binding to the active site of an enzyme. Molecules whose structures are similar may bind to the same regulatory site, but molecules whose structures are dissimilar likely require separate binding sites. Thus, the structures of the diverse regulatory metabolites require different binding sites on the glutamine synthase and are required to explain cumulative inhibition.

3. In the absence of cellular serine hydroxymethyltransferase, the mutant would be unable to synthesize glycine from serine and would be dependent on glycine supplied in the medium. Glycine is not only required for protein biosynthesis but is also an important donor to the one-carbon pool. A number of biosynthetic pathways (e.g., purine and methionine pathways) depend on an adequate pool of one-carbon metabolites.

   Methionine may be synthesized by methylation of homocysteine. In a mutant organism dependent on glycine supplied by the medium, the one-carbon pool may restrict growth rate. Addition of methionine would ease the demand on the one-carbon pool and should increase the growth rate of the mutant.

4. Oxaloacetate is the precursor from which aspartate and asparagine are synthesized, and α-ketoglutarate is the precursor of glutamate and glutamine. Oxaloacetate and α-ketoglutarate are TCA cycle intermediates whose concentrations can be maintained by glucose metabolism. Glucose is metabolized to pyruvate in the glycolytic pathway, and the pyruvate is oxidatively decarboxylated by the pyruvate dehydrogenase, forming acetyl-CoA. Pyruvate may also be carboxylated by the acetyl-CoA-dependent pyruvate carboxylase. Citrate, formed from the condensation of oxaloacetate and acetyl-CoA, is isomerized by aconitase to form isocitrate, and the isocitrate is oxidatively decarboxylated to α-ketoglutarate by the NAD^+-dependent isocitrate dehydrogenase. The α-ketoglutarate may be transaminated to form glutamate, which is subsequently amidated to form glutamine. Proline is formed by the reduction of the γ-carboxyl group of glutamate to γ-glutamylsemialdehyde and sequential internal condensation and reduction. (See solution 9 for details.)

   Oxaloacetate formed from the carboxylation of pyruvate is transaminated to form aspartate, and the aspartate subsequently amidated to form asparagine.

5. Synthesis of aspartate by transamination of oxaloacetate and of glutamate by transamination of α-ketoglutarate depletes the concentration of oxaloacetate and α-ketoglutarate in the TCA cycle. Were these TCA cycle intermediates not replenished, the rate of acetyl-CoA oxidation and subsequent ATP production would be markedly diminished. Pyruvate, derived from the glycolytic metabolism of glucose, is used to replenish the concentration of each of the cycle intermediates. As noted in the previous solution, oxaloacetate is formed from the carboxylation of pyruvate, catalyzed by pyruvate carboxylase. Synthesis of citrate, from the condensation of oxaloacetate and acetyl-CoA, in turn replenishes the concentration of the other intermediates.

6. Formation of the amide linkage to both glutamate and aspartate requires activation of the carboxyl group. In each case, amidation of glutamate or aspartate requires the hydrolysis of ATP, but the products of the hydrolysis are different. Amidation of glutamate is activated by transfer of phosphate from ATP, yielding γ-

glutamylphosphate (a mixed anhydride) and ADP. Amidation is driven by the displacement of phosphate, a good leaving group, from the enzyme-bound γ-glutamylphosphate.

The β-carboxyl group of aspartate is also activated by formation of a mixed anhydride between the carboxylate and the AMP portion of ATP. Pyrophosphate is released and is hydrolyzed to orthophosphate by pyrophosphatase. AMP is the leaving group displaced by the amine group donated by glutamine. Whereas ADP is rephosphorylated to ATP by addition of one high-energy phosphate linkage, regeneration of ATP from AMP requires the input of two high-energy phosphate groups.

7. Serine is synthesized from 3-phosphoglycerate, a glycolytic pathway intermediate. In glycolysis glucose-6-phosphate is metabolized to fructose-1,6-bisphosphate and cleaved by FBP aldolase yielding glyceraldehyde-3-phosphate and dihydroxyacetone phosphate. The numbers by the structures indicate the corresponding carbon from glucose-6-phosphate.

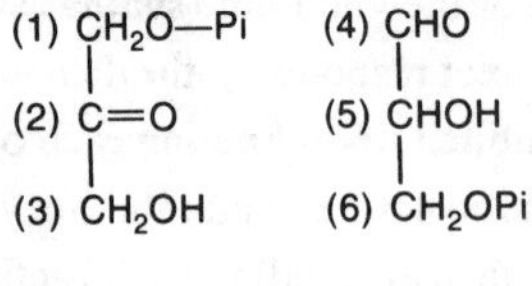

DHAP	Ga3P
(1) CH_2O—Pi	(4) CHO
(2) C=O	(5) CHOH
(3) CH_2OH	(6) CH_2OPi
(DHAP)	(Ga3P)

Triose phosphate isomerase interconverts dihydroxyacetone phosphate and glyceraldehyde-3-phosphate without scrambling the carbon chain, so labeling of glyceraldehyde-3-phosphate will be

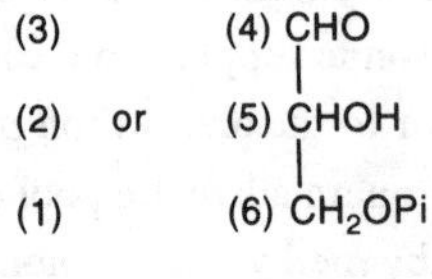

(3)		(4) CHO
(2)	or	(5) CHOH
(1)		(6) CH_2OPi

Oxidation of glyceraldehyde-3-phosphate to 3-phosphoglycerate occurs without rearranging the carbon chain, and serine will contain the carbons of glucose-6-phosphate as shown.

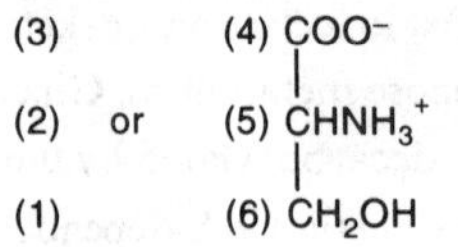

(3)		(4) COO^-
(2)	or	(5) $CHNH_3^+$
(1)		(6) CH_2OH

8. Cystathionine-β-synthase is a key enzyme in the synthesis of cysteine and catabolism of methionine in mammals. Methionine, the source of sulfhydryl group for cysteine biosynthesis, is converted to *S*-adenosylmethionine, a methyl donor. Following methyl transfer, the *S*-adenosylhomocysteine is hydrolyzed to homocysteine and adenosine. Homocysteine is condensed with L-serine to form cystathionine. Sulfur transfer occurs in the γ-cystathionase-catalyzed elimination of cysteine from cystathionine.

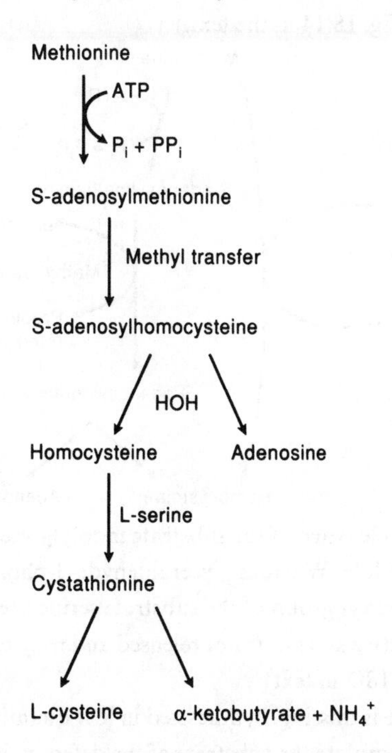

However, a genetic deficiency of cystathionine-β-synthase would inhibit the formation of cystathionine with a subsequent accumulation of homocysteine. The homocysteine accumulates in the liver and subsequently is found in blood and is excreted as the oxidized product homocystine. (See fig. 19.30 in text.)

9. Glutamate is the precursor for proline biosynthesis in humans and other animals. The γ-carboxyl group is activated by phosphoryl transfer from ATP, and the γ-glutamylphosphate is reduced to the γ-glutamylsemialdehyde. γ-Glutamylsemialdehyde forms an internal Schiff's base between the γ aldehyde and the α-amino group. The Δ^1pyrroline-5-carboxylate is reduced, forming proline. The Schiff's base formation can be considered an "internal amination" because the carbonyl group and amino group are constituents of the same molecule. Transamination is the process of aminotransfer from a donor α-amino acid to a different α-keto acid catalyzed by pyridoxal phosphate-dependent aminotransferases.

10. Although an important amino acid constituent of proteins, methionine has a second role as a source of methyl groups in a number of methyltransfer reactions. The methyl group is activated for transfer upon formation of *S*-adenosylmethionine. *S*-adenosylhomocysteine, remaining after methyl transfer, is hydrolyzed to homocysteine and adenosine (see problem 8). Homocysteine may also be methylated via catalytic transfer of CH_3- from N^5-methyl tetrahydrofolate. The catalytic role of homocysteine in transferring methyl groups from the one-carbon pool to methionine is shown. (See fig. 18.14 in the text.)

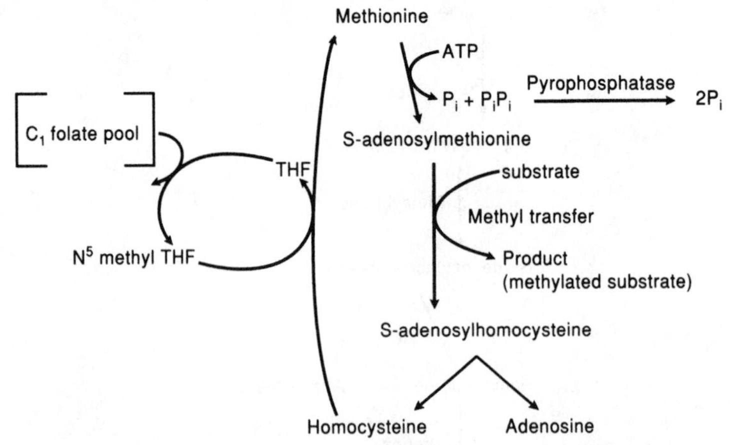

11. Tryptophan synthase catalyzes the cleavage of the substrate indolylglycerol phosphate, yielding the products glyceraldehyde-3-phosphate and indole. Whereas glyceraldehyde-3-phosphate is released, indole is retained at the active site and replaces the hydroxyl group of the substrate serine used to form tryptophan. Indole is both formed and used at the synthase active site but is not released and may be considered an enzyme-bound intermediate in synthesis. (See box 18D in text)

12. Mutants in which a specific enzyme is missing, synthesized in low quantity, or synthesized with altered kinetic properties, in principle should accumulate the substrate of the defective enzyme. If the mutant cells were supplied with the end product of the biosynthetic pathway, the committed step to the pathway likely would be inhibited, or synthesis of enzymes of the pathway would be repressed. No accumulation of substrate for the altered enzyme would be observed. Upon consumption of the end product of the pathway, feedback inhibition is alleviated and synthesis of enzymes required in the biosynthetic pathway is derepressed. The activity of each of the enzymes in the pathway is needed to supply the required end product. Were the activity of an enzyme missing or severely curtailed, we would anticipate accumulation of metabolite (or some derivative of the metabolite) on the substrate side of the altered enzyme in the metabolic pathway.

13. Hydroxylation of proline in collagen is a posttranslational modification. Proline, rather than 4-hydroxyproline, is incorporated into collagen precursors and is subsequently hydroxylated by proline-4-hydroxylase. Neither free proline nor proline bound to the prolyl-t-RNA are hydroxylated. See Borstein, P. *Annu. Rev. Bioch.* 43:567—603, 1974. The hydroxylated peptides are then assembled and processed into collagen.

14. Phenylalanine-4-monooxygenase, an enzyme present in the human liver, catalyzes the hydroxylation of phenylalanine to form tyrosine. Although phenylalanine is not synthesized by humans and is therefore an essential amino acid, tyrosine is readily synthesized from phenylalanine, except in individuals with phenylketonuria who are deficient in phenylalanine 4-monoxygenase.

19 The Metabolic Fate of Amino Acids

Problems

1. What consequences would a severe deficiency in pyridoxal phosphate have on amino acid metabolism?

2. Differentiate between transamination and oxidative deamination as mechanisms to deaminate amino acids.

3. Compare the mechanisms for release of ammonia from the amine and amide groups of glutamine used in urea formation.

4. Would you anticipate elevated arginase activity in the liver of an untreated diabetic animal? Why or why not?

5. Explain the central role of pyruvate in amino acid catabolism in oxygen-limited muscle tissue.

6. Glutamic acid, valine, and aspartic acid are called glycogenic amino acids, whereas lysine and leucine are called ketogenic amino acids. Explain. Why are tyrosine and phenylalanine considered both ketogenic and glycogenic amino acids?

7. Glutamate and aspartate are deaminated to α-keto acids that are TCA cycle intermediates, yet cannot be completely oxidized by the TCA cycle alone. Explain. Propose a pathway for the complete oxidation of the deamination products of glutamate and aspartate.

8. Predict the effect of glycine cleavage enzyme deficiency on (a) a one-carbon pool and (b) energy production from glycine catabolism.

9. In what way is arginine catabolism uniquely associated with the "high-energy phosphate" reservoir in the skeletal muscle?

10. (a) The branched-chain α-keto acid dehydrogenase catalyzes an oxidative decarboxylation that requires thiamine pyrophosphate, lipoamide, and FAD as cofactors. Review the chemistry of the pyruvate dehydrogenase (an enzyme that also catalyzes an oxidative decarboxylation) and suggest a mechanism for the branched-chain α-keto acid dehydrogenase.

10. (b) In principle, why is it possible for individuals with decreased activity of dihydrolipoamide dehydrogenase to also have increased serum levels of the branched-chain α-keto acids?

11. Isoleucine, valine, and methionine are all glycogenic amino acids. Genetic defects that affect catabolism of these amino acids include deficiency in propionyl-CoA carboxylase and methylmalonyl-CoA mutase activities. Given this information, propose a common pathway to glucose-6-phosphate from these amino acids.

12. Some patients with the deficiencies described in problem 11 accumulate propionate, whereas methylmalonic acidemia is observed in other patients. In each case, explain why the accumulated metabolite is predicted to result from the genetic deficiency.

13. A patient is found to have a defect in the enzyme 4-hydroxyphenylpyruvate dioxygenase. What diet would you recommend as treatment for this condition? What is the basis for your recommendation?

14. The concentration of phenylalanine in the blood of neonates is used to screen for phenylketonuria (PKU). Explain the biochemical basis for the correlation of elevated blood phenylalanine concentration and PKU. Explain why restriction of dietary phenylalanine is critically important for youngsters with PKU.

15. (a) L-Glutathione is not a primary gene product as are proteins. What "information" is used to direct the synthesis of L-glutathione?

(b) Predict the effect of a glutathione synthase inhibitor on cells exposed to oxidative stress.

Solutions

1. Pyridoxal phosphate is the cofactor used by aminotransferases (transaminases) to catalyze transfer of α-amino groups from donor amino acids to acceptor α-keto acids. The broad specificities of the aminotransferases for amino acids allow funneling of the amino groups from a range of amino acids to a few keto acid acceptors, primarily α-ketoglutarate, oxaloacetate, and pyruvate. α-Ketoglutarate is transaminated to glutamate and may then be amidated to glutamine. Oxaloacetate and pyruvate are transaminated to aspartate and alanine, respectively. Glutamate (and glutamine) and aspartate are sources of amino groups used in urea biosynthesis in the liver.

   The α-keto acids remaining from transamination of the various amino acids are oxidized to provide energy, used in gluconeogenesis, and are substrates for numerous anabolic pathways. A serious deficiency of pyridoxal phosphate would curtail aminotransferase activity and decrease the flux of α-keto acids and their metabolites through the energy-producing and biosynthetic pathways.

2. Transamination is the process of removal of the α-amino group from a donor amino acid, forming an α-keto acid, and addition of the amino group to an acceptor α-keto acid, forming a different α-amino acid. The amino group is retained as an organically bound functional group with no net loss of the amine as ammonium ion. There is no net change in oxidation state between the substrates and products.

   Transamination is used in catabolism to channel the α-amino groups from the various amino acids provided by diet or protein catabolism to a limited number of α-keto acid acceptors (i.e., α-ketoglutarate, oxaloacetate, and pyruvate) used directly or indirectly in urea formation. In anabolic (biosynthetic) pathways, transamination supplies amino groups for the formation of amino acids.

   Deamination is an oxidative process wherein the α-amino group is oxidized to the α-imino function, with subsequent hydrolysis of the iminoacid. The products of oxidative deamination are ammonium ion, α-keto acid, and reduced electron-transfer cofactor (flavin or pyridine nucleotide). The ammonium ion is used in the synthesis of carbamoyl phosphate but can be toxic to the cell if not maintained in low concentration.

   Transamination and deamination are linked in ureogenesis as shown in these reactions.

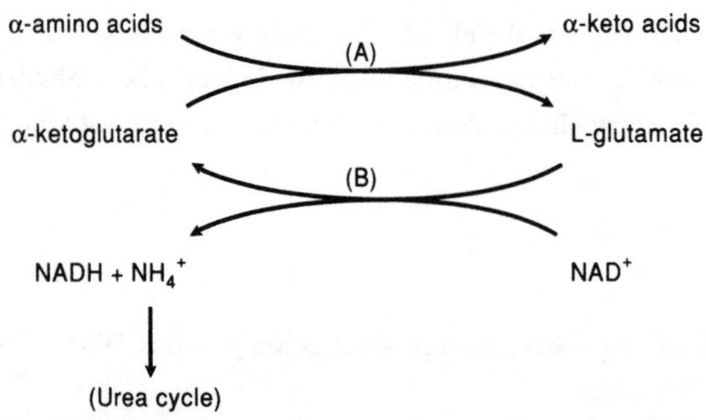

Reaction (A) is catalyzed by PLP-dependent amino transferases. Reaction (B) is an oxidative deamination catalyzed by glutamate dehydrogenase.

3. Deamination of glutamine provides two molar equivalents of ammonium ion for urea biosynthesis. Both the α-amino group and the amide group are released in the liver and incorporated into urea. The amide nitrogen is released as ammonium ion after hydrolysis catalyzed by glutaminase. Glutamate is the other product. Glutamate dehydrogenase oxidatively deaminates glutamate to ammonium ion plus α-ketoglutarate. The α-ketoglutarate may be used in another cycle of transamination and deamination, and the NADH is reoxidized by the electron-transport system.

4. The untreated diabetic animal synthesizes glucose primarily by hepatic gluconeogenesis utilizing amino acids derived from protein catabolism as the carbon source. Recall that the ATP required for gluconeogenesis is generated by fatty acid β oxidation.

   The amino acids are transaminated with α-ketoglutarate and oxaloacetate, forming glutamate and aspartate to be used in ureogenesis. The α-keto acids (those from glycogenic amino acids) then enter gluconeogenesis as pyruvate or as TCA cycle intermediates.

   The urea cycle activity must increase to accommodate the increased flux of amino groups removed from the amino acids. Arginase catalyzes the hydrolysis of arginine yielding urea plus ornithine and is the rate-limiting step in the urea cycle. Hence you would predict an increased arginase activity in the liver of the untreated diabetic animal. (Bond, J. S., M. L. Failla, and D. F. Unger, *J. Biol. Chem.* 258:8004–8009 (1983).)

5. A portion of the skeletal muscle protein or amino acid pool may be catabolized when glucose supply to muscle is limited. The amino acids are transaminated with pyruvate, yielding alanine plus α-keto acids used subsequently in energy metabolism. Alanine transports the ammonium ion equivalents from the muscle to the liver as a nontoxic metabolite. In the liver, the amino group of alanine enters the urea cycle, whereas pyruvate is a substrate for glucose synthesis (gluconeogenesis). Glucose is thus resupplied to the bloodstream. The glucose-alanine cycle transports amino groups from the muscle to the liver and glucose from liver to skeletal muscle. In this metabolic scheme, alanine (via pyruvate) rather than lactate is the gluconeogenic substrate.

6. Glutamic acid, valine, and aspartate are catabolized to α-ketoglutarate, succinyl-CoA, and oxaloacetate, respectively. Each of these metabolites is a TCA cycle intermediate convertible to L-malate. L-Malate is transported to the cytoplasm, oxidized to oxaloacetate, and used in gluconeogenesis. The carbon skeletons of Glu, Val, and Asp may be incorporated into glucose (or glycogen), hence the designation glycogenic. Leucine and lysine are catabolized to acetyl-CoA and acetoacetyl-CoA. Acetoacetate (or the reduced form β-hydroxybutyrate) are ketone bodies released from the liver for metabolism by nonhepatic tissue. Acetyl-CoA is condensed with oxaloacetate to form citrate. The citrate may either be oxidized via the TCA cycle or be exported to the cytosol where it is cleaved by ATP-citrate lyase to acetyl-CoA and oxaloacetate. Cytoplasmic acetyl-CoA is used to synthesize palmitate. The carbon skeletons of Leu and Lys are metabolized to ketone bodies or fatty acid but not glucose. Tyrosine and phenylalanine are catabolized to acetoacetate (ketogenic substrate) and fumarate (glycogenic substrate).

7. Glutamate and aspartate are catabolized to TCA cycle intermediates. Glutamate is oxidatively deaminated to α-ketoglutarate, which may be oxidized to oxaloacetate. Aspartate is transaminated directly to oxaloacetate. The TCA cycle regenerates oxaloacetate during the oxidation of acetyl-CoA so there can be no direct net oxidation

of oxaloacetate. One pathway leading to the oxidation of oxaloacetate is reduction of OAA to L-malate in the mitochondrial matrix, translocation of the L-malate to the cytoplasm, and oxidative decarboxylation of the malate to pyruvate. The latter reaction is catalyzed by the cytoplasmic $NADP^+$-malic enzyme. The pyruvate enters the mitochondrial matrix where it is oxidized by the pyruvate dehydrogenase to acetyl-CoA. The acetyl-CoA is then oxidized to two CO_2 equivalents.

8. Glycine may be catabolized by either of two routes. Glycine may be oxidized by the glycine cleavage enzyme in the mitochondrial matrix, yielding NADH and N^5,N^{10}-methylene tetrahydrofolate, or, glycine may be transhydroxymethylated to form serine and converted to pyruvate by β-elimination of the serine —OH group.

   Catabolism of glycine by the glycine cleavage enzyme augments the C-1 pool (1 mole N^5,N^{10}-methylene THF per mole glycine) but provides only approximately 2.5 moles ATP per mole glycine if the NADH is oxidized via the electron-transfer system. If glycine is converted to serine and subsequently to pyruvate, the C-1 pool is depleted by one mole per mole serine formed. However, complete oxidation of the pyruvate by pyruvate dehydrogenase, TCA cycle, and electron-transfer system provides approximately 12.5 moles ATP per mole pyruvate.

   Deficiency of glycine cleavage enzyme would result in decreased contribution of glycine to the C-1 pool or, at worst, a drain on the pool. Net ATP yield would be increased because of the pyruvate oxidation.

9. Oxidation of amino acids whose carbon skeletons are substrates for gluconeogenesis, or provide acetyl-CoA, acetoacetate, or β-hydroxybutyrate may contribute to the "high-energy" pool in the muscle mitochondria. Catabolism of arginine, after conversion to ornithine, may also contribute to the energy reserves. However, arginine is also a substrate for synthesis of creatine. The amidino group of arginine is transferred to glycine as a step in creatine biosynthesis. Creatine is phosphorylated by ATP forming a phosphoramide, creatine phosphate. Creatine phosphate in turn transfers the phosphate to ADP generated concomitantly with muscle contraction. Each of the phosphoryl transfer steps is catalyzed by creatine kinase, an enzyme that exists as both cytosolic and mitochondrial isoforms. Bessman and Carpenter have reviewed the role of creatine and creatine kinases in shuttling phosphoryl groups from the mitochondria to the muscle fibrils. (Bessman, S. P., and C. L. Carpenter, *Annu. Rev. Biochem.* 54:931–62 (1985).)

10a. TPP, lipoamide, and CoASH reactions yielding acyl-CoA are shown here.

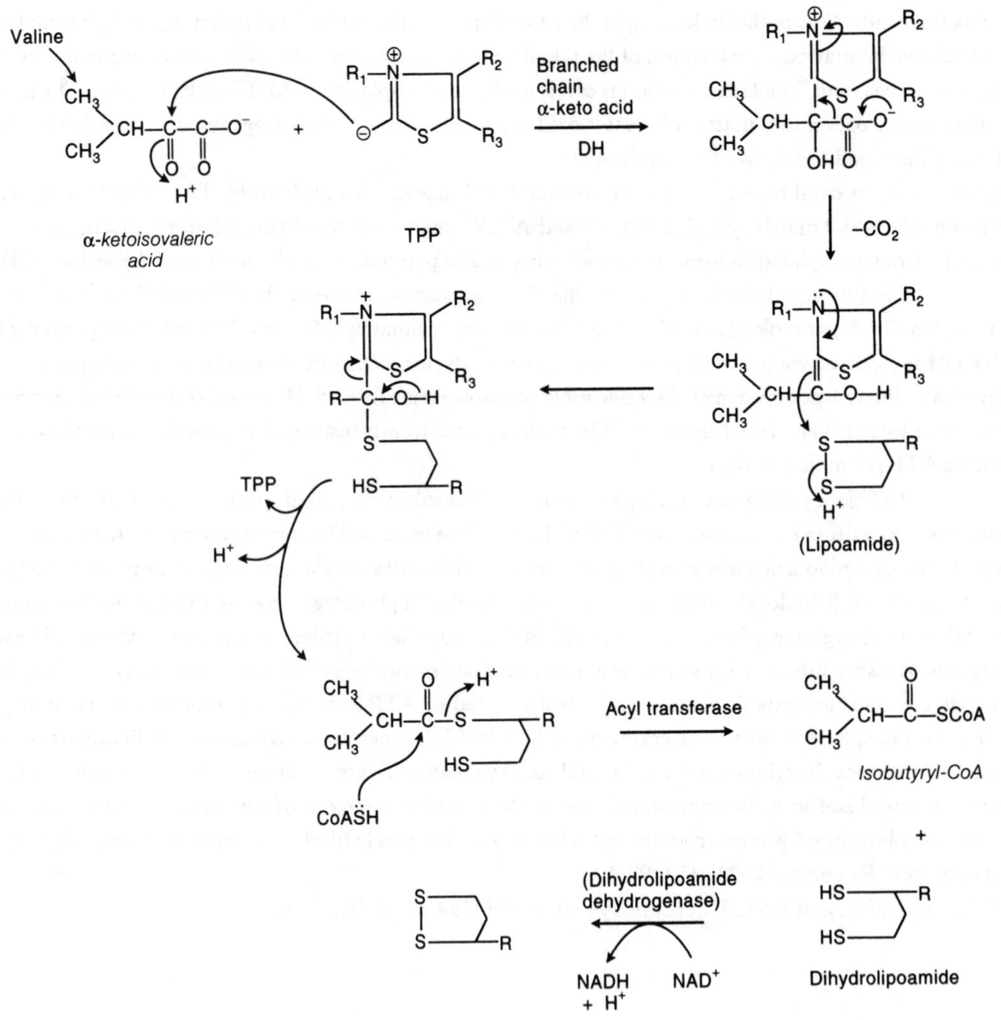

10b. The branched-chain **α**-keto acid dehydrogenase oxidizes the **α**-carbonyl to a carboxyl group that is trapped as a thioester with coenzyme A. The lipoamide is reduced to dihydrolipoamide in the process and is reoxidized by the flavoprotein lipoamide dehydrogenase with transfer of reducing equivalents to NAD^+. Were the lipoamide dehydrogenase activity limited, the rate of oxidation of the branched-chain **α**-keto acids would be diminished, potentially causing an increased concentration of the metabolite in the blood plasma.

11. Isoleucine, valine, and methionine are catabolized to the common intermediate propionyl-CoA. Metabolism of propionyl-CoA derived from catabolism of these amino acids (or by **β** oxidation of odd-chain fatty acids and pristanoyl-CoA) requires the biotin-dependent carboxylation to D-methylmalonyl-CoA, inversion to the L-isomer, and rearrangement to succinyl-CoA. The latter reaction is catalyzed by the vitamin B_{12}-dependent methylmalonyl-CoA mutase. Succinyl-CoA, a TCA cycle intermediate, is metabolized to succinate by succinate thiokinase, and oxidized to L-malate. L-Malate, transported to the cytoplasm, is oxidized to oxaloacetate, a substrate for gluconeogenesis.

12. Propionyl-CoA, the common catabolite of isoleucine, valine, and methionine, must be metabolized to succinyl-CoA for entry into the TCA cycle. Propionyl-CoA carboxylase, methylmalonyl-CoA racemase, and L-methylmalonyl-CoA mutase are required for the synthesis of succinyl-CoA from propionyl-CoA. The first of these enzymes, propionyl-CoA carboxylase, requires biotin for the activation of CO_2 and subsequent transfer of

the carboxyl group to propionyl-CoA. If the propionyl-CoA carboxylase were not active, propenyl-CoA would accumulate and the hydrolytic product, propionate, would likely be excreted in the urine.

The conversion of L-methylmalonyl-CoA to succinyl-CoA is a rearrangement requiring vitamin B_{12}, catalyzed by the methylmalonyl-CoA mutase. Were this enzyme deficient, methylmalonyl-CoA and the hydrolysis product, methylmalonate, would accumulate. Methylmalonate would eventually be excreted in the urine.

13. 4-Hydroxyphenylpyruvate is a metabolite common to the catabolism of tyrosine, and of phenylalanine, which is hydroxylated to tyrosine. Tyrosine is transaminated to 4-hydroxyphenylpyruvate and oxidized to homogentisate by the 4-hydroxyphenylpyruvate dioxygenase. Deficiency in the latter enzyme causes accumulation of the substrate, 4-hydroxyphenylpyruvate, with subsequent (presumably) deleterious effects. A diet containing only maintenance levels of phenylalanine (an essential amino acid) and deficient in tyrosine is recommended to prevent accumulation of 4-hydroxyphenylpyruvate.

14. The phenylalanine supplied by the diet is used in the biosynthesis of proteins or is catabolized in a pathway dependent on the formation of tyrosine. The formation of tyrosine is catalyzed by phenylalanine 4-monooxygenase, an enzyme whose activity is deficient in phenylketonurics. As observed in other metabolic pathways, concentrations of metabolites on the substrate (supply) side of a metabolic block will accumulate to above normal levels, whereas concentrations of metabolites on the product side will be markedly diminished. Hence, phenylalanine, the substrate of phenylalanine monooxygenase, will accumulate, an event reflected in the blood concentration of phenylalanine. Tourian and Sidbury report that blood levels in excess of 20 mg phenylalanine per 100 ml blood of individuals on a nonrestricted diet are considered diagnostic for phenylketonuria or PKU-like disorders. (Tourian, A. Y., and J. B. Sidbury, in *The Metabolic Basis of Inherited Disease.* eds. J. B. Stanbury, J. B. Wyngaarden, and D. S. Fredrickson. New York: McGraw-Hill, 1978, 240–255.)

Accumulation of phenylalanine or products of the minor secondary pathway for phenylalanine metabolism (phenylpyruvate, phenyllactate) apparently adversely affect neuronal development, leading to severe mental retardation. Accumulation of the products may be prevented by management of dietary intake of phenylalanine.

15a. L-Glutathione is synthesized in successive steps catalyzed by γ-glutamylcysteine synthase and glutathione synthase. The information for the synthesis is dictated by the specificity of each enzyme. These enzymes are primary gene products (proteins), whereas L-glutathione is a secondary gene product. The formation of γ-glutamylglycine or cysteinylglycine is prevented because of the lack of the appropriate enzymes.

15b. Glutathione, a substrate for glutathione peroxidase, is used catalytically in the cell as an antioxidant but is also a substrate consumed stoichiometrically by glutathione transferases. Loss of glutathione by conjugation with certain xenobiotics slated for excretion could compromise the cell's antioxidant defenses. Hence the cell requires an adequate supply of glutathione. One role of glutathione as an antioxidant is shown

$$2\ GSH + H_2O_2 \rightarrow G\text{-}S\text{-}S\text{-}G + 2\ H_2O \quad (Rxn\ 1)$$

where GSH and G-S-S-G are reduced and oxidized glutathione, respectively.

$$G\text{-}S\text{-}S\text{-}G + NADPH + H^+ \rightarrow 2\ GSH + NADP^+ \quad (Rxn\ 2)$$

Reaction 1 is catalyzed by glutathione peroxidase and reaction 2 by glutathione reductase. Thus, an inhibition of glutathione synthesis would lead to depletion of cellular GSH and an increased likelihood of oxidative damage, particularly if the cell was oxidatively stressed.

20 Nucleotides

Problems

1. When phosphoribosylpyrophosphate (PRPP) is incubated in alkali, 5-phosphoribose-1,2-cyclic phosphate is formed with the release of phosphate. Draw a chemical reaction mechanism for this reaction. (Hint: The reaction is similar to the mechanism for the base hydrolysis of RNA.)

2. Using the information in problem 1, design an experiment to prove that ribose-5-phosphate pyrophosphokinase (rib-5-P + ATP → PRPP + AMP) transfers a pyrophosphate group, making it an unusual kinase. (In most cases a pyrophosphate group is constructed in two steps. For example, in the formation of mevalonate pyrophosphate, two phosphates are *separately* transferred to form the pyrophosphate group.) Also, could you prove that the pyrophosphate group in PRPP is in the **α** position? (Hint: Use [γ-^{32}P] and [β-^{32}P] ATP.) Draw a chemical reaction mechanism for this pyrophosphokinase.

3. In bacteria, pyrimidine biosynthesis is regulated at aspartate carbamoyl transferase, while most of the regulation in humans is at carbamoyl phosphate synthase. Why does this observation make biochemical sense?

4. If mammalian cells in tissue culture are treated with increasing concentrations of *N*-(phosphonacetyl)-L-aspartate (PALA)—a transition-state analog inhibitor of aspartate carbamolytransferase—cells resistant to this toxin can be isolated. When the enzyme aspartate carbamoyltransferase was assayed, its activity was elevated 100-fold in such cells. Also, activities of carbamoyl phosphate synthase and dihydrooratase were elevated about 100-fold. No other pyrimidine pathway enzyme activities were elevated. Explain these observations.

5. Genetic defects in adenosine deaminase (ADA) in humans leads to severe defects in the immune system. This disease has been treated by injecting the enzyme adenosine deaminase and more recently by gene therapy. Discuss how this defect could lead to toxic effects on lymphocytes. (Hint: Deoxyadenosine is metabolized by ADA.)

6. What would happen if you gave allopurinol to a chicken? (Give a biochemical explanation.)

7. Explain why Lesch-Nyhan patients suffer from severe gout. Although these patients can be treated with allopurinol to relieve the symptoms of gout, this treatment has no effect on the severe mental retardation. Suggest a possible explanation.

8. Propose a chemical reaction mechanism for the second enzyme in purine biosynthesis, GAR synthase (phosphoribosylamine + ATP + glycine → GAR + ADP + P_i).

9. Explain how antifolates like methotrexate selectively kill cancer cells. Why do these patients when treated lose their hair, the intestinal mucosa, cells of the immune system, etc.?

10. Which atoms of nucleotide bases isolated from a hydrolysate of DNA would be labeled by the following precursors?
    (a) [3-^{14}C] serine.
    (b) [2-^{14}C] glucose.
    (c) [^{14}C] CO_2.

11. Compare the number of high-energy phosphate bonds required for the synthesis of GTP and the synthesis of CTP. Assume that PRPP and folate one-carbon derivatives are available.

12. The pathway for the *de novo* synthesis of dTTP is more complex than the pathway for synthesis of the other deoxyribonucleotides. Illustrate this difference by reference to known inhibitors of DNA synthesis.

13. Allopurinol administered together with 6-mercaptopurine under certain conditions enhances the anticancer effectiveness of the latter. How can the known site of action of allopurinol explain this effect?

14. The growth of many bacteria is inhibited by sulfanilamide and other sulfa drugs. The toxic effect of sulfanilamide can be reversed by *p*-aminobenzoate. How do sulfa drugs work and why are they not very toxic to people?

Solutions

1.

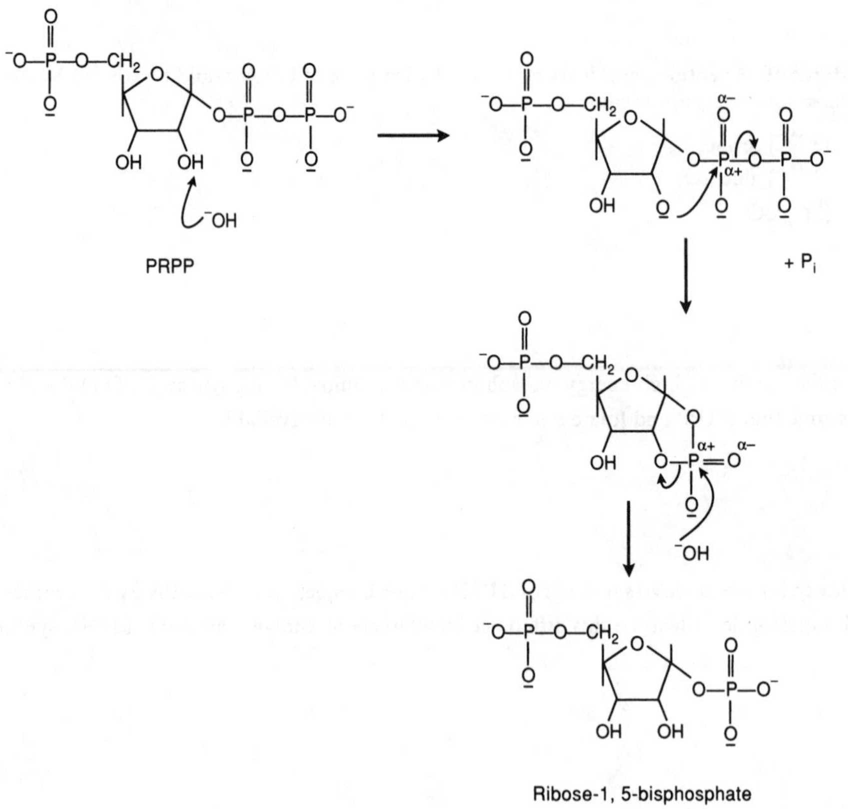

PRPP

+ P_i

Ribose-1, 5-bisphosphate

2. The mechanism of the pyrophosphokinase is shown here.

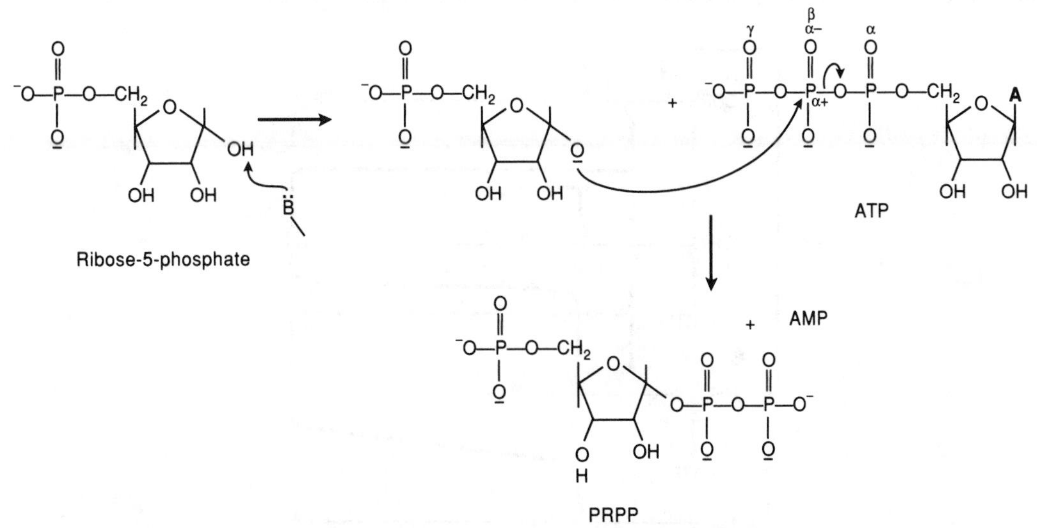

Ribose-5-phosphate

ATP

AMP

PRPP

If the γ phosphate of ATP is labeled with ^{32}P, then the phosphate group labeled in the illustration would be labeled in PRPP. When the radioactive PRPP* is incubated with base, radioactive inorganic phosphate would be released (PRPP* $\rightarrow$ PRP + P*) and the phosphoribosylphosphate (PRP) would not be radioactive. If the β phosphate was labeled in ^{32}P-ATP, then the PRPP formed would be labeled in the phosphate attached to the 1′-carbon (PRP*P), and when the PRPP was treated with base the radioactive phosphate would remain attached to the phosphoribosylphosphate (PRP*P $\rightarrow$ PRP* + P). If the pyrophosphate group in PRPP was attached in the β configuration, then it would not be close enough for the deprotonated 2′-OH to carry out a nucleophilic attack on the phosphorous group, because the 2′-OH group and the 1′-PP group would be *trans* (see the mechanism in problem 1). Therefore, although pyrophosphate groups are very common in biochemical systems, the transfer of an intact pyrophosphate group such as seen with PRPP synthetase (pyrophosphokinase) is rare.

3. Most biochemical pathways are regulated at the first committed step (enzyme unique to that pathway). Since humans have two carbamoyl synthetases, one in the mitochondria (CPase I), which is part of the pathway for the formation of urea or arginine, and another in the cytosol (CPase II), which is part of the pyrimidine biosynthetic pathway, the cytosolic CPase II is the first committed step in the pyrimidine pathway in humans and is highly regulated. In *E. coli* there is only one CPase that makes product used in pyrimidine as well as arginine biosynthesis. Therefore, in these bacteria the first committed step in pyrimidine biosynthesis is ATCase, which is highly regulated. The different points of regulation in the pyrimidine biosynthetic pathways in humans and bacteria make good biochemical sense, i.e., regulation at the first committed step in the pathway.

4. When mammalian cells in culture are treated with *N*-(phosphonacetyl)-L-aspartate (PALA), cells resistant to this toxic compound can be isolated. This drug binds very tightly to the active site of ATCase because it is a transition-state analog. By a mechanism of DNA amplification, more genes are produced in order to generate a large enough gene dose to produce enough enzyme to bind the PALA and have enough excess enzyme to allow pyrimidine biosynthesis to continue. The fact that carbamoyl phosphate synthase and dihydrooratase were also elevated about 100-fold suggests that these three enzymes in pyrimidine biosynthesis existed as a multienzyme (three active sites on one polypeptide). Therefore, one gene coding for three enzyme activities on one polypeptide was amplified.

5. If a person lacked the enzyme adenosine deaminase (ADA), the nucleoside deoxyadenosine would accumulate to high levels in some cells. The deoxyadenosine would then be phosphorylated to dATP, which inhibits the enzyme ribonucleotide reductase, leading to the inhibition of DNA replication (see the illustration). This genetic defect has been treated by injecting patients with the enzyme adenosine deaminase, which is apparently taken up by the

cells, reducing deoxyadenosine levels. The newest approach to treatment, however, is to use gene therapy to insert a gene for ADA into lymphocytes. This is the first genetic defect to be treated by gene therapy.

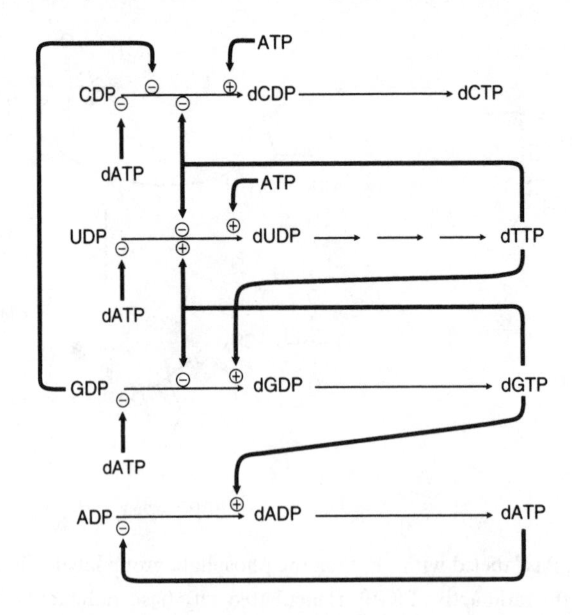

Source: L. Thelander and P. Reichard, "Reduction of Rebonucleotides," *Annual Review of Biochemistry,* 48:133, 1979, Annual Reviews Inc., Palo Alto, CA.

6. Allopurinol would be very toxic to chickens and probably would lead to rapid death. Chickens (birds) use the purine biosynthetic pathway and the synthesis of uric acid to remove their nitrogen waste. If the enzyme that produces uric acid, xanthine oxidase, was inhibited, then more water-soluble purines would be excreted, leading to a large water loss and dehydration. Also, purine synthesis would be inhibited and birds would not be able to rid themselves of excess nitrogen. Toxic levels of ammonia would result. The combination of dehydration and ammonia toxicity would lead to rapid death of the birds.
7. Patients with Lesch-Nyhan syndrome have a deficiency in hypoxanthine-guanine phosphoribosyl transferase, an important enzyme in the salvage pathway for purines. When this enzyme is missing, the levels of phosphoribosylpyrophosphate (PRPP) become elevated and stimulate the synthesis of purines. These excess purines are degraded to uric acid, leading to severe gout. The gout can be treated with drugs like allopurinol, but this treatment has no effect on the severe mental retardation seen in these patients. The brain may lack a *de novo* purine biosynthetic pathway; it depends on the salvage pathway to produce purines, which are necessary for DNA replication. When a human is born, the rapid brain growth and development that normally occurs is inhibited if adequate levels of purine nucleotides are not produced. This genetic defect may eventually be treated with gene therapy.

8. The chemical reaction mechanism for GAR synthase is

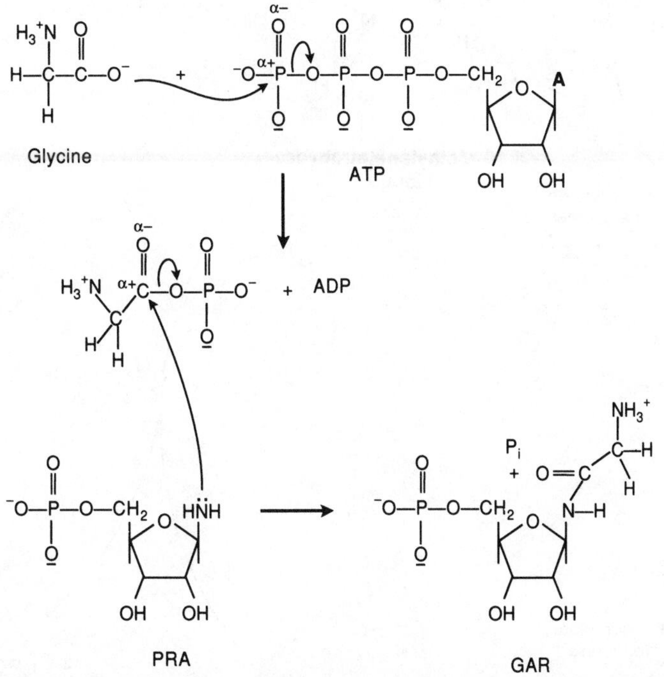

Glycine + ATP → H_3^+N–CH_2–C(=O)–O–PO_3^{2-} + ADP

PRA → GAR + P_i

9. Anticancer drugs like the antifolates have greater toxicity toward rapidly growing cells because of the cells' demands for DNA replication. For example, methotrexate is a potent inhibitor of dihydrofolate reductase. The lack of synthesis of tetrahydrofolate prevents the formation of the various one-carbon tetrahydrofolate compounds (see figure). This limited one-carbon pool inhibits purine biosynthesis and, more importantly, the formation of methylene-THF, which is used in thymidylate synthase (see figure). The lack of dTTP synthesis in rapidly growing cells leads to death from lack of thymine. However, not all normal cells in human adults are nondividing. Some are growing rapidly, leading to the side effects observed in cancer patients treated with these compounds. Stem cells in bone marrow, epithelial cells of the intestine, and hair follicle cells are sensitive to the toxic effects of many anticancer drugs. This also explains why chemotherapeutic drugs are more toxic to children, especially when they are in a growth spurt.

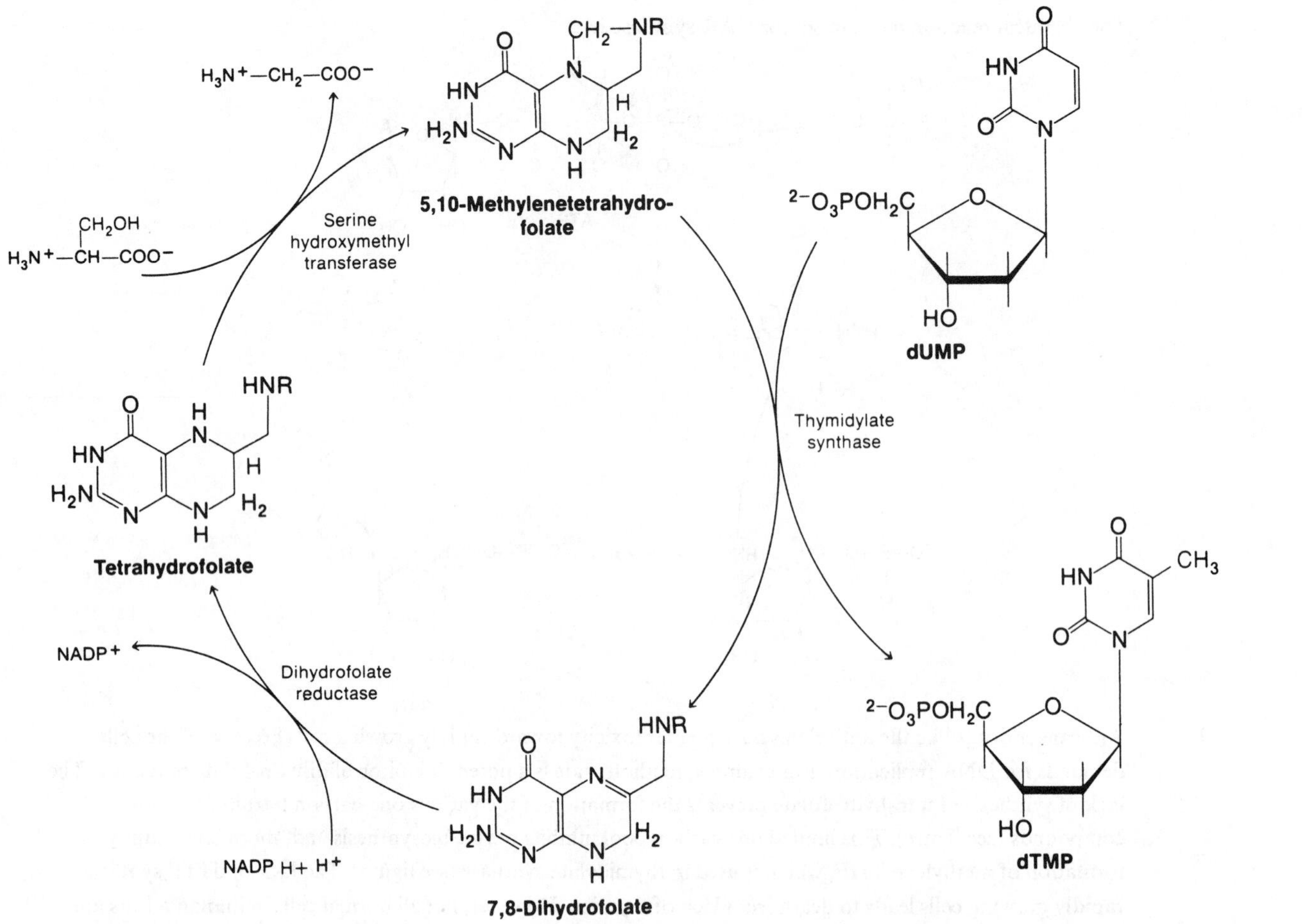

5,10-Methylenetetrahydrofolate

Serine hydroxymethyl transferase

dUMP

Thymidylate synthase

Tetrahydrofolate

Dihydrofolate reductase

$NADP^+$

NADP H+ H^+

7,8-Dihydrofolate

dTMP

10. (a) Serine contributes its side chain (C-3) to the formation of three tetrahydrofolate derivatives; 5,10-methenyltetrahydrofolate, which contributes the formyl group that becomes C-8 in the purine ring; 10-formyltetrahydrofolate, which contributes the formyl group that becomes C-2 of the purine ring (see figure 20.13, steps 3 and 9); and 5,10-methylene tetrahydrofolate, which contributes the methyl group of thymidylate (see the figure for solution 9). Thus ^{14}C would be detected at the C-2 and C-8 positions of adenine and guanine, and the methyl group of thymine.

(b) Glucose is rapidly metabolized to 3-phosphoglyceraldehyde, which is the major precursor of serine synthesis; therefore, the C-2 of serine would be labeled. Conversion of serine to glycine would put the ^{14}C at C-2 of glycine, which would then end up as C-5 in the purine ring as a result of the condensation of glycine with phosphoribosylamine (see figure 20.13, step 2). Thus the ^{14}C would be found at C-5 in adenine and guanine.

(c) CO_2 is used directly in the synthesis of both purines and pyrimidines. CO_2 is added to the imidazole ring as C-6 of the purine ring (see figure 20.13, step 6). CO_2 is also used to combine with the amine side group of glutamine to form carbamoylphosphate, the first step in pyrimidine ring synthesis (see figure 20.18). The ^{14}C from CO_2 would therefore be found at C-6 in adenine and guanine and at C-2 in cytosine and thymine.

11. The synthesis of GTP requires the hydrolysis of eight high-energy phosphate bonds (8~P) (see figures 20.13 and 20.16). The formation of phosphoribosylamine from glutamine and PRPP releases PP_i, which is subsequently hydrolyzed to P_i (2~P used). The formation and closure of the imidazole ring requires the hydrolysis of 3 ATP to

3 ADP + 3 P_i (3~P used). The addition of N-7 of the purine ring from aspartate requires the hydrolysis of 1 ATP to ADP and P_i (1~P used). The conversion of xanthine monophosphate to GMP requires the hydrolysis of 1 ATP to AMP + PP_i, of which PP_i is subsequently hydrolyzed (2~P used).

The synthesis of CTP requires the hydrolysis of five high-energy phosphate bonds (5~P used) (see figures 20.18 and 20.19). Two ATP are used in the synthesis of carbamoyl aspartate, the first two steps in the formation of orotate (2~P used). The transfer of orotate to PRPP to form OMP releases 1 pyrophosphate (PP_i), which is subsequently hydrolyzed (2~P used). The amination of UTP to CTP requires the hydrolysis of 1 ATP to ADP + P_i (1~P used).

12. The pathways used in the *de novo* synthesis of the deoxyribonucleotides dGDP, dCDP, and dADP are relatively simple, involving replacement of the 2′-OH group of the ribose-5′-phosphate of the corresponding ribonucleotides with hydrogen in a reaction catalyzed by ribonucleotide reductase. The pathway by which dTMP is synthesized is more complex for the reason that there is no corresponding ribonucleotide to be used as a substrate by ribonucleotide reductase. Instead, dTMP must be synthesized from dUMP, which in turn is generated ultimately by the reduction of UDP to dUDP. The conversion of dUMP to dTMP, catalyzed by thymidylate synthase, involves transferring a methyl group and reducing equivalents from 5,10-methylene FH_4 to dUMP (see the figure in solution 9). In the process, this reaction generates dihydrofolate as a product, which must be converted back to 5,10-methylene FH_4 via tetrahydrofolate if dTMP is to continue to be synthesized.

    A number of specific inhibitors of dTMP synthesis have been described that prevent the necessary regeneration of 5,10-methylene FH_4 from dihydrofolate. These inhibitors, the principal of which are methotrexate and trimethoprim, act to competitively inhibit the conversion of dihydrofolate to tetrahydrofolate, catalyzed by dihydrofolate reductase. Several other inhibitors of dTMP synthesis have also been described, most notably 5-fluorouracil, that act directly by inhibiting the reaction catalyzed by thymidylate synthase.

13. Allopurinol is an analog of hypoxanthine that inhibits the activity of xanthine oxidase. Xanthine oxidase, which catalyzes the conversion of xanthine to uric acid (see figure 20.27), plays an important role in the degradative pathway for 6-mercaptopurine, oxidizing 6-thioxanthine to 6-thiouric acid, which is then excreted. Inhibition of xanthine oxidase would therefore lead to elevated intracellular levels of 6-thio IMP (6-mercaptopurine is converted to 6-thio IMP by HGPRTase). 6-Thio IMP inhibits key steps in the conversion of IMP to AMP and GMP, and the formation of phosphoribosylamine from PRPP and glutamine, which is the first committed step in the *de novo* pathway for purine nucleotide synthesis. The resulting depletion of purine nucleotides would therefore kill fast-growing cells, such as cancer cells, far more effectively than 6-mercaptopurine alone.

14. Sulfa drugs are analogs of *p*-aminobenzoic acid (see structures), a precursor in folic acid biosynthesis. If excess *p*-aminobenzoic acid (PABA) is available, the competition is overcome and folic acid can be made. (You may be familiar with the use of PABA or its esters in sunscreens, as this compound absorbs ultraviolet light.) Humans do not make folic acid but require it in their diet. When the bacterial infection is treated with sulfanilamide, the bacteria are killed because they cannot make folic acid (important in one-carbon metabolism). There is little toxic effect on the patient since humans do not make folic acid but require it in their diet.

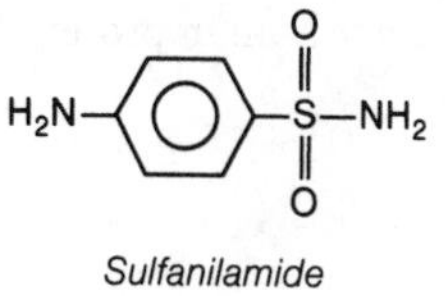

$H_2N-C_6H_4-SO_2-NH_2$

Sulfanilamide

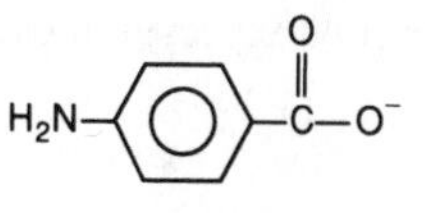

$H_2N-C_6H_4-C(=O)-O^-$

p-Aminobenzoic acid (PABA)

21 Biosynthesis of Complex Carbohydrates

Problems

1. How is lactose made in mothers' milk? What is unusual about the subunit structure of lactose synthase?

2. The synthesis of dextran does not use nucleotide sugars as seen in the synthesis of other glucose polymers. What is the driving force for this reaction, i.e., how would you analyze the thermodynamics?

3. How can lectins be used to identify complex carbohydrate structures on glycolipids or glycoproteins? (Hint: You could label the lectins with radioactive iodine [^{125}I] or use lectins bound to a column matrix.)

4. A great variety of different oligosaccharides result from a limited number of sugars. Explain.

5. How can two different genes for different glycosyltransferases determine ABO blood group types in humans? Explain why blood group O is considered a universal donor. If you have AB type blood, why can you accept any blood type? A small number of people lack the H antigen (the glycosyltransferase that adds Fucα 1) and have Bombay type blood. What blood type could you give to a person with Bombay type blood and why?

6. There is a lot of interest in expressing human genes in bacteria to produce a useful human protein in large amounts. Why would this not be expected to work with some human proteins?

7. A mutant cell line that does not synthesize Dol-P-man does not express glypiated proteins on its cell surface. Why?

8. Mutants with a glycosylation defect are used to follow transport between cellular compartments. Using one of these mutants, design an assay to detect transit between Golgi compartments that lie beyond the medial Golgi.

9. How would you provide biochemical evidence that a particular glycoprotein is a resident protein of the endoplasmic reticulum?

10. Explain why fibroblasts from a patient with I-cell disease (discussed in chapter 6) will secrete lysosomal enzymes when grown in tissue culture.

11. Why does penicillin kill susceptible bacteria only when they are growing?

12. What complications have to be overcome to synthesize complex carbohydrates (such as cell-wall components and O-antigens) outside the cell?

Solutions

1. Lactose is formed in the mammary gland from glucose and UDP-galactose by the interesting enzyme lactose synthase. Lactose synthase is a complex of two proteins, galactosyltransferase and α-lactalbumin. Galactosyltransferase is found in all body tissues and normally catalyzes the reaction [UDP-galactose + *N*-acetyl-D-glucosamine → UDP + *N*-acetyllactosamine]. However, when α-lactalbumin binds to galactosyltransferase, the specificity of the galactosyltransferase is changed so that D-glucose is used as the galactose acceptor. Thus, the α-lactalbumin allows the milk sugar lactose to be synthesized in the mammary gland during lactation.
2. Some bacteria synthesize the glucose polymer dextran without the use of nucleotide sugars such as UDP-glucose or ADP-glucose. The (1,6) linkage is catalyzed by dextran sucrase using the disaccharide sucrose as the substrate [*n* sucrose → dextran + *n* fructose]. The energy for this reaction comes from the glycosidic bond between glucose and fructose, which is broken to form the linkages between the glucose monomers (this could be called a glycosidic bond exchange reaction).
3. Lectins are proteins that bind specific carbohydrate structures (for example, concanavalin A binds oligomannosyl N-linked sugars but not O-linked or branched mannose). Glycolipids could be separated on thin

layer chromatography (TLC) and a specific radioactive lectin could be used to locate specific structures. The location of these structures could be detected by exposing the TLC to x-ray film (radioautography). The lectin could also be linked to a column matrix and the compounds of interest could be applied to the column. Only structures recognized by the lectin would bind. Lectin affinity columns are very useful in purifying complex carbohydrates (see figure 6.26).

4. The large variety of different oligosaccharides results from a limited number of sugars. Sugars have a large number of hydroxyl groups that form glycosidic bonds with the anomeric carbons of other sugars; these anomeric hydroxyl groups can be either α or β conformation. Over eighty different glycosidic linkages are known. With different sugars, number of residues, branching, and other combinations, the number of different oligosaccharides possible is quite large.

5. Only two genes determine the ABO blood group scheme (figure 21.2). The A gene encodes a glycosyltransferase that adds a terminal *N*-acetylgalactosamine residue onto a core oligosaccharide, and the B gene encodes a glycosyltransferase that adds a galactose residue (figure 21.15). If both genes are present, then both antigens are present, resulting in blood type AB. If both genes are missing, then the core oligosaccharide is not substituted and the blood type is O. People with O type blood are considered universal donors because they have only the core oligosaccharide, which will not generate antibodies when their blood is given to people with A or B type blood. If you have the AB blood type, all types of blood are compatible because you already have all the different blood antigens; antigens in the transfused blood would not be considered foreign. People who lack the H antigen cannot add the fucose residue to the core oligosaccharide. These people can receive blood only from a person with Bombay type blood (lacking the H antigen) but can donate blood to persons with Bombay type, A, B, or O.

6. Many proteins are glycosylated and require that proper oligosaccharides be added for proper folding and translocation. If one of these glycoproteins was expressed in *E. coli,* it would be improperly modified, thus generating a nonfunctional protein. There is much interest in expressing these human glycoproteins in mammalian systems, such as in the milk of transgenic cows.

7. In mutant cells lacking Dol-P-mannose synthase a complete glycophosphitadyl inositol (GPI) is not made, and the glycoproteins with GPI anchors, which are normally found in the plasma membrane, are secreted from the cell because they are no longer attached by their anchors to the membrane.

8. Mutants that lack a specific transferase can be used to assay specific transport reactions between various Golgi membranes. For example, donor Golgi membranes lacking GlcNAc-TI enzyme cannot modify the oligosaccharides of marker glycoproteins. However, if Golgi membranes from wild-type cells are mixed with the mutant Golgi membranes containing the marker glycoprotein in the presence of radioactive UDP-GlcNAc, the transfer of GlcNAc to the glycoprotein can be assayed. The location of the radioactive glycoprotein can be detected with electron microscopic autoradiography. Similar assays can be used to follow transport between different medial and *trans* compartments. (See box 21D in the text.)

9. A number of proteins, such as the P_{450} cytochromes, are considered resident proteins of the endoplasmic reticulum (ER). These proteins are anchored to the membrane by a hydrophobic sequence of amino acids and are generally not glycosylated. These proteins are not transported to the Golgi but remain in the ER.

10. When fibroblasts from patients with I-cell disease are grown in culture, many lysosomal enzymes are excreted outside the cells. Normally, a lysosomal enzyme is glycosylated and mannose is phosphorylated. There is a mannose-6-phosphate receptor that binds the phosphorylated lysosomal protein and targets it to a transport vesicle that carries the lysosomal enzymes to the lysosome. The genetic defect in I-cell patients is a lack of the enzyme required to phosphorylate mannose. The enzyme thus becomes targeted for secretion outside the cell instead of targeting to the lysosome. If normal lysosomal enzymes (mannose phosphorylated) are placed in the media, these enzymes are taken up by the fibroblast and are targeted to the lysosome, restoring normal lysosome function.

11. The major effect of penicillin is to act as a substrate analog of the D-alanyl-D-alanine moiety of the cell wall's peptidoglycan strands and inhibit, irreversibly, the transpeptidase that cross-links the strands together. Thus

cell-wall synthesis is inhibited. Cell-wall synthesis is concomitant with bacterial cell growth and multiplication. Stationary, nongrowing cells already have complete cell walls and obviously do not require the transpeptidase reaction (no cell-wall synthesis is occurring). Therefore, only growing bacterial cells that are actively synthesizing cell walls are subject to penicillin's inhibitory effects. Without a complete cell wall, bacterial cell lysis occurs.

12. The major problem in the synthesis of complex carbohydrates outside the cell is the lack of an external energy source such as ATP outside the cell. The biosynthesis of a bacterial cell wall such as the peptidoglycan occurs mainly inside the cell, the final step being the cross-linking of the peptidoglycan strands outside the cell. This reaction is a transpeptidation, which does not require any energy source. A peptide bond is broken and another one is formed with the release of alanine. The final step of the synthesis of O-antigens in *Salmonella typhimurium* also occurs outside the cell. In this case the activated forms of the precursors are synthesized inside the cell on a lipid carrier (the same one used in peptidoglycan) and transferred to the outside surface of the cell (held in place by the lipid carrier). The activated galactose of a lipid-linked polymer attacks the mannose of the tetrasaccharide, which is also attached to undecaprenol phosphate (the lipid carrier). In both of these complex reactions to produce extracellular complex carbohydrates a lipid carrier is used to keep intermediates on the surface of the cell. The polymerization is either a transpeptidation (equal bonds broken and formed) or an activated sugar using the lipid carrier.

22 Biosynthesis of Membrane Lipids and Related Substances

Problems

1. Why are human genetic defects in phospholipid biosynthesis not observed?

2. Phosphatidylcholine biosynthesis appears to be regulated principally at the step catalyzed by CTP: phosphocholine cytidylyltransferase. Does the type of regulation observed make biochemical sense? Draw a chemical reaction mechanism for this enzyme.

3. The majority of phospholipids contain an unsaturated fatty acid at the C-2 position. Give an example of a phospholipid that has a saturated fatty acid in this position. How is it synthesized?

4. Explain why the reactions from choline to phosphatidylcholine in eukaryotic cells are thermodynamically feasible.

5. Where are ether-linked phospholipids found and what are some of their functions?

6. A patient accumulated the lipid Gal-β(1,4)-Glc-β(1,4)-ceramide. What enzymatic reaction might be defective?

7. Arachidonic acid is the major precursor of prostaglandins and thromboxanes. If a person were unable to absorb arachidonic acid from the diet but could absorb linoleic acid, could that person still make PGE_2?

8. Low doses of aspirin (one aspirin every other day) are recommended to prevent heart attacks and strokes. Why would three or four tablets per day not work better? (Hint: Remember that TXA_2 is made in platelets and that PGI_2 is made in the arterial walls.)

9. Snake venoms contain many types of lipases, including phospholipase A_2. Why would small amounts of this enzyme contribute to some of the toxic effects of snake venom? (Bee venom contains a protein that stimulates phospholipase A_2.)

Solutions

1. Phospholipids are very important to cells because they are the major components of membrane structure. The demand for phospholipids is highest in growing cells, such as those in a developing fetus. Any genetic defects in the biosynthesis of phospholipids would be lethal in the early stages of fetal development.

2. The CTP:phosphocholine cytidylyltransferase is found free in the cytosol and membrane-bound in the endoplasmic reticulum (ER); however, only the membrane-bound fraction is active. When phosphatidylcholine levels decrease in the membrane, there is increased binding of cytidylyltransferase to the ER where the enzyme is active. After phosphatidylcholine levels increase, the cytidylyltransferase is released in the cytosol where it is inactive. Phosphorylation of the enzyme also releases it from the membrane. Fatty acids and diacylglycerol will increase binding of the cytidylyltransferase. When fatty acid levels are high, diacylglycerol levels will increase, providing substrate for phosphatidylcholine biosynthesis, and activate the rate-limiting enzyme CTP:phosphocholine cytidylyltransferase by increasing its binding to the ER membrane.

The chemical reaction mechanism for the cytidylyltransferase is

$$\text{C-ribose-}CH_2\text{-O-}\overset{\alpha}{P}(O)(O^-)\text{-O-}\overset{\beta}{P}(O)(O^-)\text{-O-}\overset{\gamma}{P}(O)(O^-)\text{-O}^- \;(\text{CTP}) + {}^-O\text{-}P(O)(O^-)\text{-O-}CH_2\text{-}CH_2\text{-}\overset{+}{N}(CH_3)_3 \;(\text{Phosphocholine})$$

$$\downarrow$$

$$\text{C-ribose-}CH_2\text{-O-}P(O)(O^-)\text{-O-}P(O)(O^-)\text{-O-}CH_2\text{-}CH_2\text{-}\overset{+}{N}(CH_3)_3 \;(\text{CDP-choline}) + {}^-O\text{-}P(O)(O^-)\text{-O-}P(O)(O^-)\text{-O}^- \;(\text{Pyrophosphate})$$

3. Deacylation-reacylation of phospholipids occurs in various tissues at both the SN-1 and SN-2 positions. This remodeling allows alteration of the phospholipid without having to completely synthesize the entire molecule. In lung tissue the surfactant dipalmitoylphosphatidylcholine is synthesized by removing the fatty acid in the SN-2

position and then acylating with palmitoyl-CoA. Another example of remodeling occurs when the fatty acid at the SN-2 position is replaced with arachidonic acid. The arachidonic acid is stored there until it is needed for eicosanoid biosynthesis.

4. The synthesis of phosphatidylcholine is thermodynamically feasible because of the ATP used to phosphorylate choline and the CTP used to form CDP-choline (see the mechanism in solution 2). The cytidylyltransferase reacts phosphocholine with CTP to form CDP-choline and pyrophosphate. The hydrolysis of this pyrophosphate by pyrophosphatase is the major driving force for these reactions.

5. Ether-linked phospholipids are widely found in nature. Ether-linked lipids are the major component in archaebacteria cell membranes. Eukaryotic membranes contain significant amounts of ether-linked glycerolphospholipids. For example, 50% of all phospholipids in heart tissue are ether-linked in the SN-1 position. Plasmalogens are the major form of ether-linked phospholipids found in the heart and are characterized by a vinyl ether linkage. Alkenyl-ether-containing phospholipids protect cells against singlet oxygen, which can kill cells. Some ether-linked phospholipids are bioactive, such as 1-alkyl-2-acetylglycerolphosphocholine (platelet-activating factor), which reduces blood pressure and causes blood platelets to aggregate.

6. The degradation of glycosphingolipids proceeds in an essentially stepwise fashion. Each step involves the removal of the exposed terminal glycosyl residue. In all but a few cases, diseases in which there is an abnormal accumulation of one of the intermediates of glycosphingolipid catabolism (sphingolipidoses) are caused by a deficiency in an activity of one of the enzymes of this pathway. In the case described, in which Gal-β(1,4)-Glc-β(1,4)-ceramide has been found to accumulate, it is likely that the defect lies in the activity responsible for conversion of this intermediate to Glc-β(1,4)-ceramide (see figure 22.19). The enzyme that catalyzes this reaction is lactosylceramide-β-galactosidase.

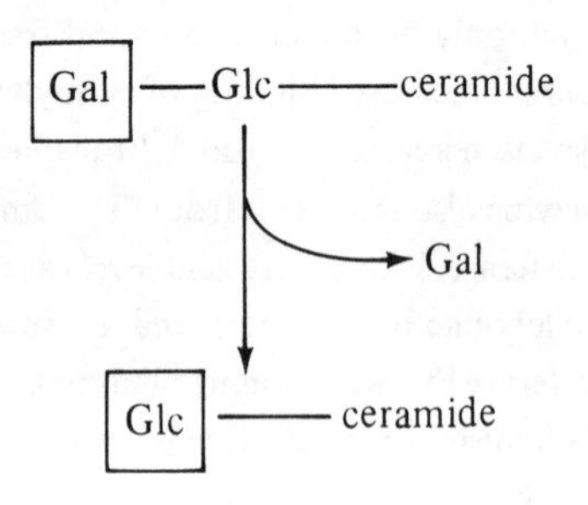

7. In animal cells, the C_{20} polyunsaturated fatty acid arachidonic acid, which is the precursor for synthesis of PG_2 and TX_2 prostaglandins and thromboxanes, may be derived from the diet or from desaturation and elongation of linoleic acid. The scheme for the synthesis of arachidonic acid ($20{:}4^{\Delta 5,8,11,14}$) from linoleic acid ($18{:}2^{\Delta 9,11}$) is:

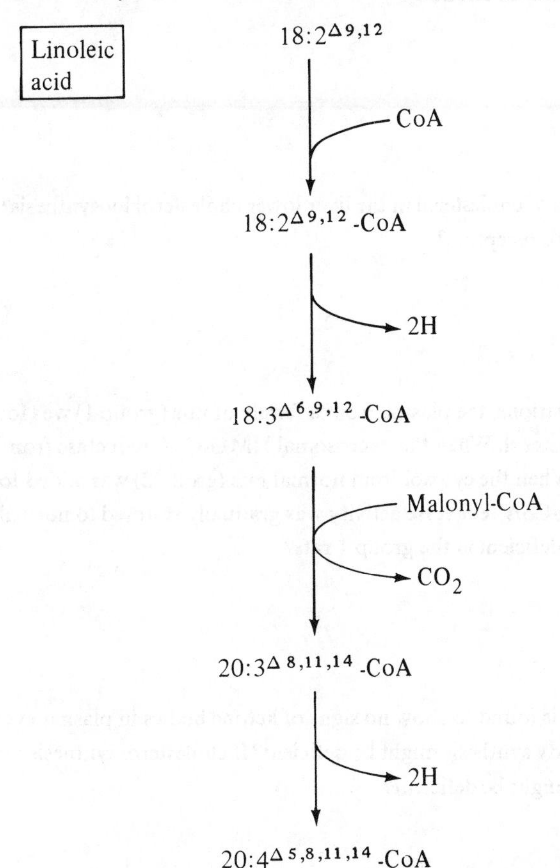

8. Low doses of aspirin appear to help prevent heart attacks and strokes in people over 40 years old. TXA_2 is made in blood platelets and causes platelet aggregation; it is a vasoconstrictor. If the cyclooxygenase is inactivated by aspirin, no new enzyme can be made because these cells have no nucleus. New platelets would have to be made to regain the synthesis of TXA_2. This would take days. Vascular endothelial cells make PGI_2, a vasodilator that inhibits platelet aggregation. These cells in the arterial walls have a nucleus and can make more cyclooxygenase in a few hours. If high dose aspirin is taken, the synthesis of PGI_2 does not recover rapidly. Thus, PGI_2 helps prevent heart attacks and strokes. A small amount of aspirin is certainly better in this case.

9. Phospholipase A_2 selectively removes fatty acids from the SN-2 position of phospholipids. As discussed in solution 3, arachidonic acid is stored in membranes by linkage at the SN-2 position in phospholipids. If a lot of arachidonic acid were released, it would stimulate the synthesis of prostaglandins, which then would induce inflammation. The release of arachidonic acid from phospholipids is believed to be the rate-limiting step in eicosanoid biosynthesis.

23 Metabolism of Cholesterol

Problems

1. How can elevated levels of cholesterol in the liver lower cholesterol biosynthesis? What effect will elevated cholesterol have on LDL receptors?

2. During routine investigations, the plasma from a family of rats (group 1) was found to have very low concentrations of cholesterol. When the microsomal HMG-CoA reductase from liver was assayed, extremely low activities were found. When the cytosol from normal rats (group 2) was added to the microsomal fraction from group 1 rats, the HMG-CoA reductase activity was gradually restored to normal values. What enzyme activity (or activities) might be deficient in the group 1 rats?

3. A person with diabetes is found to show no signs of ketone bodies in plasma even when in diabetic shock. Which enzyme(s) of ketone body synthesis might be deficient? If cholesterol synthesis were normal, would this be a clue as to which enzyme(s) might be deficient?

4. A patient homozygous for familial hypercholesterolemia (FH) was treated with lovastatin to lower LDL levels in the blood. This treatment did not have any effect on LDL levels. Why? After a number of heart attacks, a heart and liver transplant were done and LDL levels were dramatically lowered. Why were both organs replaced?

5. An FH heterozygote had plasma LDL-cholesterol levels about twice the normal levels. How could this patient be treated to lower LDL levels? What is the biochemical rationale for these treatments?

6. Liver cells in culture are given 2-[^{14}C]-acetate. Where would this label appear in HMG-CoA?

7. A deficiency of apoprotein C-II results in the disease hyperlipoproteinemia type I, in which there is a massive increase in the concentration of plasma triacylglycerol. Provide an explanation for this clinical finding.

8. What is the hereditary defect in Wolman's disease? Would you expect HMG-CoA reductase activity to be high or low in skin fibroblasts cultured from patients with this disorder? Would the number of LDL receptors be high or low in these fibroblasts?

9. There is an inherited disease in which lecithin: cholesterol acyltransferase (LCAT) is deficient. What effect would you expect this deficiency to have on the composition of HDL and other lipoproteins in plasma?

Solutions

1. The major regulation of cholesterol biosynthesis occurs with the first committed step in the pathway, HMG-CoA reductase. This enzyme is located on the outer surface of the endoplasmic reticulum (ER). It has two structural domains: One region of the protein has seven hydrophobic segments, which cross the membrane of the ER, while the other region protrudes into the cytosol and has the catalytic site. The activity of the reductase is regulated by three mechanisms. When cholesterol is high, the gene for the reductase is turned off and less mRNA for HMG-CoA reductase is produced. High cholesterol levels will also reduce the half-life by affecting the membrane domain of the enzyme by dissolving into the ER membrane. The third mechanism involves the phosphorylation/dephosphorylation of the reductase. When the enzyme is phosphorylated, the reductase is less active. High levels of cholesterol will also reduce the number of LDL receptors synthesized, thus reducing the amount of cholesterol brought into the liver from the blood.
2. The low plasma cholesterol level exhibited by the defective rats may have any of a number of possible explanations. The low activity of HMG-CoA reductase (a microsomal enzyme) in microsomal fractions prepared from the livers of these rats indicates that a defect in this enzyme is responsible. HMG-CoA reductase catalyzes the rate-limiting step in cholesterol biosynthesis (see the solution to problem 1), and a defect in the activity of this enzyme would have severe consequences for the plasma cholesterol level.

   The low activity of HMG-CoA reductase observed for the unusual rats might result from a defect in either the amount or activity of this enzyme. The restoration of activity observed upon addition of normal rat liver cytosolic fractions to the inactive microsomal fractions prepared from the defective rats indicates that the deficiency derives from inactivation of an otherwise normal enzyme. The activity of HMG-CoA reductase is known to be subject to regulation by a phosphorylation-dephosphorylation system, as shown. The aberrant inactivation of reductase and its reactivation by a normal cytosolic fraction may be explained by a deficiency in this system in either reductase kinase or reductase phosphatase activities (both cytosolic), preventing dephosphorylation and reactivation of inactive HMG-CoA reductase-Ⓟ.

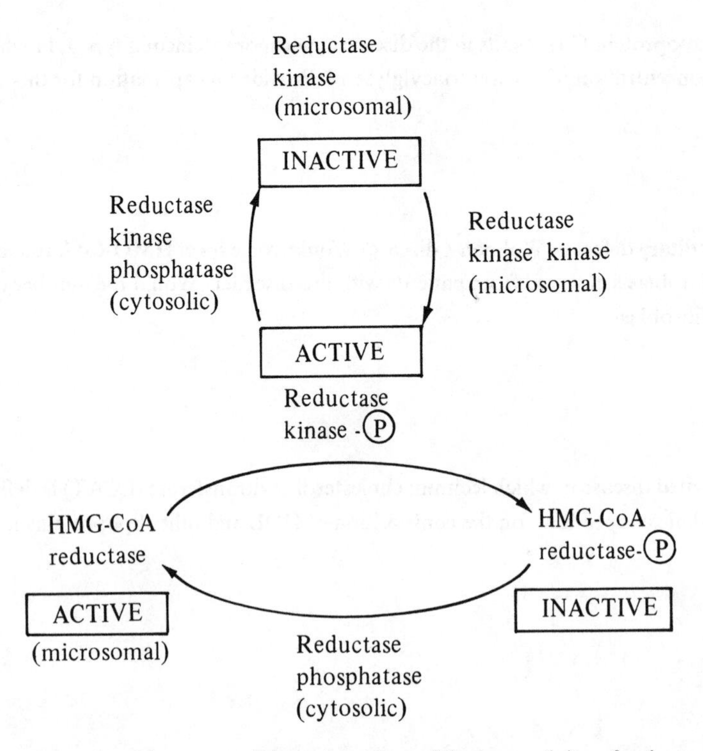

Further support for this as a possible explanation of the low activity of reductase in the unusual rats might be obtained by determining the effect on activity of inactive microsomal fractions, of addition of pure phosphorylase a-phosphatase, the enzyme in rat liver believed to function in these dephosphorylation reactions.

3. Individuals in an acute diabetic condition generally accumulate a considerable quantity of the ketone bodies acetoacetate and β-hydroxybutyrate in plasma (ketonemia) owing to overproduction of acetyl-CoA as a product of fatty acid oxidation in the liver. The synthesis of these ketone bodies, which in the liver occurs by the pathway shown, requires the activity of four principal enzymes: acetoacetyl-CoA thiolase, HMG-CoA synthase, HMG-CoA lyase, and D-β-hydroxybutyrate dehydrogenase.

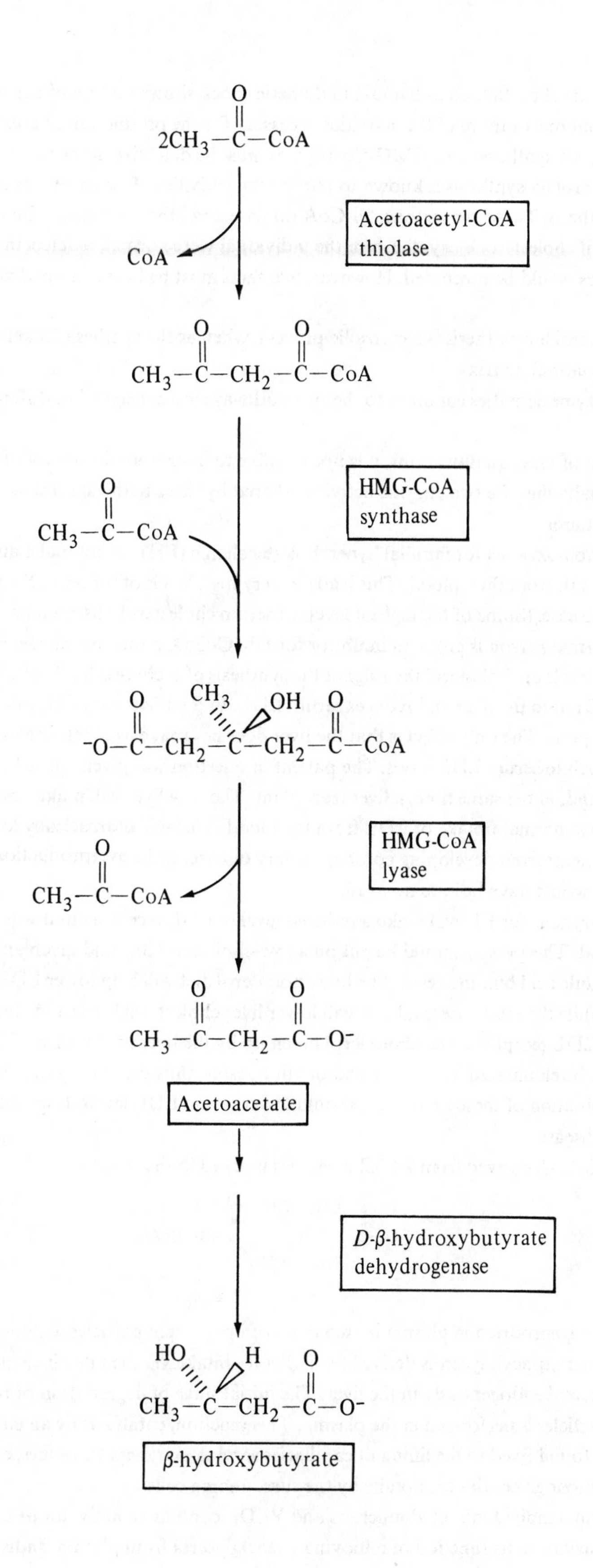

$2CH_3-C(=O)-CoA$
Acetoacetyl-CoA thiolase
CoA
$CH_3-C(=O)-CH_2-C(=O)-CoA$
$CH_3-C(=O)-CoA$
HMG-CoA synthase
$^{-}O-C(=O)-CH_2-C(CH_3)(OH)-CH_2-C(=O)-CoA$
HMG-CoA lyase
$CH_3-C(=O)-CoA$
$CH_3-C(=O)-CH_2-C(=O)-O^-$
Acetoacetate
D-β-hydroxybutyrate dehydrogenase
$CH_3-C(HO)(H)-CH_2-C(=O)-O^-$
β-hydroxybutyrate

The observation that an individual in diabetic shock shows no sign of ketone bodies in the plasma would indicate that one or more of the activities necessary for the production of acetoacetate (acetoacetyl-CoA thiolase, HMG-CoA synthase, and HMG-CoA lyase) must be defective in the individual.

Cholesterol biosynthesis is known to require the activities of two enzymes common to the pathway for ketone body synthesis. These are acetoacetyl-CoA thiolase and HMG synthase. On the face of it, therefore, it would seem that if cholesterol biosynthesis in the individual were normal, a defect in the activities of either of these two enzymes would be precluded. However, two facts must be borne in mind when considering these observations:

(1) Cholesterol biosynthesis is a ctyosolic process, whereas the synthesis of ketone bodies is confined to the mitochondrial matrix.

(2) The enzyme activities common to the two pathways are consigned to different isozymes.

Because of these qualifications, it is not possible to infer from the ability of the individual to synthesize cholesterol normally that the two enzyme activities shared by these pathways are not the site of the defect in ketone body synthesis.

4. People who are homozygous for familial hypercholesterolemia (FH) do not make any receptors for LDL and cannot remove LDL from their blood. This leads to very high levels of serum LDL, which results in cardiovascular disease. (Some of the highest levels of serum cholesterol observed have been found in FH patients.) If a normal person is given an inhibitor for HMG-CoA reductase, cholesterol synthesis is inhibited in the liver. Lower levels of cholesterol then signal the synthesis of increased levels of LDL receptors. This increases the uptake of LDL into the liver and reduces serum LDL. In a patient with FH, this has little effect,since there are no LDL receptors. The only effect is that the liver does not make as much cholesterol and does not contribute as much to serum LDL levels. The patient in question was given a heart transplant to replace the damaged heart and, at the same time, a liver transplant. The new liver will make normal amounts of LDL receptors and have normal uptake of LDL from the blood. This will dramatically lower serum LDL levels and prevent the new heart from developing coronary artery disease. If the liver transplant had not been done, the heart transplant would have been to no avail.

5. A patient heterozygous for FH will make a reduced level of LDL receptors that will lead to elevated levels of LDL in the blood. This person should be put on a low-cholesterol diet and given an inhibitor of HMG-CoA reductase and a bile acid binding resin. The low-cholesterol diet will help lower LDL because of the lower intake of cholesterol, while the reductase inhibitor will lower liver cholesterol levels and stimulate the synthesis of higher levels of LDL receptors. The cholesterol taken up by the liver in the form of LDL can be secreted as bile acids, which can be eliminated by the bile acid binding resins, thus preventing the absorption of the bile acids by the gut. A combination of these treatments should lower serum LDL levels dramatically and reduce the risk of cardiovascular disease.

6. The ^{14}C in HMG-CoA derived from 2-[^{14}C]-acetate is marked in the structure.

$$^{-}O-\overset{O}{\overset{\|}{C}}-\overset{*}{C}H_2-\overset{\overset{*}{C}H_3}{\underset{OH}{C}}-\overset{*}{C}H_2-\overset{O}{\overset{\|}{C}}-S-CoA$$

$* = {}^{14}C$

7. Triacylglycerol is transported in plasma in two types of lipoprotein particles, chylomicrons and VLDL. Chylomicrons carry triacylglycerols derived from dietary intake via the intestine, and VLDL triacylglycerols are stored or synthesized endogenously in the liver. The initial stage of degradation of triacylglycerols contained by both types of particles is performed in the plasma. This reaction, catalyzed by an enzyme, lipoprotein lipase (LPL), which is found fixed to the lining of capillaries serving extrahepatic tissues, converts triacylglycerol to free fatty acids and monoglycerides for uptake by the surrounding cells.

In normal individuals, chylomicrons and VLDL contain an activator of LPL, apoprotein C-II, which stimulates the enzyme in its function of removing triacylglycerol from plasma. Individuals suffering from the condition hyperlipoproteinemia type I, however, lack this activating activity and, as a result, accumulate triacylglycerol in plasma to abnormally high levels.

8. Wolman's disease is caused by a complete lack of lysosomal acid lipase, the activity responsible for normal catabolism of cholesterol esters and triacylglycerols in lysosomes. In individuals suffering from this condition, the uptake and accumulation of these materials within extrahepatic cells is severely affected.

The major part of cholesterol transport in plasma is achieved through the medium of low-density lipoprotein (LDL) particles. These particles ferry cholesterol esters to extrahepatic cells, such as skin fibroblasts, from a variety of sources, including the liver and the extrahepatic cells themselves. In the extrahepatic tissues, LDLs are first taken up in an endocytotic process mediated by specific LDL receptors; they are then exposed to lysosomal degradation by fusion of endocytotic and lysosomal vesicles. In normal individuals this degradation includes removal of the acyl group from cholesterol esters to give cholesterol, a reaction catalyzed by lysosomal acid lipase. This free cholesterol released into the cytosol provokes reductions in both HMG-CoA reductase activity, reducing the cell's own ability to synthesize cholesterol, and LDL receptor synthesis, reducing further uptake of LDL.

In individuals suffering from Wolman's disease, the enzyme catalyzing the hydrolysis of cholesterol esters is defective. As a result, the activity of HMG-CoA reductase and the synthesis of LDL receptors remain high, leading to accumulation in extrahepatic tissues of cholesterol esters and triacylglycerols (LDL particles contain approximately 10% triacylglycerol and 46% cholesterol ester by weight).

9. The enzyme lecithin:cholesterol acyltransferase (LCAT) plays an extremely important role in the transport of cholesterol in plasma. It performs this role in association with one of the longer-lived lipoprotein carriers, HDL. In normal individuals, dietary cholesterol transferred from chylomicrons, and waste cholesterol released into plasma from extrahepatic cells, adsorb to HDL and are converted by the action of LCAT to cholesterol esters. These esters are then shuttled to other lipoprotein carriers, principally VLDLs and LDLs, destined for conversion to LDLs and subsequent uptake by extrahepatic tissues. Much of the cholesterol ester in human LDLs is cholesterol that has been recycled from tissues in this way.

In LCAT deficiency, the flow of cholesterol from dietary and waste endogenous sources to extrahepatic tissues is severely impaired. HDL particles, which leave the liver and intestine essentially devoid of cholesterol esters, remain so throughout their lifetime. Other lipoprotein carriers, notably VLDLs and LDLs, suffer a similarly reduced content of cholesterol esters, while plasma exhibits an elevated level of free cholesterol.

24 Integration of Metabolism and Hormone Action

Problems

1. Most metabolic conversions can occur in either direction by using different pathways. Eukaryotic cells frequently take advantage of subcellular compartments to separate oppositely directed pathways. Use fatty acid synthesis and degradation as an example and discuss the design of the paths from the point of view that they are thermodynamically favorable and kinetically regulated (be sure to consider subcellular compartments). Discuss the design of glycolysis and gluconeogenesis in the liver.

2. If a starving person (one who has gone a number of weeks with no food) is given a shot of insulin, what will happen? Explain your answer.

3. Why are all known hormone receptors proteins? Could other macromolecules serve as receptors?

4. The contraceptive pill contains synthetic progestin or progestin plus estrogen. How do you suppose the pill works?

5. List several reasons why polypeptide hormones are synthesized as precursors.

6. A patient has a hypothyroid condition. He has low serum T_3 and T_4 levels and elevated serum TSH. Upon injection of TRH his serum TSH goes even higher. Is his defect primary (thyroid), secondary (pituitary), or tertiary (hypothalamus)?

7. Activation of most membrane associated hormone receptors generates a second messenger. What is a second messenger and what are the five second messengers currently known? What role do G proteins play in second messenger formation?

8. If cells are repeatedly exposed to hormones, the secondary response is lower or does not occur. Explain the phenomenon of desensitization (how does it happen) and why this is useful to the cell.

9. Is vitamin D a hormone or a vitamin? Explain your answer.

10. Inhibitors of protein synthesis have been shown to block both the rapid and slow auxin-mediated growth responses. How could you explain these observations?

Solutions

1. Pathways are designed so that the equilibrium is very favorable for the conversion. The same pathway cannot be used for conversion in either direction because if it were thermodynamically favorable in one direction, it would be unfavorable by the same amount in the opposite direction. Also, the simultaneous occurrence of conversions in both directions would serve no purpose, as it would result in a futile cycle and waste energy. Regulatory enzymes are designed so that pathways never operate simultaneously in both directions.

Fatty acid synthesis and degradation demonstrate these principles. In a liver cell, fatty acid synthesis takes place in the cytosol using acetyl-CoA carboxylase and a large, multifunctional polypeptide fatty acid synthase. The first committed step is acetyl-CoA carboxylase, which is highly regulated by hormonal control; this results in the phosphorylation of the enzyme. Citrate is transported from the mitochondria and is used to generate acetyl-CoA and reducing power in the form of NADPH (NADPH is used in biosynthetic reactions instead of NADH). Citrate activates the carboxylase while the end product, palmitate, inhibits the reaction. Fatty acid degradation occurs in the matrix of the mitochondria. A key point of regulation is on the uptake of the fatty acid into the matrix of the mitochondria. Malonyl-CoA, the product of the acetyl-CoA carboxylase, inhibits uptake and prevents the newly made palmitic acid from being degraded.

Gluconeogenesis utilizes many of the glycolytic enzymes, yet three of these enzymes in glycolysis have large negative free energy changes in the direction of pyruvate formation. These reactions must be replaced in gluconeogenesis to make glucose formation thermodynamically favorable. This allows both glycolysis and gluconeogenesis to be thermodynamically favorable and at the same time permits the pathways to be independently regulated so that a futile cycle does not exist. The first step in gluconeogenesis involves the movement of pyruvate into the matrix of the mitochondria, where it is converted to phosphoenolpyruvate (PEP). PEP is transported back into the cytosol where it is converted by the glycolytic enzymes back to fructose-1,6-bisphosphate. Then two additional enzymes unique to gluconeogenesis allow glucose to be made. One of the most important allosteric effectors is fructose-2,6-bisphosphate, which activates phosphofructokinase, stimulating glycolysis, while at the same time inhibiting fructose bisphosphatase, inhibiting gluconeogenesis. The levels of fructose-2,6-bisophosphate are under hormonal regulation in response to blood glucose levels.

2. A starving person is metabolizing fats for energy and generating large amounts of acetyl-CoA, which is converted to ketone bodies. The brain has switched from depending entirely on glucose for energy to depending on ketone bodies. When this starving person is given a dose of insulin that normally would be large enough to generate insulin shock (a rapid drop in blood glucose levels that causes a person to pass out because the brain is not getting enough glucose), nothing happens because the brain is getting its energy from ketone bodies, not glucose.

3. A hormone receptor must do two things if it is to function properly:
   (1) Distinguish the hormone from all other surrounding chemical signals and bind it with a very high affinity (K_d ranges from 1×10^{-7} M to 1×10^{-12} M).
   (2) Upon binding the hormone, undergo a conformational change into an active form that can then interact with other molecules that initiate the molecular events leading to the hormone's elicited response.

   Proteins are the only macromolecules that can exhibit this kind of behavior (specific binding and conformation change).

   Acceptors are macromolecules that react to a receptor's conformational change by mediating enzyme activation (or inactivation). This is done by specific phosphorylation of proteins or production of regulatory molecules, both enzymatic events. Therefore, acceptors must be proteins also. The only exception to this (in a loose sense) is a DNA sequence to which a hormone-receptor complex binds and whose perturbation affects a distant promoter and thereby alters the transcriptional activity of that gene (steroid hormone-receptor function; catabolite activator protein; see chapter 29 in the text).

4. The pill works via a feedback inhibition mechanism (see figure 24.20 in the text). Taken together, progestin and estrogen inhibit secretion of FSH and LH by the pituitary (probably by suppressing GnRH release by the hypothalamus) and thus prevent follicular growth and ovulation.

5. The precursor-product relationship serves the cause of hormone function in a variety of contexts, some of which are listed here.
   (1) A polypeptide signal sequence must be present if the protein is to be transported into the endoplasmic reticulum and subsequently secreted.
   (2) Additional polypeptide sequences are necessary for proper peptide chain folding (e.g., C peptide of insulin).
   (3) Cleavage allows control of hormones from inactive to active form (e.g., thyroxine).
   (4) Production of a number of different hormones from the same precursor allows coordinate production of several hormones. Specific cleavage by the cell allows control of which peptides are produced (e.g., cleavage of prepro-opiocortin to corticotropin, β-lipotropin, γ-lipotropin, α-MSH, β-MSH, γ-MSH, endorphin, and enkephalin).
   (5) A large precursor of the hormone can serve as a storage form (e.g., thyroglobulin).

6. This patient most likely has a primary defect, that is, of the thyroid. Since his thyroid cannot produce T_3 and T_4, his TSH levels are elevated due to lack of feedback inhibition. His TRH (from the hypothalamus) cannot be the cause, since under such conditions TSH (from the pituitary) would be reduced or eliminated instead of elevated.

7. Most membrane associated hormone receptors generate a diffusible intracellular signal called a second messenger. The five currently known second messengers are cyclic AMP, cyclic GMP, inositol triphosphate, diacylglyerol, and calcium. Part of the transmembrane signaling that occurs when a ligand binds to the membrane associated receptor is the activation of a membrane-bound G protein (the receptor interaction

stimulates the exchange of GTP for GDP) and the resulting G protein-GTP complex, which can activate or inhibit an intracellular enzyme associated with the inner cell membrane. For example, adenylate cyclase can be activated to produce cAMP (a second messenger that activities protein kinases). Another example would be the activation of phospholipase C by a G protein-GTP complex, which leads to the generation of inositol triphosphate (the inositol triphosphate causes the release of calcium from the lumen of the endoplasmic reticulum into the cytosol) and diacyglycerol, both of which stimulate the phosphorylation of proteins leading to a cellular response.

8. The desensitization of cells to a hormonelike epinephrine prevents chronic stimulation, which results in transient activation of cellular events. The reduced response is due to a decrease in the number of accessible receptors and the uncoupling of receptors from adenylate cyclase activation. The receptor will become phosphorylated and lose its ability to activate its G protein. Polypeptide hormones such as insulin also show a diminished response by a loss of receptors from the cell surface in a process called down-regulation. The hormone–receptor complex is taken up by the cell and transported to the lysosome where it is degraded (this process may take hours). New receptors must then be made by protein synthesis so that the cell can again be hormone responsive.

9. Vitamin D can be considered both a hormone *and* a vitamin. Its mode of action is like many other steroid hormones (forming a receptor–hormone complex that activates transcription of specific genes in the nucleus), and it is synthesized in the body where it acts at a distant location. Vitamin D_3 is formed in the skin of animals through the action of ultraviolet light on 7-dehydrocholesterol. Vitamin D can also be taken in the diet (commonly as a vitamin supplement in milk) and is then considered a vitamin.

10. Both the rapid and slow growth response initiated by auxin binding are mediated by protein synthesis. The rapid response is due to an increase in proton transport out of the cell by a pump coupled with a membrane ATPase. Synthesis of polypeptide factors that stimulate proton transport (and the associated ATPase) may occur in the first few minutes after auxin binding. These polypeptides and the events they trigger probably modulate mechanisms controlling cellular growth. The slow response is due to an increase in the synthesis of proteins and nucleic acids necessary for sustained growth. This dependency of both rapid and slow auxin-mediated growth response upon *de novo* synthesis of proteins accounts for their sensitivity to the effects of inhibitors of protein synthesis.

25 Structures of Nucleic Acids and Nucleoproteins

Problems

1. **Briefly describe how Avery was able to show that DNA was the genetic material in cells.**

2. **Summarize the evidence that RNA is the genetic material in tobacco mosaic virus (TMV).**

3. **Describe two physical methods that could be used to estimate the base composition of DNA. What would the data look like with two DNA samples, one with high G-C content and another with high A-T content? (Assume that the concentration of the samples is equal.)**

4. **Why is DNA denatured at either low pH (pH 2) or high pH (pH 11) and why is DNA stable at pH 7? (Hint: See p*K* values in table 20.2.)**

5. **Standard conditions for hydrolyzing RNA to necleotides are 0.3-*N* NaOH, 37°, for 16h. Draw a chemical reaction mechanism for this hydrolysis. Why is DNA not hydrolyzed under these conditions?**

6. **What effect would the following reagents have on the T_m of duplex DNA: 7-M urea, 90% formamide, higher concentrations of NaCl, pure water, and T4 gene 32-encoded protein? Explain how these chemicals act to affect DNA duplex stability.**

7. **Why can't RNA duplexes or RNA-DNA hybrids adopt the B conformation?**

8. There are DNA-binding proteins that specifically bind to Z DNA. How could these proteins help stabilize DNA in the Z configuration? (Hint: How do single-stranded DNA-binding proteins destabilize duplex structures?)

9. Linear duplex DNA can bind more ethidium bromide than covalently closed circular DNA of the same molecular weight. Why? How could ethidium bromide be used to separate these two forms of DNA on CsCl gradients? (Hint: The ethidium cation has a density less than that of water.)

10. What is the structure of the nucleic acid called A, given the clues listed below?

(a) Nucleic acid A sediments as a single species in a neutral sucrose density gradient.

(b) In an alkaline sucrose gradient, one half (structure B) of the mass of A sediments down the tube; the other half remains at the meniscus (top of the tube).

(c) The buoyant density of A is much higher than that of duplex DNA of the same G + C content.

(d) The thermal transition profile of B is broad when A is melted.

(e) When A is heated and rapidly cooled it gives rise to B and C. The buoyant density of C is heavier than that of B. (Hint: Single-stranded DNA is more dense than duplex DNA, and RNA is more dense than DNA.)

11. Renaturation of randomly sheared denatured DNA can be measured by the hypochromic shift at 260 nm, by S1-nuclease resistance, or by retention on hydroxyapatite. Which method is likely to give an overestimate for the extent of renaturation?

12. Give the relative times for 50% renaturation of the following pairs of denatured DNAs, starting with the same initial DNA concentrations.

 (a) T4 DNA and *E. coli* DNA, each sheared to an average single-strand length of 400 nucleotides.

 (b) Unsheared T4 DNA and sheared T4 DNA.

13. When histone proteins are isolated from chromatin their mass is equal to the DNA, and the ratio of four of the histones is 1:1:1:1 (H2a:H2b:H3:H4), while H1 is found in half the yield (0.5). Discuss whether or not these data fit the bead-and-string model for nucleosomes.

14. You are given a sample of nucleic acid. How would you determine whether: (a) it is DNA or RNA and (b) whether it is single or double stranded?

Solutions

1. Avery, working with two different strains of pneumococcus, was able to show that a fraction isolated from the pathogenic S strain that transformed the nonpathogenic R strain was DNA. This transforming activity was not affected by RNase, proteases, or enzymes that degrade capsular polysaccharides but was destroyed by treatment with DNase. Purified DNA from S cells was able to transform R cells into S cells *in vitro.*
2. Purified RNA isolated from tobacco mosaic virus (TMV) was able to produce lesions on tobacco plants similar to those produced by the intact virus. Other researchers reconstituted two different strains of the virus, HR and TMV, in all possible combinations (HR-protein + TMV-RNA, TMV-protein + HR-RNA, etc.). These reconstituted viruses were able to infect tobacco plants and produce viral progeny. The type of viral protein found in the progeny was determined by the type of RNA. This proved that the nucleic acid carries the genetic determinants of the viral protein.
3. One method that could be used to estimate the base composition of DNA is CsCl gradients. The density of the duplex DNA increases with higher G-C content, therefore the density of the high G-C DNA would be higher than the A-T-rich DNA and could be determined on CsCl gradients (see figure). The melting temperature (T_m) of DNA is also affected by the base composition, with the G-C-rich DNA having a higher T_m than the A-T-rich DNA. The DNA samples could be heated in a spectrophotometer and the increase in absorbance of ultraviolet light (hyperchromism) could be plotted against temperature. The A-T-rich DNA would have a lower T_m than the G-C-rich DNA. Also, if the DNA melting was monitored at 260 nm, the A-T-rich DNA would show a larger

hyperchromic increase in absorbance than an equal amount of G-C-rich DNA. If the DNA denaturation was followed at 280 nm, the G-C-rich DNA would have a larger hyperchromic increase.

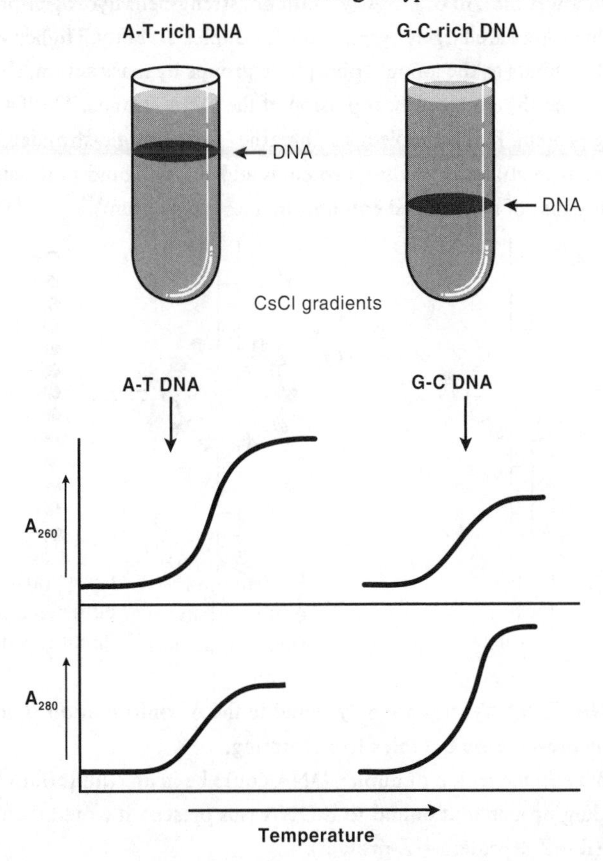

4. At high pH the thymine base and the guanine base would be deprotonated and have a negative charge, while at low pH the adenine, guanine, and cytosine bases would be protonated and have a positive charge. These charges in the hydrophobic interior of the duplex would disrupt the structure. At neutral pH the bases would not have any charge and the duplex would have its greatest stability. Living systems maintain their pH around 7, which preserves the duplex structures found in their nucleic acids (DNA and RNA).
5. The chemical reaction mechanism for the base hydrolysis of RNA is shown here:

Obviously, DNA could not be hydrolyzed by this mechanism because DNA lacks a 2′-OH group.

6. The following reagents would lower the melting temperature (*Tm*) of duplex DNA: 7-M urea, 90% formamide, pure water (low salt), and the T4 gene 32-encoded protein. Increasing the salt concentration would increase the *Tm*. Urea and formamide disrupt the hydrophobic forces that hold the duplex together. Urea does not have the hydrogen-bonding potential that water has, and a 7-M urea solution has about 50M water. This would suggest that urea is not disrupting the hydrogen bonding, leading to a lower *Tm*, but that it is disrupting the

hydrophobic forces. Urea also disrupts protein structure by disrupting hydrophobic interactions. The nonpolar solvent ethanol will also lower the *Tm* of DNA, yet ethanol strengthens hydrogen bonds. These observations suggest that hydrophobic forces are largely responsible for duplex stability. Higher salt will increase the stability of the duplex because Na^+ binds to the anionic phosphate groups by mass action, shielding the phosphate groups from each other and limiting the electrostatic repulsion of the DNA strands. The T4 gene 32-encoded protein is a single-stranded binding protein. Duplex molecules "breathe" (small single-stranded bubbles form and close in equilibrium). When the single-stranded binding protein is added it will bind to the single-stranded regions and destabilize the duplex because of the coupled equilibrium (see the diagram).

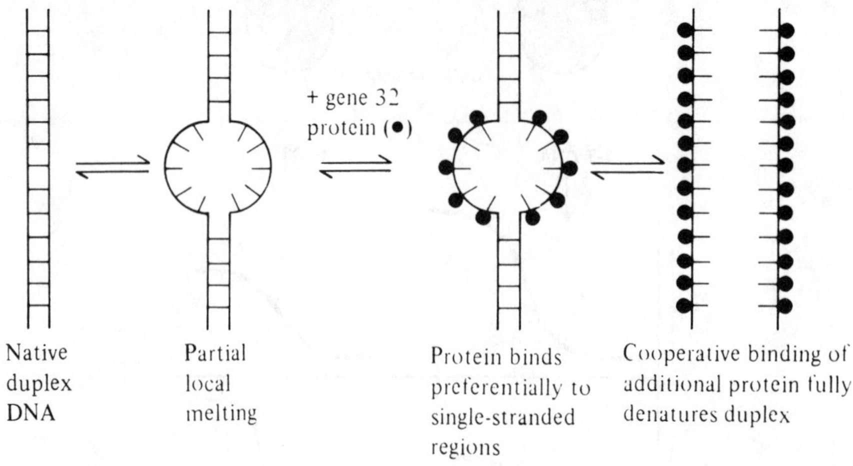

7. RNA duplexes and RNA–DNA hybrids are only found in the A conformation. The 2′-OH group found on the ribose in RNA sterically prevents the B duplex from forming.
8. The duplex structure in a specific region of duplex DNA could be in an equilibrium between the Z and B forms ($Z \rightleftharpoons B$). If a DNA binding protein that bound to Z DNA was present, it would then stabilize the Z structure by a coupled equilibrium ($B \rightleftharpoons Z$ + protein $\rightleftharpoons$ Z-protein).
9. The differences in Ethidium (Et) binding capacities between linear duplex DNA and covalently closed circular DNA (cccDNA) can be understood in terms of the differences in topological constraints imposed on the two molecules. By intercalating between two adjacent base pairs, Et unwinds the double helix, which results in an increase in the length of the helix (pitch). For cccDNA, the conformational stress introduced by unwinding is compensated for by a change in tertiary structure of the molecule, i.e., supercoiling. The winding and unwinding (twist) of a cccDNA molecule is related to supercoiling (writhe) in the following way:

$$\Delta L = \text{twist}\,(T) + \text{writhe}\,(W)$$

The linking number L remains constant provided that no covalent bonds are broken and re-formed, as in the present case. Since unwinding reduces T, W must assume more positive values in order to satisfy the relationship. Simply stated, unwinding the helix results in the introduction of positive supercoils.

At some point, the torsional stress caused by the positive supercoils will become energetically unfavorable and the tendency of the molecule will be toward winding, thus preventing further binding of Et. The tendency toward winding in positively supercoiled DNA is driven by the excess free energy of the supercoiled conformation resulting from torsional stress (as evidenced by the fact that introducing a nick into a supercoiled molecule completely relaxes the DNA). Linear duplex DNA and nicked circular DNA do not experience this torsional stress, since they are not covalently closed, and would thus be expected to have a greater binding capacity for Et.

It should be noted that if one starts with negatively supercoiled DNA (as opposed to *relaxed*, covalently closed circular DNA) and adds increasing concentrations of Et, the negatively supercoiled DNA will initially bind more Et than will duplex linear DNA. This is so because the negatively supercoiled conformation energetically favors unwinding of the helix and thus intercalation of Et. Ultimately, however, when Et is present in excess concentrations, linear duplex DNA has a greater capacity for Et binding as a result of the topological constraints imposed on cccDNA discussed earlier.

The difference in Et binding capacities is exploited to separate the two DNAs by CsCl density gradient ultracentrifugation as follows. Et has a lower density than water, which has a lower density than nucleic acids. Thus intercalation of Et into a DNA molecule lowers its buoyant density in a CsCl solution. Because linear duplex DNA ultimately binds more Et than cccDNA, it will have a lower buoyant density. This permits the separation of the two species in a CsCl density gradient, since each species will band in the gradient where its respective buoyant density equals the CsCl solution density.

10. Structure A is a duplex molecule composed of complementary DNA and RNA strands:

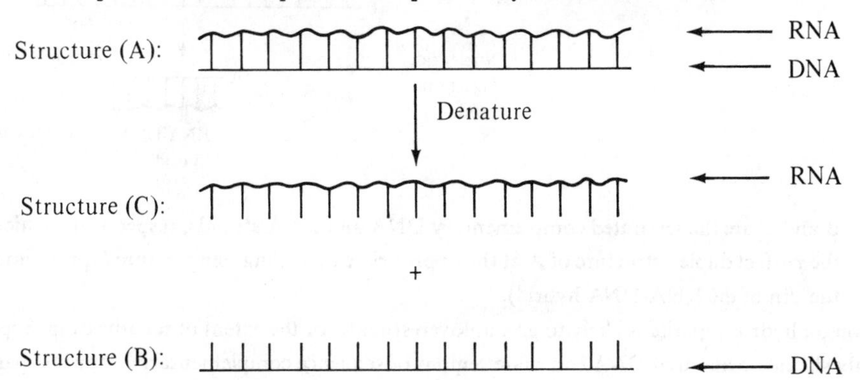

The clues in the problem provide the following information:

(a) Neutral sucrose density gradients separate molecules by size and are non-denaturing. The narrow sedimentation profile of structure A indicates that the RNA-DNA hybrids are homogeneous in size:

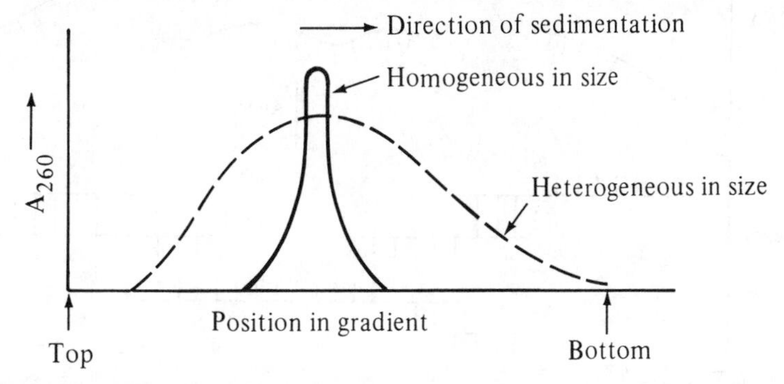

(b) The denaturing effect of alkali is exploited in alkaline sucrose gradients to separate single-stranded DNA molecules by size. Structure B, which sediments down the tube, is single-stranded DNA, which comprises approximately half of the mass of the RNA-DNA hybrids. RNA, however, is unstable in alkali, and is hydrolyzed to nucleoside monophosphates, which remain at the meniscus.

(c) RNA is denser than DNA, and RNA-DNA hybrids are denser than DNA-DNA hybrids.

(d) The sharp thermal transition profile indicates regular *inter*molecular base pairing of the complementary RNA and DNA strands in structure A; i.e., structure A is a perfectly base-paired heteroduplex. The broad melting curve of B, on the other hand, indicates the formation of irregular *intra*molecular base pairing in the DNA strands. Note that structure C, the complement of structure B, would also be expected to have a broad melting curve, also due to irregular *intra*molecular base pairing. A possible mixture of molecules with irregular *intra*molecular base-pairing is illustrated:

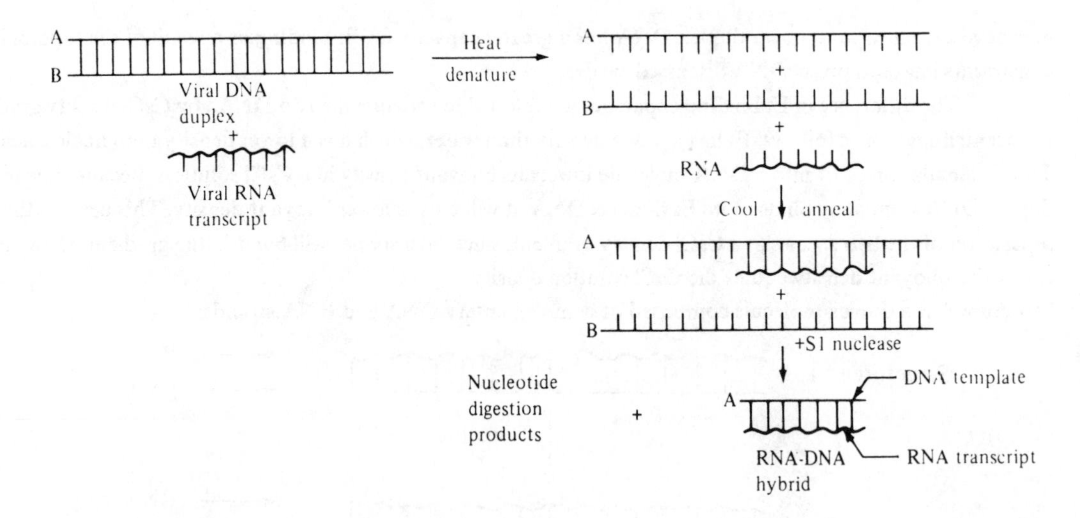

(e) B and C are the separated complementary DNA and RNA strands, respectively, which will renature to the perfect duplex structure of A at the appropriate annealing temperature (approximately 20° C below the *Tm* of the RNA-DNA hybrid).

11. Retention on hydroxyapatite is likely to give an overestimate for the extent of renaturation. Separated strands of randomly sheared denatured DNA can share regions of sequence complementarily over just parts of the strands:

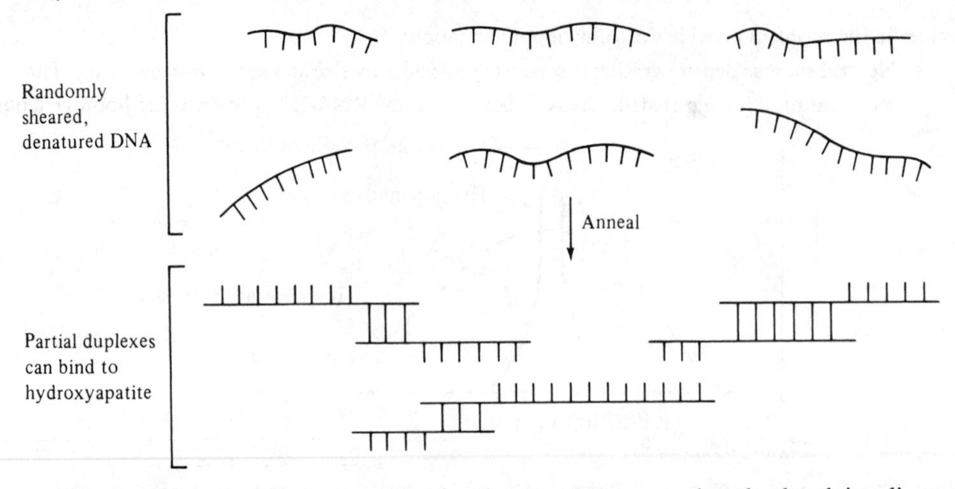

The hypochromic shift, measured by absorbance at 260 nm, on the other hand, is a direct physical measurement of the extent of base stacking characteristic of duplex regions and is thus a more accurate measure of renaturation. S1 nuclease is also more accurate, since the nuclease will digest single-stranded regions of partial duplexes, while duples regions are protected from digestion. Note, however, that none of the preceding methods can distinguish intramolecular duplex regions from intermolecular duplex regions:

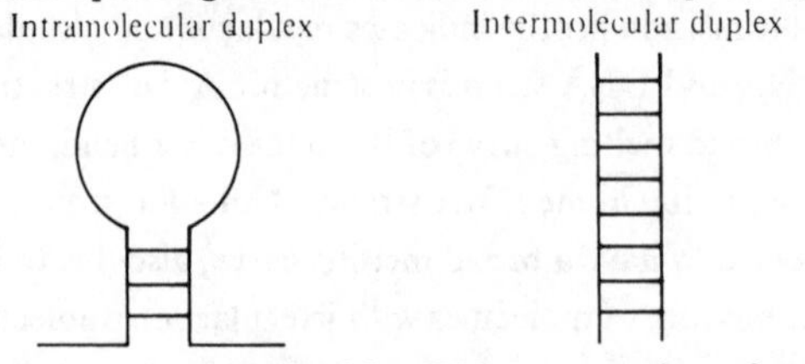

12. (a) The Cot ½ values give the time required for 50% renaturation of complementary single strands sheared to a specified average length, starting with an initial total DNA concentration of Co at time $t = 0$. The

Cot ½ values for renaturation of T4 and *E. coli* DNAs sheared to an average single-strand length of 400 nucleotides can be determined directly from figure 25.25 in the text:

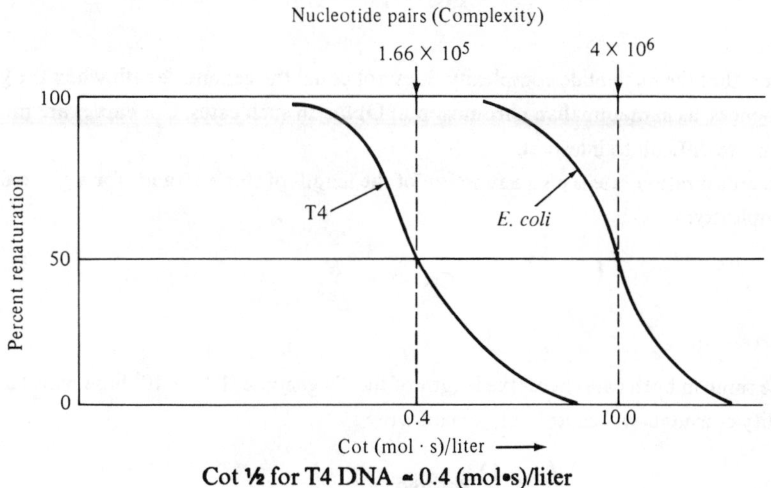

Cot ½ for T4 DNA = 0.4 (mol•s)/liter

Cot ½ for *E. coli* DNA = 10.0 (mol•s)/liter

If we designate the Cot ½ for T4 DNA at Cot′ ½, we can set up the following proportionality:

$$\frac{\text{Cot}'\frac{1}{2}}{\text{Cot}\frac{1}{2}} = \frac{0.4}{10.0}$$

Since Co′ = Co in this case, we then have

$$\frac{t'\frac{1}{2}}{t\frac{1}{2}} = 0.04$$

Rearranging, we obtain

$$25\left(t'\frac{1}{2}\right) = t\frac{1}{2}$$

Therefore, T4 DNA will renature about 25 times faster than *E. coli* DNA under the given conditions. This reflects the fact that the Cot ½ is proportional to the nucleotide complexity *N*. This is so because for a given total molar concentration of single strands of the same size, the molar concentration of complementary strands increases with decreasing nucleotide complexity, thus increasing the probability of two complementary strands "finding" each other.

In the examples given here, in which the DNAs consist predominantly of unique sequences, the nucleotide complexity is equal to the genome sizes of T4 and *E. coli* in nucleotide pairs. In other words, we could have arrived at the same answer as above simply by dividing the genome size of *E. coli* by that of T4 DNA.

$$\frac{4 \times 10^6}{1.66 \times 10^5} = 25$$

Note, however, that the nucleotide complexity does not equal the genome length when the genome contains repeated sequences, as in mammalian chromosomal DNA. In such cases, Cot curves are more complex and, accordingly, more difficult to interpret.

(b) The renaturation rate is also a function of the length of single strands for a given nucleotide complexity:

$$\text{Cot}\ \frac{1}{2} \propto \frac{N}{L^{0.5}}$$

Since N is the same in both cases here (the length of the T4 genome, 1.66×10^5 base pairs) and other proportionality constants will cancel out, we can write

$$\frac{\left(\text{Cot}\ \frac{1}{2}\right)\text{sheared}}{\left(\text{Cot}\ \frac{1}{2}\right)\text{unsheared}} = \frac{(L^{0.5})\ \text{unsheared}}{(L^{0.5})\ \text{sheared}}$$

Therefore,

$$\frac{\left(\text{Cot}\frac{1}{2}\right)\text{sheared}}{\left(\text{Cot}\frac{1}{2}\right)\text{unsheared}} = \frac{(1.66 \times 10^5)0.5}{(400)^{0.5}} = 20$$

Thus starting with the same initial total DNA concentration, unsheared T4 DNA will renature approximately 20 times faster than T4 DNA sheared to an average length of 400 nucleotides.

The basis for this dependence of renaturation rate on nucleotide length can be understood in molecular terms as follows: The observed second-order kinetics of renaturation is consistent with a two-step process, generally thought to involve an initial nucleation step followed by zippering:

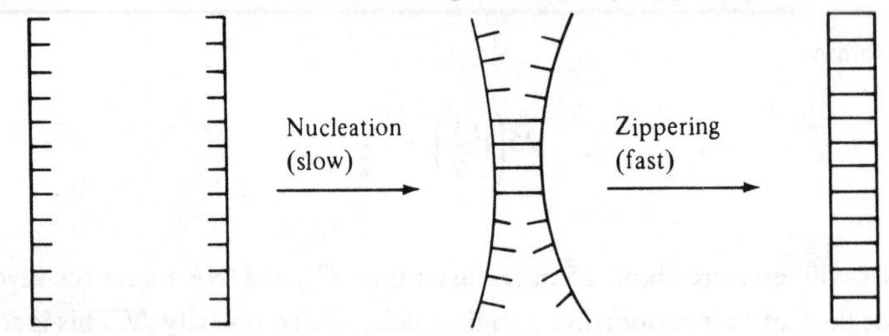

The rate-limiting step is nucleation, and the renaturation are increases with increasing fragment length because more nucleation sites are available per strand for reaction. The dependence on less than the first power of fragment length is due predominantly to polymer excluded-volume effects or steric hindrance effects. With increasing fragment length, interpenetration of two strands becomes more difficult, interfering with the ability of complementary base pairs to find each other.

13. The ratio of the histones in chromatin supports the model proposed for nucleosome structure. The core of the nucleosome is made up of an octamer of two molecules each of H2A, H2B, H3, and H4. The H1 histone seals off

the nucleosome, i.e., only one H1 per nucleosome. Thus, the ratio of histones fits the model proposed for nucleosome structure.

14. DNA and RNA can be differentiated by the use of colorimeteric reactions such as orcinol, which gives a blue-green color with ribose but does not react with deoxyribose (only RNA). The diphenylamine test will react with deoxyribose and not ribose, and thus will give a color reaction with DNA. Another method to determine the type of nucleic acid is to treat the sample with base and determine whether the nucleic acid is hydrolyzed. RNA would be hydrolyzed by high pH conditions, while DNA is resistant to base hydrolysis. Another similar approach would be to see if nucleases such as DNase or RNase would digest the sample.

    To determine whether the sample is single or double stranded, it could be heated. A large hyperchromic increase (30% to 40% increase in absorbance), would be characteristic of duplex structure. Single-stranded structure could have significant duplex structure, such as is seen in tRNA with hairpin loops, and would have significant increases in absorbance when heated. The melting curve, however, would be very broad and not show the sharp transition seen with double-stranded duplex structure. Double-stranded nucleic acids would also have a lighter density in CsCl gradients (about 1.7 g/cc), while single-stranded molecules would have a density greater than 1.9 g/cc.

26 DNA Replication, Repair, and Recombination

Problems

1. Explain how Meselson and Stahl were able to demonstrate that *E. coli* replicates its DNA in a semiconservative mode. How would the data appear for the first and second doublings if the mode of replication was dispersive? Does the dispersive mode of DNA synthesis occur in cells? (Explain your answer.)

2. The genetic map below represents the distribution of eight genes (*a–h*) as they occur on a bacterial chromosome. The plot is a graphic representation of the average number of copies for each of the eight genes, for a cell that grows with a doubling time equal to the time required for a complete round of DNA replication.
   (a) Estimate the location of the origin of replication.

   (b) Infer whether replication is bidirectional or unidirectional.

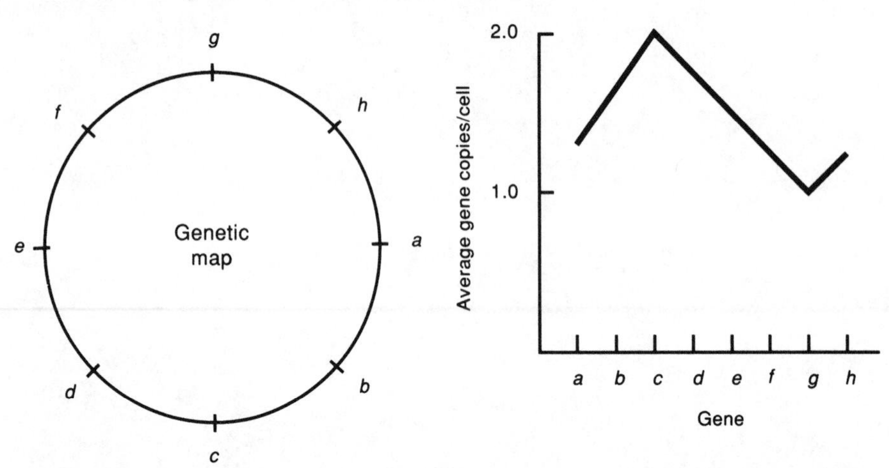

3. Draw the chemical reaction mechanism for the formation of a phosphodiester linkage during DNA synthesis. Discuss the significance of the pyrophosphate product that is formed. What is the significance of the Mg^{2+} requirement?

4. Would you expect the replication of either ϕX174 DNA or the *E. coli* chromosome to be sensitive to inhibition by the antibiotic rifampicin? If inhibition occurs, at what stage during replication does it occur?

5. Draw the chemical reaction mechanism for DNA ligase in *E. coli* (uses NAD^+ as the source of energy and forms a covalent intermediate with an ε-amino group of lysine). Why are ligation reactions that require ATP more thermodynamically favorable?

6. *E. coli* has a genomic complexity of about 4×10^6 base pairs (bp) and each replication fork can move at about 10^3 bp per second. How long would it take to replicate the *E. coli* chromosome? With an ample carbon source and ideal growth conditions, cells of *E. coli* can divide in about 20 min. How can this shorter division time occur if the rate of fork migration remains constant at 10^3 bp per second?

7. Humans have about 3×10^9 base pairs of DNA in their genome and the replication forks migrate much slower than in bacteria (about 30 bp per second). How long would it take to replicate the entire genome if it was a single continuous piece of DNA (one chromosome)? How many replication origins would be required to replicate this DNA in an hour?

8. In the graph below, *E. coli* was labeled with radioactive thymidine for a short pulse (10 s) followed by a chase with an excess of nonradioactive thymidine. The DNA was extracted and centrifuged in alkaline sucrose gradients (under high pH conditions the DNA denatures). Explain what these data imply, and interpret these results in light of our current model for DNA replication.

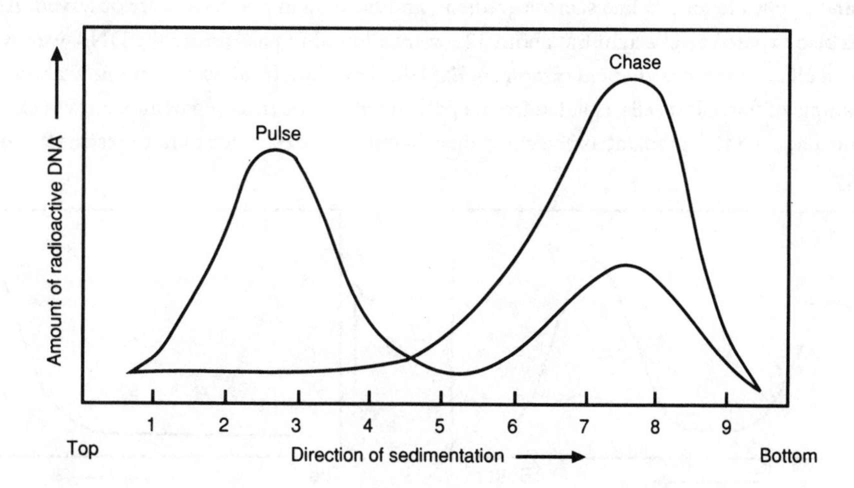

9. Cairns and De Lucia isolated a mutant strain of *E. coli* that had only about 1% of the DNA Pol I activity found in wild-type cells, yet the strain replicated its DNA at a normal rate. Explain how this discovery was important in understanding the role of the different DNA polymerases in replication and repair.

10. All enzymes that make DNA in a template-dependent fashion require a primer. Why does a primer increase the fidelity of DNA synthesis and why is this primer usually RNA?

11. What is the novel priming system for adenovirus DNA replication? Why is this system an advantage to the virus? What other viruses have similar priming systems?

12. Outline in general terms how a retrovirus replicates its RNA. What unique viral protein is used in this replication process? What serves as the primer for retrovirus replication?

13. Eukaryotic DNA is replicated at a slower rate than prokaryotic DNA. One reason may be the requirement for the deposition of histone proteins on DNA (histone synthesis and DNA replication are coupled). Describe a model for the replication of eukaryotic DNA and nucleosome formation.

14. Normal human fibroblasts were grown in culture and then exposed to UV light. A short time later the DNA was extracted and applied to an alkaline sucrose gradient, and the data in graph (*a*) were observed. Another sample of cells was also exposed to UV light but about 12 h were allowed to pass before the DNA was extracted and applied to an alkaline sucrose gradient (graph *b*). Explain these data from what you know about DNA repair. Another sample of fibroblast cells, isolated from a patient with xeroderma pigmentosum was exposed to UV light and then applied to a gradient after a short time. Would you expect the data to resemble those in graph (*a*) or (*b*)? Why?

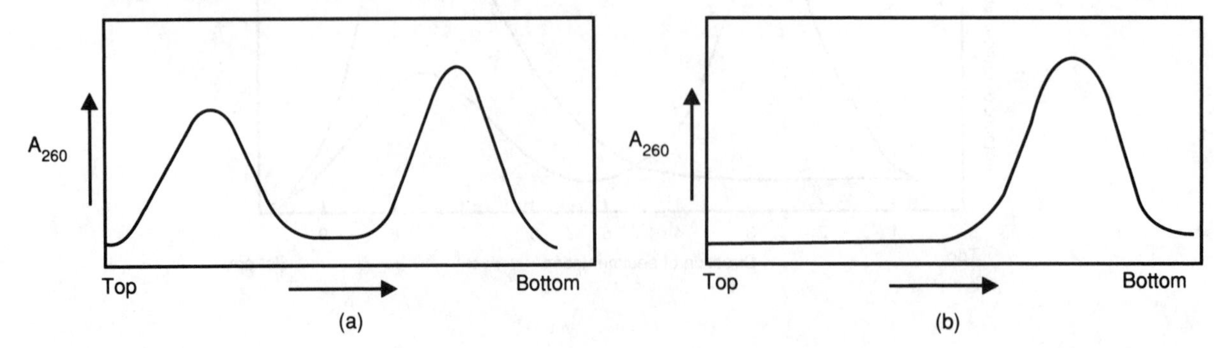

15. Why is the uracil-DNA glycohydrolase very important in DNA repair? Why is thymidine-containing and not uracil-containing DNA the modern product of evolution?

16. What is the role of the recA protein in *E. coli* DNA repair and in recombination? What use are *recA* mutants in biochemical and genetic research?

Solutions

1. Meselson and Stahl were able to show that DNA replicated by a semiconservative mode of replication using the heavy isotope of nitrogen, ^{15}N, and CsCl gradients. *E. coli* was grown for many generations in medium containing $^{15}NH_4Cl$. Then the cells were transferred to a medium containing a ^{14}N-nitrogen source, and the density of their DNA was determined on CsCl gradients at different generation times. After one generation, the DNA had a density halfway between densities of fully ^{15}N-labeled DNA and unlabeled DNA (^{15}N-^{14}N-DNA). After the second generation, one half of the DNA was unlabeled with ^{15}N (^{14}N-^{14}N-DNA), while the other half had the hybrid density (^{15}N-^{14}N-DNA). These results could only be explained by a semiconservative mode of replication. If the mode of replication was dispersive, the first doubling would generate DNA of a hybrid density (both semiconservative and dispersive replication would give identical results at the first generation). After the second doubling there would still be only a single band of DNA; the density, however, would be closer to unlabeled DNA as it would be 75%^{14}N-25% ^{15}N-DNA. Dispersive replication does not occur, but repair synthesis does occur in a dispersive manner (such as repair synthesis after UV light exposure to remove thymine dimers).

2. (a) The location of the origin of replication must be closest to the gene with the highest average number of copies/cell, in this case gene *c*.

   (b) Taking into account the plot of gene frequency and the physical map of the chromosome described, we can see that the average number of copies/cell of genes on both sides of the origin (gene *c*) declines gradually from 2 to 1. This pattern would be expected if replication were bidirectional. If replication were instead unidirectional, the pattern of decline in gene frequency (assuming the same physical map), would be

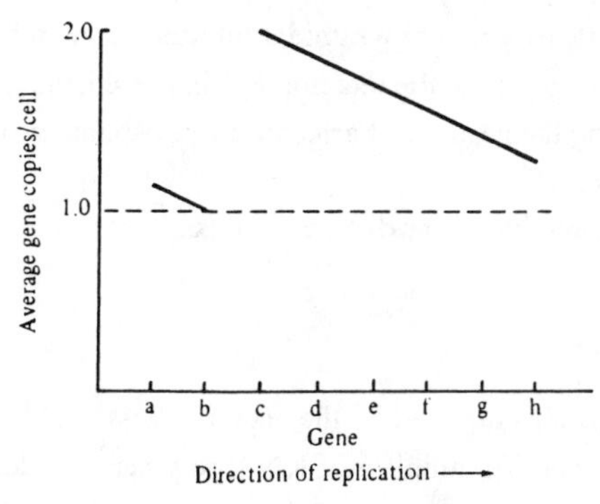

Note, in this case, in contrast to the plot referred to in the problem, the gene immediately counterclockwise of the origin (gene *b*), is the last to replicate and so is present at the lowest frequency. For bidirectional replication, assuming the rate of replication is the same in both directions, the gene present at the lowest frequency is that opposite the origin on the circular map (gene *g*).

3. The mechanism for the formation of a phosphodiester linkage during DNA synthesis is shown:

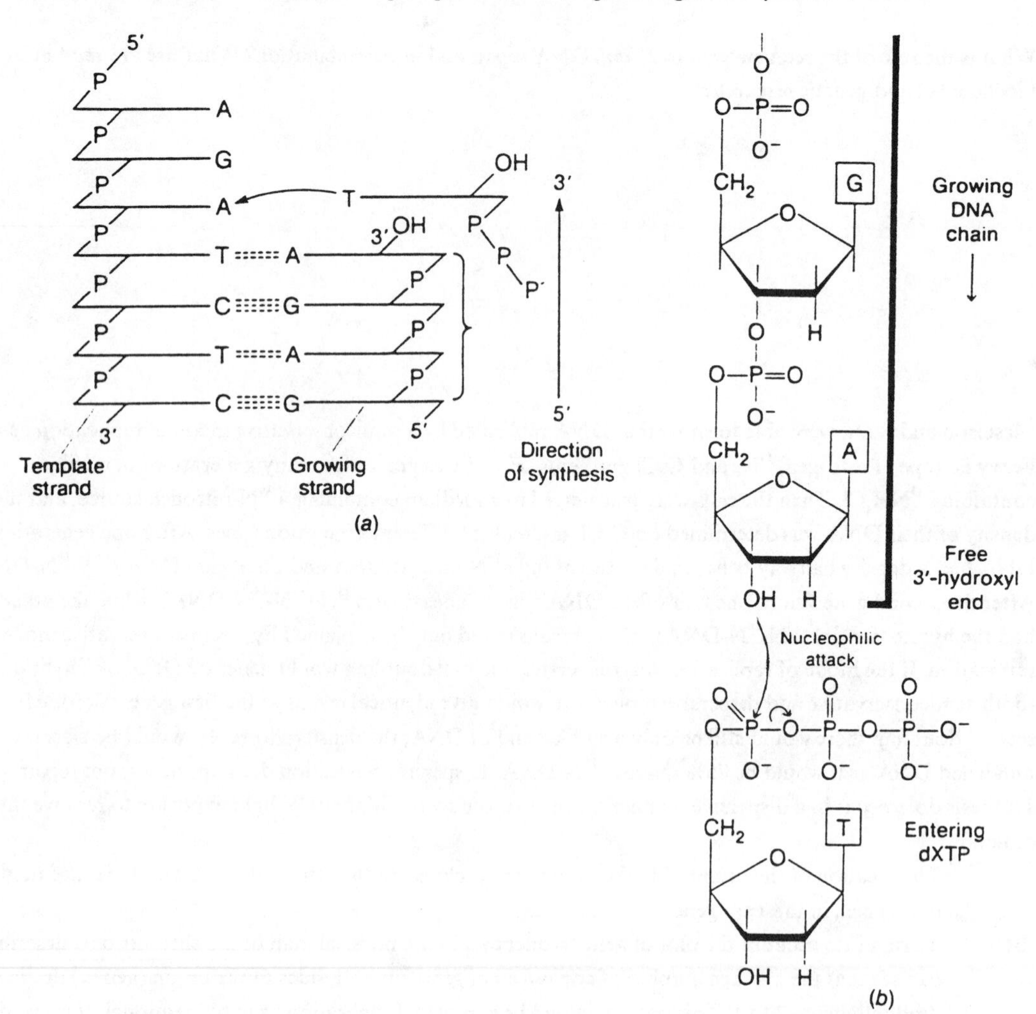

As has been seen in many other biochemical reactions in the cell, the generation of pyrophosphate coupled with its hydrolysis by pyrophosphatase is the major driving force for DNA synthesis. The Mg^{2+} can bind to the transition state intermediate (trigonal bipyramidal intermediate) and stabilize it, lowering the energy of activation. The metal ion can also promote the reaction by charge shielding; the Mg^{2+}NTP complex is the actual substrate, with the metal reducing the negative charge on the phosphate groups so as not to repel the electron pair of the attacking nucleophile.

4. The replication of ϕX174 DNA involves three distinct phases:

(1) + ssDNA → RF.

(2) RF → RF.

(3) RF → + ssDNA.

Step (1) is accomplished with proteins supplied by the bacterial host, and is primed by primase. Thus it would not be inhibited by rifampicin, a specific inhibitor of RNA polymerase, added at the time of infection. In step (2), synthesis of the viral + strand is primed using a 3′OH generated by cleavage by the viral gene A protein. As the production of this protein requires transcription of the viral A gene by RNA polymerase, rifampicin, added at the time of infection, would inhibit replication of ϕX174 RNA at this point.

Replication of the *E. coli* chromosome includes three phases:

(1) Initiation.

(2) Elongation.

(3) Termination.

The initiation of replication at oriC requires RNA polymerase both for transcription and priming. Therefore, rifampicin added prior to initiation would inhibit replication of the *E. coli* chromosome at this stage.

5. The chemical reaction mechanism for DNA ligase is shown here:

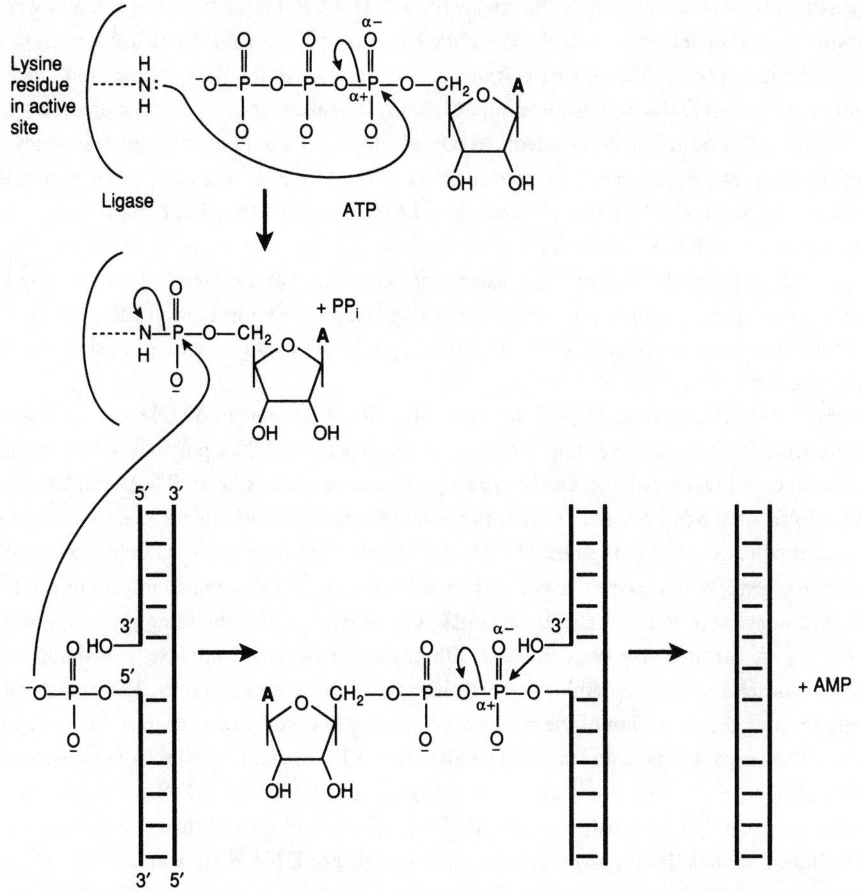

Lysine residue in active site

Ligase

ATP

$+ PP_i$

5′ 3′

3′ HO

5′

3′ 5′

3′ HO

+ AMP

Ligation reactions that require ATP are more thermodynamically favorable because they produce pyrophosphate as a product. The pyrophosphate is hydrolyzed by pyrophosphatase, and because of coupled equilibrium the reaction is pulled to the formation of the ligated DNA (see also the explanation in solution 3).

6. It would take *E. coli* about 33 min to replicate its genome ($4 \times 10^6/2 \times 10^3 = 2 \times 10^3$ s or about 33 min). Since a replication fork moves at 10^3 bp per second and because there are two replication forks, the DNA is being replicated at 2×10^3 bp per second. Cells would divide faster if one round of replication started before the other finished, generating multiple replication forks (multifork replication) with division times of about 20 min.

7. The human genome is much larger than bacteria and the rate of migration of the replication fork is much slower. If there was a single point of replication and the human DNA was contained in one chromosome, the time for replication would be very long (3×10^9 bp/60 bp per second = 5×10^7 s or about 580 days!). To make the time of replication an hour, about 14,000 origins of replication would be needed.

8. These data support the model in which one strand is made in a continuous model (leading strand) while the other strand is made discontinuous (lagging strand). Small pieces (Okazaki fragments) are made first and then become incorporated into large segments of DNA. When the DNA is applied to the alkaline sucrose gradient the DNA is denatured to single-stranded DNA. The small single-strand fragments would not sediment as fast in the sucrose gradient as would the large segments of DNA. The pulse "looks" only at initial DNA made, while the chase allows one to observe the product (small pieces → large segments).
9. The small amount of DNA polymerase I in the mutant could not account for the rapid rates of synthesis found in the mutant (normal rate of replication). This suggests that another polymerase must be the primary enzyme in replication (DNA Pol III). However, this mutant was deficient in repair synthesis (could not repair UV damage, etc.), suggesting that this must be one of the major roles of DNA Pol I. Cairns and De Lucia were lucky, because if the mutant was completely lacking DNA Pol I, the bacteria could not survive (a lethal mutation). The Pol I is important in the maturation of the Okazaki fragments by removal of the RNA primer and "nick translation."
10. The requirement that all DNA polymerases initiate chain elongation with a primer contributes greatly to the fidelity of DNA replication. The first nucleotides coupled together would have a high frequency of mismatches because of the cooperative nature of base pairing interactions. However, if a short RNA primer is made (ribo-oligos form a more stable duplex), the RNA can later be removed by DNA Pol I, permitting accurate base pairing by replacing the RNA with DNA.
11. Adenoviruses prime the replication of DNA using a protein primer at the 5′ end of the growing DNA strand. This unusual priming system allows the virus to specifically replicate without replicating the genomic DNA of the host. Two other animal viruses, polio virus and encephalomyocarditis virus, also prime their DNA replication with a protein.
12. Retroviruses (like the AIDS virus, HIV) first convert their RNA genome into DNA using the unique enzyme reverse transcriptase (converts RNA into DNA, an RNA-dependent DNA polymerase) and using a cellular tRNA as a primer (all DNA polymerases require a primer, even ones using an RNA template). This DNA is made into a double-stranded circular DNA which enters the nucleus and integrates into the host genomic DNA where it can remain latent for a long time (called a provirus). The integrated viral sequence is similar to any other eukaryotic gene. When the provirus is activated (turned on) it produces an infectious mRNA that is packaged with the viral proteins (coded by the mRNA) and released by a budding process from the cell surface.
13. DNA in eukaryotic chromosomes is complexed with histone proteins in complexes called nucleosomes. These DNA-protein complexes are disassembled directly in front of the replication fork. The nucleosome disassembly may be rate-limiting the migration of the replication forks, as the rate of migration is slower in eukaryotes than prokaryotes. The length of Okazaki fragments is also similar to the size of the DNA between nucleosomes (about 200 bp). One model that would allow the synthesis of new eukaryotic DNA and nucleosome formation would be the disassembly of the histones in front of the replication fork and then the reassembly of the histones on the two duplex strands. Histone synthesis is closely coupled to DNA replication.
14. UV light causes the formation of pyrimidine dimers, which are removed by a general excision repair mechanism. Single-strand breaks occur as the first step in this process. The damaged DNA is removed and resynthesized, and the nicks are ligated. A short time after exposure to UV light, many single-strand breaks are observed in the DNA [this generates small fragments of single-strand DNA on the alkaline sucrose gradients, graph (*a*)]. After many hours the DNA is repaired and only large segments of single-strand DNA are found. The data from the xeroderma pigmentosum patient would look like graph (*b*), as this patient is deficient in excision repair. Thus, no single-strand breaks appear in the DNA and the single-strand DNA remains of high molecular weight. The inability to remove pyrimidine dimers leads to the early development of skin cancers and eventual blindness and death.
15. Cytosine spontaneously deaminates to uracil, leading to a G-U base pair (the duplex is not disrupted by mispairing). If this damage was not repaired, it would lead to a point mutation (G-C pair to an A-U pair). The uracil-DNA glycohydrolase will remove the uracil base, allowing repair to take place. If the thymine base was not found in DNA, there would be no mechanism to recognize deaminated cytosine bases and remove them.

16. The rec A protein is very important in DNA repair. An insult to the DNA leads to the activation of the protease function of rec A, which then cleaves lex A protein, turning on the genes in the SOS response. Once the DNA is repaired, the rec A protease is inactivated and new lex A protein is made, repressing the DNA repair genes again. Rec A mutations are totally deficient in homologous recombination, demonstrating the important role of the rec A protein in this process. The purified rec A protein will catalyze the exchange between duplex and single-stranded DNAs with the hydrolysis of ATP. Also, rec A protein will form a complex between two circular helices if one helix is gapped on one strand. These functions of rec A protein would place it at the hub of activities in recombination. Rec A mutants would be very useful in genetic research because mutants generated would not be repaired, and in recombinant DNA cloning recombinational events between vector recombinant DNA and host genomic DNA would not occur.

27 DNA Manipulation and Its Applications

Problems

1. What are the advantages and disadvantages of the two methods of sequencing DNA (Gilbert's chemical method and Sanger's chain termination method)?

2. What are the limitations on sequencing RNA by the indirect method of sequencing the cDNA obtained from the reverse transcription of RNA?

3. What are the major advantages of the polymerase chain reaction (PCR) method for amplifying defined segments of DNA as opposed to the use of conventional cloning methods? How might the PCR method be used to test for infection with the AIDS virus and how would this be an improvement over the antibody test currently used? (The current ELISA test is an indirect test for the presence of antibodies against the HIV proteins.)

4. You have just isolated a novel recombinant clone and purified the desired insert (a 10-kb linear duplex DNA) from the vector. Now you wish to map the recognition sequences for restriction endonucleases A and B. You cleave the DNA with these enzymes and fractionate the digestion products according to size by gel electrophoresis. You observe the following:
   (a) Digestion with A alone gives two fragments, of lengths 3 and 7 kb.
   (b) Digestion with B alone gives three fragments, of lengths 0.5, 1, and 8.5 kb.
   (c) Digestion with A plus B generates four fragments, of lengths 0.5, 1, 2, and 6.5 kb.
   Draw a restriction map for the insert, showing the relative positions of the cleavage sites with respect to one another.

5. Describe the procedure you would use to clone a DNA fragment into the *Bam*HI site of pBR322.

6. A small circular duplex viral DNA with about 5,000 bp contains a 72-bp direct repeat in tandem. Each 72-bp segment contains a cleavage site for the *Eco*RI restriction enzyme that is not present in the rest of the virus. How could you produce a modified viral chromosome with only one copy of the 72-bp segment?

7. How large a genomic library should you construct in order to detect and isolate a 15-kb gene out of a genome containing 3×10^9 bp?

8. Why isn't a primer required to promote second-strand synthesis using the single-stranded DNA produced by reverse transcriptase from an RNA template?

9. If you were interested in isolating a cDNA for human serum albumin, why would you use a cDNA library established from mRNA isolated from liver? If you wanted to isolate the gene for albumin, why would you use a genomic library established from any human tissue?

10. The analysis of DNA regulatory sequences can be undertaken via the technique of DNA-mediated gene transfer, using a gene that provides a functional assay ("reporter gene") for the effect of the putative regulatory sequence on gene expression. How could this be done?

11. Explain how the thymidine kinase gene (*tk*) can be used as a selectable genetic marker for isolating transformed mammalian cells containing DNA of interest.

12. Site-directed mutagenesis is one of the most powerful tools available to the biochemist. What are some of the applications of this technique? How can the PCR method be used to do site-directed mutagenesis, and what is the advantage of this method?

13. Briefly outline how you could map the globin gene family using recombinant DNA techniques.

14. What is the procedure called chromosome jumping? How was this procedure used to map the cystic fibrosis gene?

Solutions

1. Both methods of sequencing are useful to biochemists; however, Sanger's chain termination procedure is the most popular method for routine sequencing. Gilbert's chemical method is labor intensive (many steps, precipitations, incubations, etc., taking the better part of a day) and uses potentially dangerous chemicals such as dimethylsulfate and hydrazine. The chemical method is useful in mapping protein-binding sites on DNA and can be used to verify ambiguous sequences obtained by the Sanger method. Variations of the Gilbert method are also widely used for direct RNA sequencing. The Sanger method, however, does not use dangerous chemicals and is rapid, requiring only a short incubation time with the DNA polymerase before the samples are applied to the high-resolution sequencing gel (this is the method of choice in student teaching laboratories). The chain termination method also can use ^{35}S-labeled deoxynucleoside triphosphates instead of ^{32}P-labeled dNTPs. This makes the procedure safer and results in a higher resolution sequencing gel because the weaker beta particle emitted by ^{35}S gives sharper band patterns on the x-ray film. The half-life of the ^{35}S is longer than that of the ^{32}P (87 days as compared to 14 days), allowing more time for the investigator to work with the isotope. Also, the Sanger method can be automated using fluorescent derivatives attached to the dideoxynucleotides. This major advancement in molecular biology is the reason Gilbert and Sanger shared the Nobel prize in chemistry (Sanger's second Nobel prize).
2. The primary limitation on the indirect method of sequencing RNA using the enzyme reverse transcriptase is the inability to detect posttranscriptional modifications commonly found in many RNA species. For example, tRNAs contain a high percent of modified bases and ribose methylation. The identification of the modified nucleosides and their location in the sequence requires additional methods used in nucleic acid biochemistry.
3. The primary advantage of the PCR method over conventional recombinant DNA cloning is the speed at which a specific sequence of DNA can be cloned *in vitro.* (Instead of a week to clone a specific DNA fragment, it can now be done in an afternoon.) It is possible to start with a messenger RNA population, convert it to cDNA with reverse transcriptase, and then amplify the cDNA sequence of interest, dramatically reducing the number of steps required for cloning and the time required.

   The PCR method will probably become the method of choice for detecting many types of infections by both bacteria and viruses. This powerful technique is also being applied to detection of genetic defects and is having a major impact in medical diagnostic procedures. One of the major problems with the current test for HIV infection (infection with the AIDS virus) is that the test is indirect, testing for the presence of antibodies produced against the AIDS virus. It may take months after infection for a person to produce antibodies at high enough levels to be detected. The PCR method, however, could detect infection within five days. The DNA could be isolated from white blood cells and subjected to PCR amplification, using specific probes for HIV. The presence of the provirus could then be detected. This method could detect one infected lymphocyte out of a million.

4. From the sizes of the fragments produced, the relative positions of the restriction enzyme cleavage sites with respect to one another may be deduced as follows:

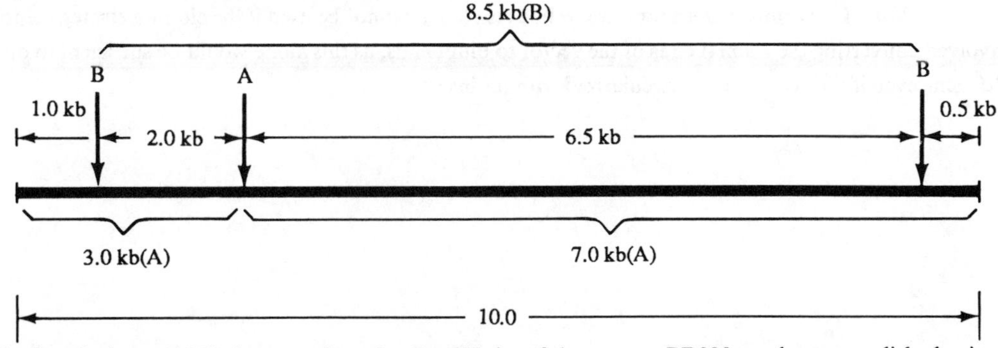

5. Cloning the insert DNA fragment into the *Bam*HI site of the vector pBR322 can be accomplished using one of the following strategies, depending upon the nature of the insert (i.e., whether or not its ends are compatible—may be ligated directly—with the *Bam*HI ends of the vector).

If the insert fragment ends are compatible with those of the linearized vector, the strategy is as follows:

Vector

1. Digest the pBR322 DNA with the restriction enzyme BamHI.
2. Treat the linearized vector DNA with phosphatase to remove the 5′terminal phosphate groups (this prevents recirculization of the vector without insert).

Insert

3. Digest the source of the insert DNA with an appropriate restriction enzyme (either BamHI or another enzyme that generates the same sticky ends).
4. Isolate the DNA fragment of interest (usually by gel electrophoresis).

Vector + Insert

5. Ligate the vector and insert DNAs (generally employing a molar ratio of 1 vector molecule to 1 insert molecule).
6. Transform the bacterial host with the ligated DNA.
7. Select for transformants containing recombinant plasmids.

If the insert fragment and linearized vector DNA ends are not compatible, they can be made so in one of several ways:

a. By treating both DNAs with *E. coli* DNA polymerase I. This enzyme possess both a 5′ to 3′ polymerizing and 3′ to 5′ degradative activities and will therefore either fill in a "sticky" end with a 5′ overhang (provided that appropriate deoxyribonucleotide triphosphates are supplied), or chew back a sticky end with a 3′ overhang. In both cases a blunt end is generated that may be ligated to any other blunt end.

b. By treating the isolated insert fragment with *E. coli* DNA polymerase I and ligating BamHI "linkers" (short double-stranded oligodeoxyribonucleotides containing the recognition/cleavage site for *Bam*HI) onto the blunt end (see the accompanying scheme).

In terms of the strategy to be employed in step 7 in selecting for transformants, it must be remembered that the *Bam*HI site contained by the vector pBR322 is located within the tetracycline resistance gene. Ligation of a DNA insert into this site will therefore disrupt the Tc^R gene and eliminate resistance to tetracycline. Selection may instead be based upon resistance to ampicillin, encoded by the Ap^R gene, and selected transformants counterscreened for sensitivity to tetracycline to identify those containing plasmids with insertions in the Tc^R gene.

Note: This counterscreen for tetracycline resistance cannot be used if the cloning strategy employed involved converting the *Bam*HI ends of the vector to blunt ends, as this alone would be sufficient to disrupt the Tc^R gene even if the vector were recircularized without insert.

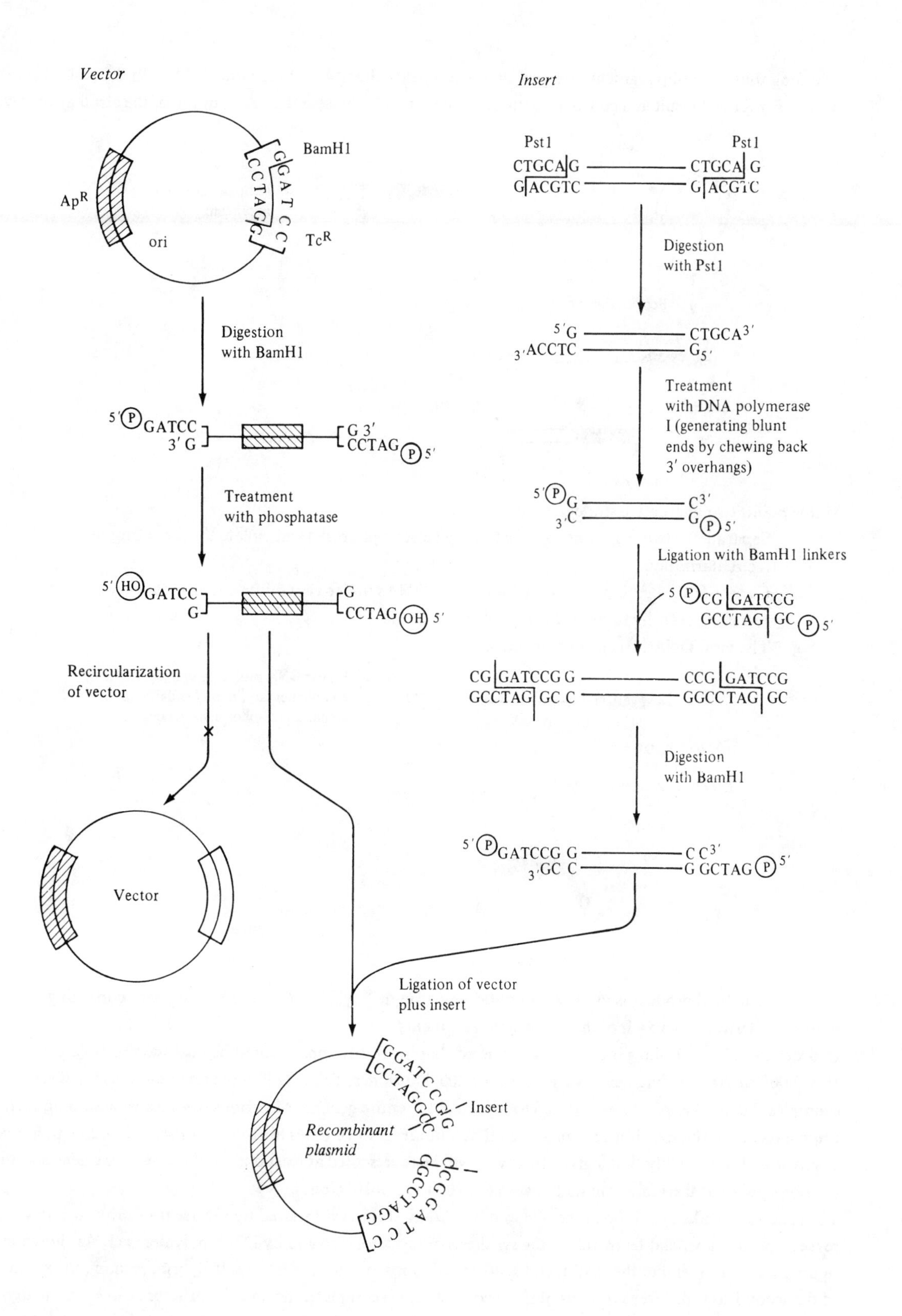
Vector
Insert
BamH1
ApR
ori
TcR
Pst1
Pst1
Digestion with Pst1
Digestion with BamH1
Treatment with DNA polymerase I (generating blunt ends by chewing back 3′ overhangs)
Treatment with phosphatase
Ligation with BamH1 linkers
Recircularization of vector
Digestion with BamH1
Vector
Ligation of vector plus insert
Insert
Recombinant plasmid

6. Assuming that the 72-bp segments are perfect direct repeats, digestion of the viral DNA with the restriction enzyme *Eco*RI will result in a construct with only one intact 72-bp segment. An outline of the strategy follows:

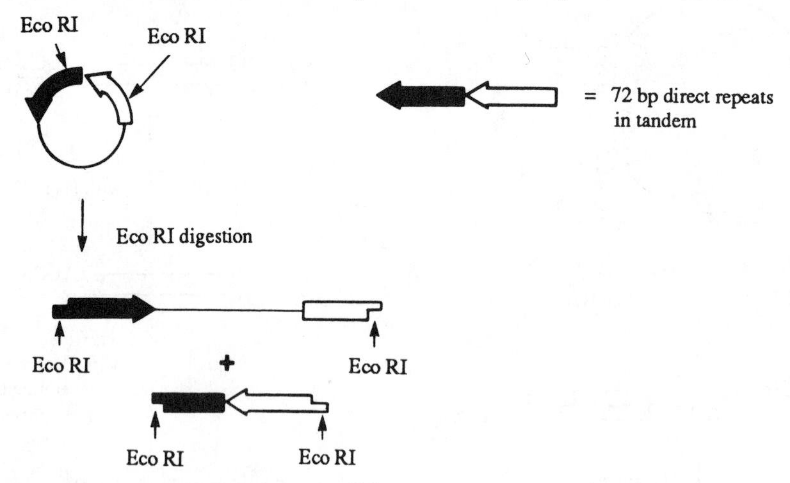

At this point, two alternative strategies are possible:

1. Separate the two fragments by size (e.g., gel electrophoresis) and isolate the large fragment for recircularization.
2. Carry out the ligation reaction with the two DNAs under very dilute conditions, which favor *intra*molecular ligation over *inter*molecular ligation.

The viral DNA is then circularized as follows:

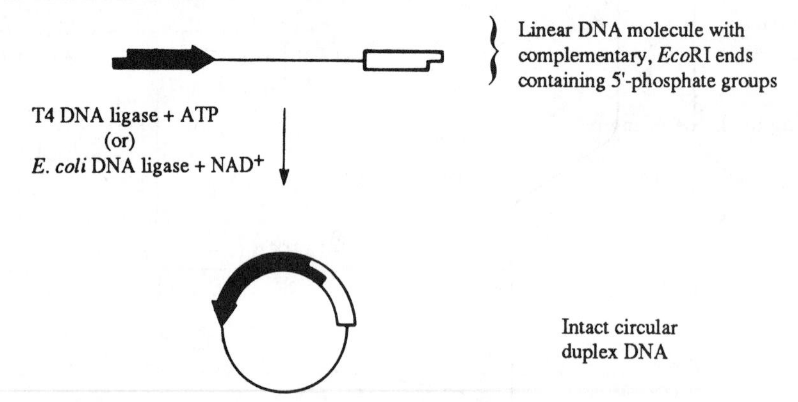

The final product is viral DNA containing a single "hybrid" 72-bp repeat segment consisting of sequences derived from each of the two original segments.

7. In order to isolate a 15-kb gene from a genome containing 3×10^9 bp, it would be necessary to isolate $3 \times 10^9/15{,}000$ or 200,000 fragments per genome, or 200,000 clones. Table 27.3 in the text describes for several examples the number of clones required to represent the entire genome of various organisms, assuming DNA fragments of a given size. It is recommended that a library 3 to 10 times the minimum size should be prepared, to ensure a high probability that a given fragment will be represented at least one. In the case of the gene specified, the library should therefore contain between 6×10^5 and 2×10^6 clones.

8. The function of a hairpin loop at the 3′ end of the DNA strand synthesized by reverse transcriptase allows it to serve as both primer and template for the synthesis of the second strand by DNA polymerase I. As shown in figure 27.17, the 3′ end of the first strand synthesized loops around to form a self-hybrid, priming the synthesis of the second strand. After synthesis of the second strand is complete, the hairpin structure can be cut with S1 nuclease, yielding a double-stranded DNA molecule with free ends. This can then be cloned into a suitable vector once appropriate ends have been generated (see figures 27.18 and 27.19).

9. The cDNA library established from a specific tissue is made up of sequences represented in the mRNA population found in that tissue. The mRNAs in the highest concentration would be represented in the cDNA library to the highest percent. Since human serum albumin is synthesized in the liver, the levels of the specific mRNA for serum albumin should be highest in that tissue. Cells of the body are not different because they contain different genes but because they *express* different genes. Thus, a genomic library established from DNA isolated from almost any human tissue could be used to isolate the gene for human serum albumin.

10. Recombinant DNA technology makes it possible to construct DNA molecules containing a gene encoding a protein whose activity is readily assayed coupled to sequences of possible regulatory significance derived from another gene.

    These hybrid constructs can be transfected into suitable recipient cells, and the activity of the assayable gene, now under the control of heterologous regulatory sequences, determined.

    One example of such a hybrid would be a construct linking the gene for chloramphenicol acetyl transferase (CAT) and regulatory sequences derived from a gene expressed in a tissue-specific manner. A hybrid molecule of this type can be transfected into the cell type in which expression is observed, and CAT expression assayed. By constructing hybrids containing increasingly smaller segments of the regulatory sequences, the precise boundaries of sequence elements conferring tissue-specificity of expression can be determined. Many regulatory sequences have been defined and characterized using variations on this strategy.

11. Selectable genetic markers have been powerful tools in isolating transformed mammalian cells containing DNA of interest. If the DNA fragment is linked to the thymidine kinase gene (*tk*), and if a cell has taken up the *tk* gene, the cell will also contain the DNA fragment that you are trying to clone. A cell line that contains a mutant *tk-* gene is used in this method. These cells will not grow in a selection media (HAT media, containing hypoxanthine, amethopterin, and thymidine), which prevents the synthesis of thymidine monophosphate from uridine monophosphate. Only cells that contain the gene for thymidine kinase and can convert thymidine to thymidine monophosphate can replicate DNA and grow. If a fragment of DNA containing the selectable marker (*tk*) is taken up by a cell, then this transformed cell can grow and divide. The investigator can then select only cells containing the DNA of interest.

12. The ability to change sequences of DNA at specific sites and observe the effects of these changes is a very powerful tool to the biochemist. Regulatory regions in genes can be changed and the effects of these changes can be monitored. This will determine what sequences are important in gene regulation. If a protein binds to a DNA region, the sequences in this region can be changed and the binding site and sequences in that site can be determined. The gene coding for the binding protein could be changed and the amino acid residues in the protein could be mutated in a specific fashion, so the region interacting with the DNA could be determined and the amino acids important to binding could be discovered. The study of enzymes (enzymology) has been helped by the ability to change amino acid residues in active sites leading to a better understanding of how the enzyme works and which residues are important for activity. The use of the PCR technique for generating site-directed mutations is outline here:

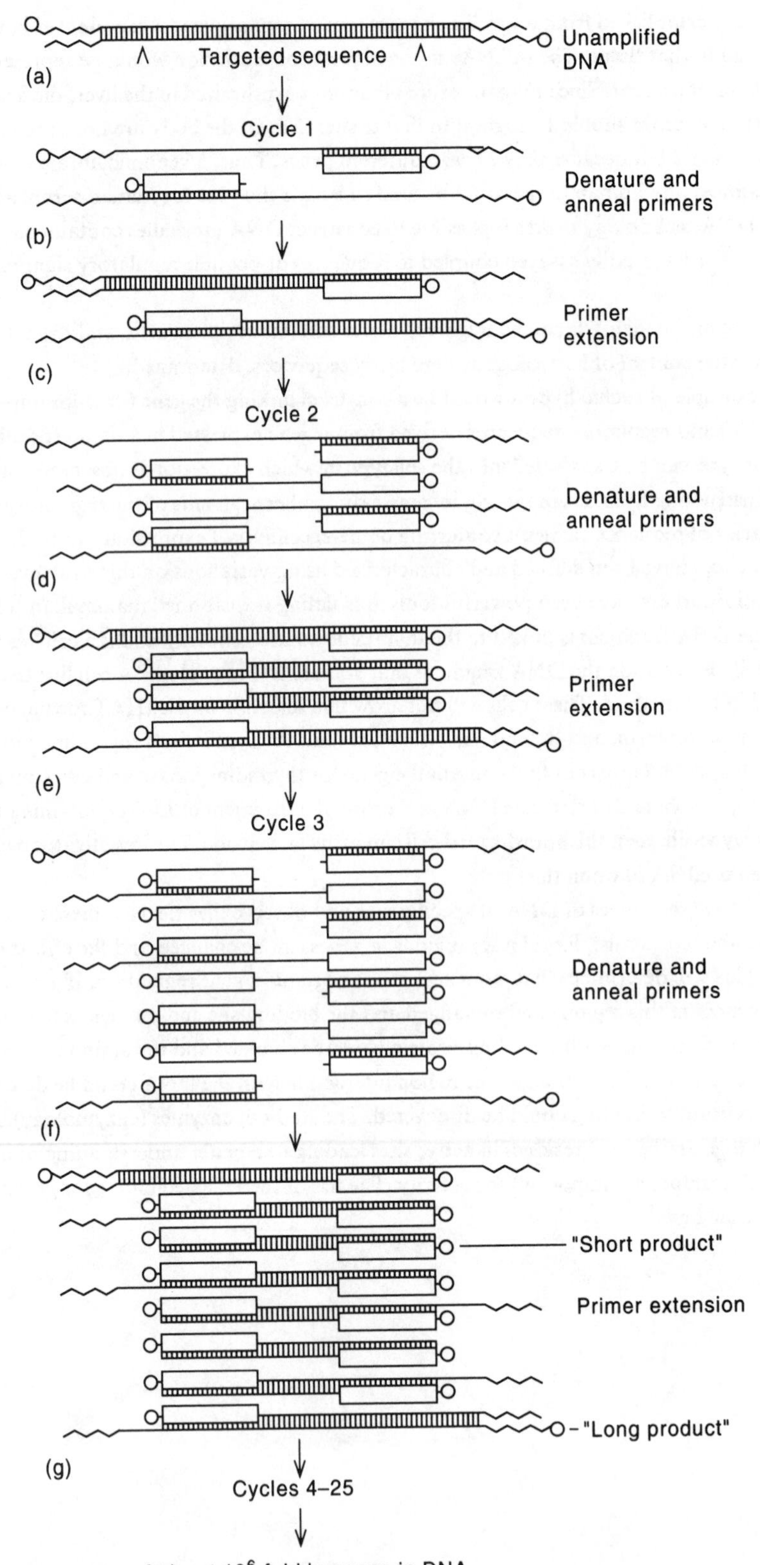
Targeted sequence
Unamplified DNA
(a)
Cycle 1
Denature and anneal primers
(b)
Primer extension
(c)
Cycle 2
Denature and anneal primers
(d)
Primer extension
(e)
Cycle 3
Denature and anneal primers
(f)
"Short product"
Primer extension
"Long product"
(g)
Cycles 4–25
At least 10^6-fold increase in DNA

This method of generating site-specific mutants is much faster than conventional cloning methods and will allow a person to produce many sequence changes in a gene of interest in a short time.

13. The α-like globin genes and the β-like globin genes are located in single clusters on chromosome 16 and chromosome 11 respectively. The β-globin will be used as an example of how to characterize a gene family. Because it was possible to isolate globin mRNA from reticulocytes, the cloning of the β-globin gene was simplified. The cDNA probe for the β-globin mRNA was used to isolate the gene from a genomic library. The restriction map analysis showed that the gene was larger than the β-globin mRNA (it contained two introns and other nontranscribed and nontranslated regions). The regions surrounding the β-globin gene region were also examined using a technique called chromosome walking. The original cDNA probe was used to find other members of the genomic library that hybridized, locating flanking sequences around the β-globin gene. The repetition of this process with other clones led to the extension of the sequence information on either side of the β-globin gene. All the other β-globinlike genes were found to be nearby (figure 27.28 in the text). This same approach was used to map the α-globin gene family.

14. Chromosome jumping is a useful procedure to traverse long distances and skip troublesome regions of the genome (repetitious sequences, etc.). One approach with this technique is to digest the genome with rare cutting restriction enzymes (*Not* I recognizes an 8-base sequence and generates an average fragment size of 500 kb of DNA). The fragments are circularized with a small marker DNA between the ends. The marker DNA contains sequences necessary for cloning in λ. These circular DNA fragments are then cleaved with another restriction enzyme that produces fragments small enough to clone in λ. With this procedure only clones that contain the marker DNA, i.e., the ends of the original fragment, are isolated. This method would permit only one jump; thus another method used with this technique is called a linking library. The same sample of DNA is digested with a restriction enzyme that gives smaller fragments. These smaller fragments are then circularized with the same marker DNA used in the jumping library. The circular DNA is then digested with *Not* I to linearlize the fragments for insertion into the vector. The linking library carries sequences on both sides of the *Not* I restriction sites while the jumping library carries sequences from one side of two adjacent *Not* I sites. These two libraries are then used as shown in the figure.

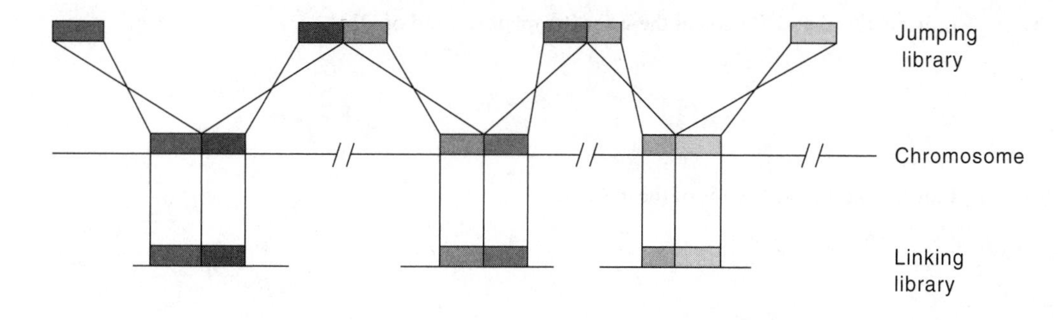

28 RNA Synthesis and Processing

Problems

1. **One strand of DNA is completely transcribed into RNA by RNA polymerase. The base composition of the DNA template strand is: G = 20%, C = 25%, A = 15%, T = 40%. What would you expect the base composition of the newly synthesized RNA to be?**

2. **The illustration is a schematic drawing representing a portion of Miller's electron microscope picture of transcription and translation in progress in *E. coli*.**

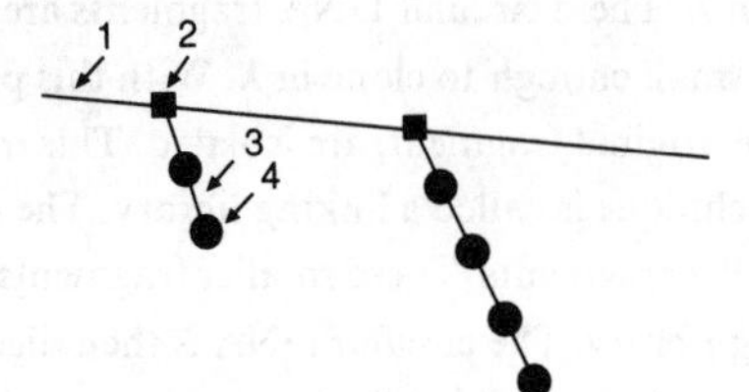

(a) **Identify 1, 2, 3, and 4.**

(b) **Indicate the 3′ and 5′ ends of the sense (template) strand of DNA.**

(c) **Indicate the 3′ and 5′ ends of the mRNA.**

(d) **Draw four peptides in the process of being synthesized, indicating relative lengths.**

(e) **Indicate the N- and C-terminal ends of the longest peptide.**

(f) Indicate with an arrow the direction in which RNA polymerase is moving.

(g) What parts of this diagram would be different in eukaryotes?

3. Although 40 to 50% of the RNA being synthesized in *E. coli* at any given time is mRNA, only about 3% of the total RNA in the cell is mRNA. Explain.

4. In *E. coli*, the mRNA fraction is heterogeneous in size, ranging from 500 to 6,000 nucleotides. The largest mRNAs have many more nucleoties than needed to make the largest proteins. Why are some of these mRNAs so large in bacteria?

5. When yeast phenylalanine tRNA is digested with a small amount of RNase (partial digest), such as T2 RNase, almost all of the cleavage sites are found in the anticodon loop. Explain this observation.

6. Speculate on the advantage of having three rRNAs (16S, 23S, and 5S) as part of the same RNA precursor (such as is found in *E. coli*)?

7. In *E. coli* the precise spacing between the -35 and -10 conserved promoter elements has been found to be a critical determinant of promoter strength. What does this suggest about the interaction between RNA polymerase and these elements? What sorts of evidence could you obtain about this interaction by doing "footprint" experiments? (Also explain how you would do these experiments.)

8. What is the maximum theoretical rate of initiation at an *E. coli* promoter, assuming that the diameter of its RNA polymerase is about 200 Å and the rate of RNA chain growth is 45 nucleotides per second?

9. Cordycepin-5′-triphosphate, 3′-deoxy ATP, is a ribonucleoside triphosphate analog that can be bound as a substrate by RNA polymerase. What effect does this analog have on transcription and how could this information be used to determine the direction of transcription?

10. The following hypothetical RNA is a precursor made in a eukaryotic nucleus and its synthesis is inhibited by low levels of α-amanitin:

pppAUUAUGCCGAUAAGGUAAGUA–(N_{100})–
AUCUCCCUGCAGGGCGUAACCAAUAAACGACGACGACGUCACC

Indicate the final processed RNA found in the cytoplasm and point out important features.

11. In early research with intact eukaryotic mRNA, the RNA appeared to have two 3′-ends and no 5′-terminus. Explain these observations.

12. What is unique about the promoter for RNA polymerase III? Diagram how this promoter works.

13. T7 and SP6 RNA polymerases are used to make RNA transcripts *in vitro* as important tools in modern molecular biology. The addition of inorganic pyrophosphatase to these reactions can increase the yield of the RNA product. Explain the increased yield.

14. Draw the chemical reaction mechanism for the formation of a "lariat" in the processing of introns from eukaryotic mRNA.

15. A particular eukaryotic DNA virus was found to code for two mRNA transcripts, one shorter than the other, from the same region on the DNA. Analysis of the translation products revealed that the two polypeptides shared the same amino acid sequence at their amino-terminal ends, but were different at their carboxyl-terminal ends. The longer polypeptide was coded by the shorter mRNA! Suggest an explanation.

16. You are a graduate student working on an *in vitro* processing system for intron removal with a precursor rRNA isolated from *Tetrahymena*. The purified RNA will process by itself without any proteins being added to the reaction. How could you prove that this RNA was autocatalytic and that the reaction was not just due to some protein contamination in your incubation?

17. Draw a chemical reaction mechanism for the first step in the self-splicing reaction of group I introns. (Group I intron splicing requires a guanosine co-factor.)

Solutions

1. The composition of the RNA must be the complement of that of the DNA template strand:

DNA template strand		RNA
%		%
G 20	→	C 20
C 25	→	G 25
A 15	→	U 15
T 40	→	A 40

2. As illustrated in the diagram, in prokaryotes, ribosomes become associated with the growing 5′ end of nascent RNA chains and translation occurs concurrently with transcription.

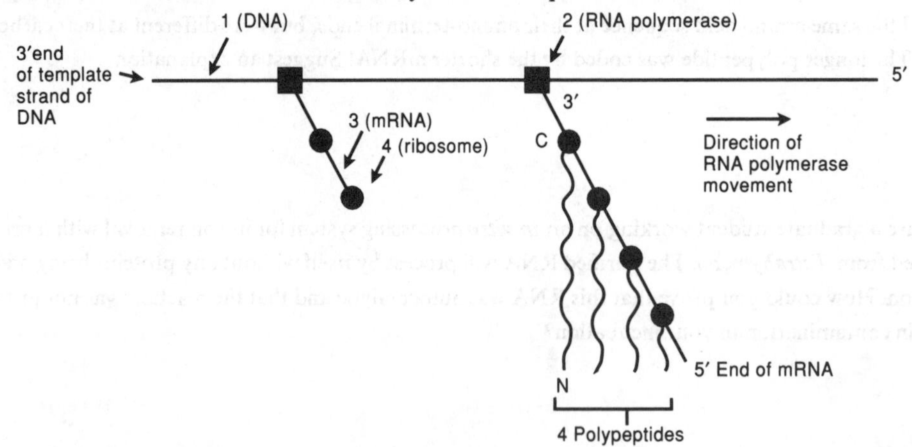

In contrast, in eukaryotes, RNA synthesis takes place in the nucleus, where the nascent transcripts are first processed into mature mRNAs. The mature mRNAs are then transported to the cytoplasm, where they become associated with ribosomes and are translated.

3. mRNA in bacteria is degraded extremely rapidly, with a turnover rate of approximately 50% in 1 to 3 min, and thus does not accumulate. The 3% of the total RNA present at any time represents that mRNA which has just been synthesized but not yet degraded. However, rRNA and tRNA are considerably more stable and accumulate. Even though these RNAs together account for roughly 50% of the total RNA synthesis, they represent greater than 95% of the total RNA present at any given time.

4. Many mRNAs in prokaryotic organisms are polycistronic, i.e., they code for a number of polypeptides. Therefore, with the noncoding regions located on the 5′ and 3′ ends of the mRNA and the multiple coding regions, the mRNA may be many times bigger than the amount of RNA needed to code for even the largest protein.

5. If yeast phenylalanine tRNA was partially digested with a RNase, the loops or single-stranded regions would be cleaved. The cloverleaf model for the secondary structure of tRNAs predicts that cleavages would occur in the D and T loops as well as the anticodon loop. However, the D and T loops interact with each other in the three-dimensional model of tRNA, leaving only the anticodon loop exposed to RNase cleavage.

6. The advantage of having a single precursor for the ribosomal RNA in prokaryotes is to make it easy to maintain a 1:1:1 ratio of the three RNAs in the ribosomes. Since you need one 16S, 23S, and 5S rRNA for the formation of the 30S and 50S ribosomal subunits, the synthesis of a balanced amount of these RNAs is easily maintained.

7. The RNA polymerase binds to the same side of the duplex at the -10 and -35 regions with about two turns of the duplex helix between these boxes. This binding site can be determined by a variety of "footprint" experiments. DNA that is 5′-labeled can be mixed with the polymerase, and regions that are protected from digestion with an enzyme like DNase I can be determined on a sequencing gel. Sequences with tight contact with the RNA polymerase are observed as a series of blank spots in the sequencing ladder (see box 28B in the text) that look like footprints.

8. At ~ 34 Å per 10 bp of B-form DNA, the polymerase covers about (200 Å/34 Å) × 10 = 60 bp of DNA. Once it has initiated, it moves away from the promoter at an average rate of about 45 nucleotides per second. Therefore, if the rate of initiation were limited only by the rate at which the polymerase clears the promoter, one could expect a maximum initiation frequency of one per 60 bp/(45 bp/s) ≈ 1.3 s. This is, in fact, very similar to the rate of initiation at rRNA promoters *in vivo*.

9. **If transcription were 5′ → 3′, the analog could be incorporated at the 3′ end of a growing chain but would lack the 3′—OH needed for further elongation:**

Growing RNA chain + Cordycepin triphosphate → Elongation blocked

If transcription were 3′ → 5′, the analog would not be incorporated at the 5′ end because it would be unable to form a 3′,5′-phosphodiester bond:

No bond formation

Since the analog could not be incorporated, RNA synthesis would continue (provided that ATP were available, along with other XTPs).

10. **Since the synthesis of this RNA is sensitive to low levels of α-amanitin, it must be transcribed by RNA polymerase II and processed into a mature mRNA before being transported to the cytoplasm for translation. The processing steps include cap formation, removal of intron sequences, and poly(A) tail addition. The consensus sequences that direct these processing events are as follows:**

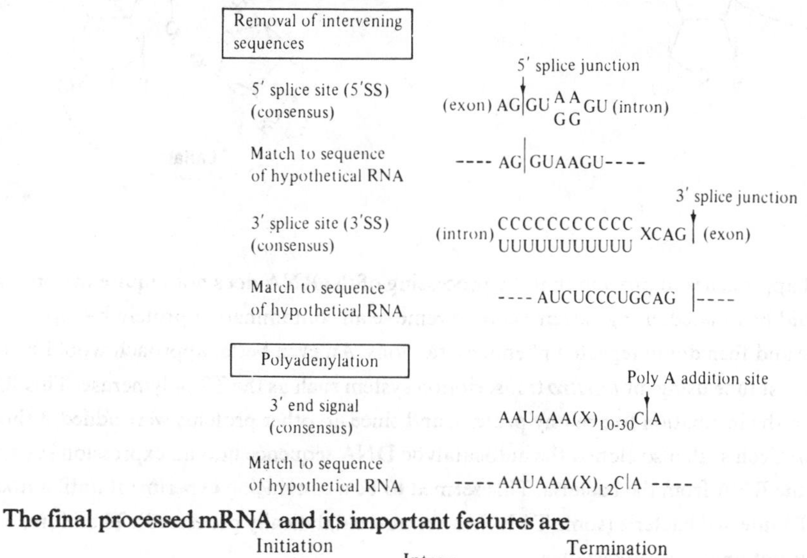

The final processed mRNA and its important features are

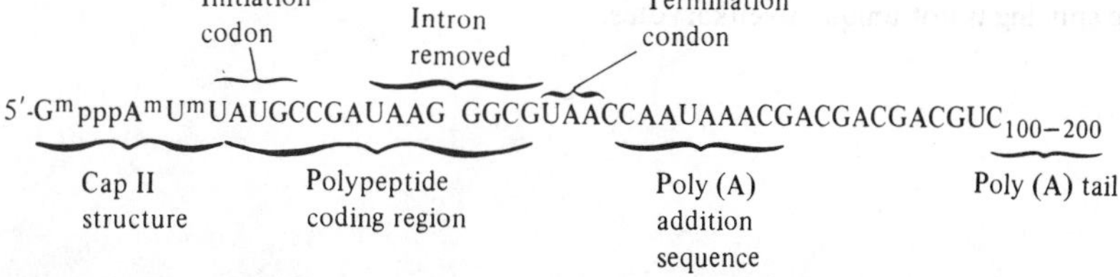

Note that capped mRNAs contain a free 3′—OH group at the 5′ terminal, since the G is attached to the penultimate residue via a 5′,5′-triphosphate link.

11. **One common way to determine the 3′-end of an RNA is to treat the RNA with Na-periodate and oxidize the cis-diols to dialdehydes, which can then be reduced to dialcohols with $^3H\text{-}BH_4$. Normally the only nucleotide in**

the sequence that has cis-diols is the 3′-terminal, which has the 2′ and 3′ hydroxyls. However, since the eukaryotic mRNAs are capped at the 5′-end (GpppGpNpN---N_{OH}) they also have a guanosine at the 5′-terminus as well as the adenosine (most eukaryotic mRNAs end in poly A) at the 3′-end with cis-diols. Also, these eukaryotic mRNAs would appear to lack a normal 5′-end. Many types of 5′-end analysis would not detect the 5′-terminus, leading to the paradox of a RNA with no 5′-end and two 3′-ends.

12. The polymerase III promoters map internal in the sequences being transcribed. For example, the region necessary for transcription of 5S rRNA is found in region +40 to +80, downstream from the transcription start site. A transcription factor binds to this site (TFIIIA) and along with two other factors initiates transcription at position +1 with RNA polymerase III (the binding of TFIIIA to this site was "footprinted" at this site using DNase I digestion; see box 28B in the text).

13. Pyrophosphate is one of the products of the reaction occurring during transcription with any of the RNA polymerases (---NpNpN + pppX $\rightarrow$ ---NpNpNpX + PP_i). Pyrophosphatase would cleave the pyrophosphate to orthophosphate, preventing product inhibition ($PP_i \rightarrow 2\ P_i$) and helping drive the reaction to produce more transcripts (the cleavage of pyrophosphate is a common theme seen in many biochemical reactions generating a negative Gibbs free energy [$-\Delta G$]).

14.

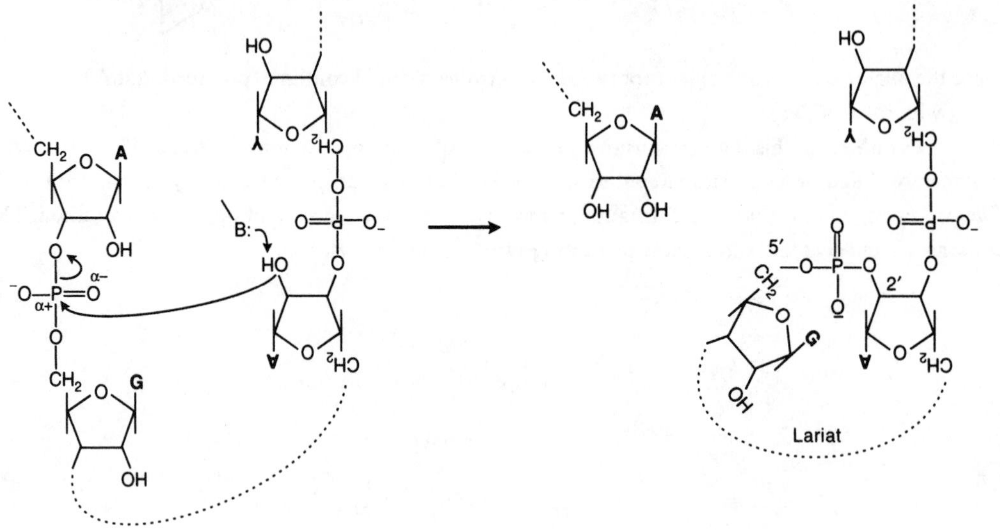

16. There are a number of approaches to proving that the processing of the RNA does not require the presence of protein. The RNA could be isolated, using extreme care to remove all contaminating protein by digesting the sample with a protease and then doing repeated phenol extractions. An even better approach would be to produce the RNA in a test tube using an *in vitro* transcription system such as the T7 polymerase. This RNA could be extracted from the incubation free of any protein, and since no other proteins were added it should be a very clean system. Tom Cech's lab also cloned the autocatalytic DNA sequence into an expression system in *E. coli* and then isolated the RNA from the bacteria. This seemed to be a convincing experiment until autocatalytic RNAs were found in T4 infected bacteria (some T4 RNAs are autocatalytically processed). Therefore, autocatalytic splicing is not unique to eukaryotes.

17.

CH2
U
O
OH
O
-O-P=O
α-
α+
O
CH2
A
OH
HO
CH2
G
3′
OH
O
H
:B
CH2
U
OH
OH
+
HO
CH2
G
O
OH
-O-P=O
O
CH2
A
OH

29 Protein Synthesis, Targeting, and Turnover

Problems

1. Compare the translation initiation signals in prokaryotic and eukaryotic systems, and describe those features of each type of mRNA that determine the frequency with which a particular message is translated. What consequences do these differences have for gene organization in the two systems?

2. Transfer RNA molecules are rather large, considering the fact that the anticodon is a trinucleotide. Why is this the case?

3. Draw the chemical reaction mechanism for the formation of an aminoacyl-tRNA.

4. The relationship between tRNAs and their synthases is sometimes called the "second genetic code." Explain.

5. Explain this statement: "The universal genetic code is not quite universal."

6. How much energy is required to synthesize a single peptide bond in protein synthesis? How does this compare with the free energy of formation of the peptide linkage, which is about 5 kcal/mole?

7. Bromouracil is a base replacement mutagen, while acridine causes single-base insertions or deletions. Mutations caused by acridine frequently can be compensated for by secondary mutations several nucleotides distant from the first, while compensation for those caused by bromouracil require changes within the same codon. Explain.

8. Assume that you have a copolymer with a random sequence containing equimolar amounts of A and U. What amino acids would be incorporated and in what ratio, when this copolymer is used as an mRNA?

9. Assuming that translation begins at the first codon, deduce the amino acid sequence of the polypeptide encoded by the following mRNA template:

AUGGUCGAAAUUCGGGACACCCAUUUGAAGAAACAGAUAGCUUUCUAGUAA

10. The effect of single-point mutations on the amino acid sequence of a protein can provide precise identification of the codon used to specify a particular residue. Assuming a single base change for each step, deduce the wild-type condon in each of the following cases.

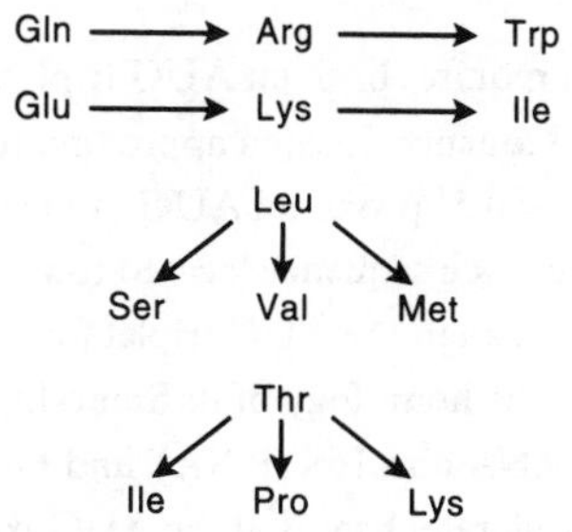

Gln → Arg → Trp

Glu → Lys → Ile

Leu → Ser, Val, Met

Thr → Ile, Pro, Lys

11. Give examples of several proteins that are not functional without posttranslational modification or processing.

12. Why are PEST sequences more likely to be found in regulatory proteins than in structural proteins?

13. Scientists have tried to isolate the peptidyl transferase from ribosomes for many years without success. It is now thought that this activity is part of the ribosome (large subunit). Discuss this point, in view of what you know about other catalytic RNP complexes (RNA-protein complexes).

14. Explain why toxins such as ricin and α-sarcin are such potent toxins, i.e., act at such low concentrations (one molecule per cell). What is unique about this group of toxic enzymes?

15. The signal recognition particle (SRP) has sometimes been called the "third ribosomal subunit." Do you think this is a valid comment? Explain.

16. Explain why normal hemoglobin is very stable, while the same protein containing the valine analog aminochlorobutyrate is very unstable.

Solutions

1. In prokaryotes, translation initiation requires both an AUG triplet, or initiation codon, and a purine-rich element, termed the Shine-Dalgarno sequence, located approximately 10 bases upstream of the AUG. The initiation codon is not generally the most 5′ proximal AUG in the mRNA. The Shine-Dalgarno sequence is thought to base-pair with a pyrimidine-rich sequence located toward the 3′ end of the 16S ribosomal RNA of the 30S ribosomal subunit, which helps align the AUG triplet for initiation. The efficiency with which a given mRNA is translated is determined by the homology of its Shine-Dalgarno sequence to the consensus (i.e., the extent of base pairing between the mRNA and 16S rRNA), and the sequence context of the initiation codon.

   In eukaryotes, translation generally begins at the AUG triplet that lies closest to the 5′ end of the mRNA. This is thought to be recognized by a scanning mechanism, in which the 40S ribosomal subunit first binds to the mRNA at the capped 5′ end, and then migrates 5′ → 3′ until it encounters the first AUG codon. (There is as yet no evidence to support the possibility of the base pairing between 18S rNA and the 5′ end of eukaryotic mRNA in initiation.) The efficiency with which a given AUG is recognized as a translation start site is determined by its sequence context; CCACC*AUG*G is the optimal context.

   The difference in the mechanism of translation initiation in prokaryotes compared to eukaryotes has profound consequences for the strategy used to coordinate expression of a set of genes in the two systems. In prokaryotes this coordination is achieved by organizing genes into transcription units which are transcribed to give polycistronic mRNAs. Each cistron within a polycistronic mRNA begins with a Shine-Dalgarno sequence and initiation AUG codon. In contrast, in eukaryotes, in which translation begins almost invariably (and exclusively) at the 5′ most AUG codon, monocistronic mRNAs are necessarily the order of the day. Genes which must be coordinately expressed are consequently not organized into transcriptional units, and coordination must be achieved in some other way.

2. The tRNAs are involved in two distinct processes: aminoacylation and the template-directed polymerization of amino acids. The demands for specific recognition of a tRNA by its aminoacyl tRNA synthase may require many regions of the tRNA structure (see solution 4). The location of the charged tRNA on the surface of the ribosome and the distance from the anticodon interaction with the codon on the mRNA and the site where the peptidyl transferase activity resides on the ribosome is relatively constant, suggesting the importance of the three-dimensional folding of the tRNAs. All of these constraints would require more structure than just the trinucleotides found in the anticodon.

3.

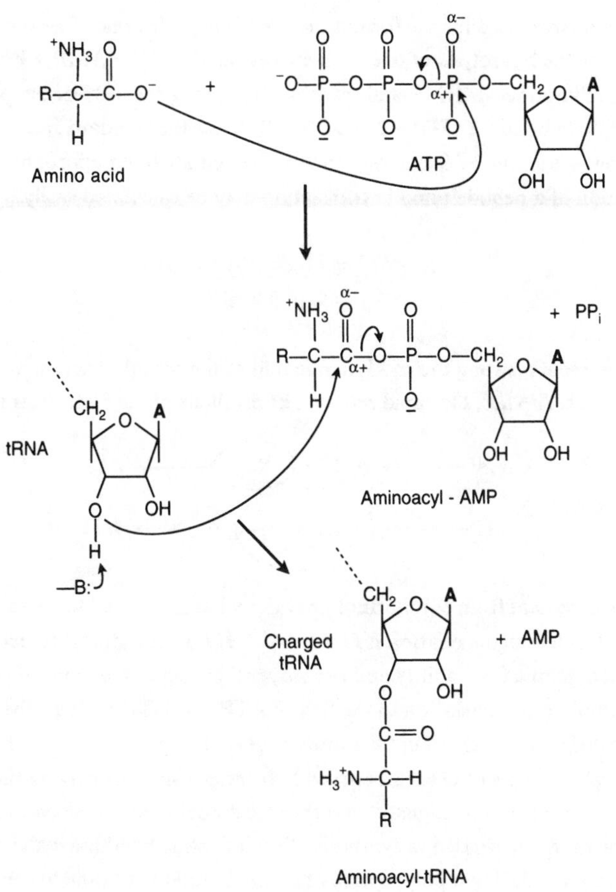

Amino acid + ATP

Aminoacyl - AMP + PP_i

tRNA

Charged tRNA + AMP

Aminoacyl-tRNA

4. The genetic code is the triplet code found in the mRNA specifying the sequence of amino acids in the protein being made. However, another important feature is the accurate charging of the tRNAs with the correct amino acids. Obviously, if the wrong amino acid is attached to the specific tRNA, a point mutation would occur. The crystal structure of the glutaminyl-tRNA synthase with its tRNA bound has been solved. This synthase requires the anticodon to be intact, thus it must bind with both ends of the L-shaped RNA. There are additional contacts between the protein and the tRNA along the inside of the L-shaped structure. The acceptor stem plunges into the active site pocket, which induces conformational changes in the 3′-terminal CCA acceptor sequence. There are many points of recognition that allow the specific charging of the glutaminyl-tRNA. Not all tRNA synthases require the anticodon for specific charging (for example, the alanine-tRNA is recognized by its synthase by only a small number of sequences in the acceptor stem, including the crucial G-U pair). These highly specific interactions found between the tRNAs and their corresponding synthases certainly could be called the "second genetic code."

5. A number of variations have been found in the universal genetic code in genes found in mitochondria and chloroplast. These variations in the meaning of some of the code words represent divergences from the standard genetic code and not an independent origin of another genetic code. These divergences probably arose in these organelles because of the limited number of genes coded and requirements for the synthesis of ribosomes and tRNA. It was clear that something was unusual when only 24 types of tRNAs were found in mitochondria. Mitochondria do not use all 61 codons (see table 29.3 in the text for the yeast mitochondria genetic code).

6. The difference between the standard free energies of the products and reactants in the synthesis of a peptide bond in translation provides a useful indication of the energy required in the process. A total of three high-

energy phosphate groups are used for the formation of each peptide bond. The synthesis of an aminoacyl-tRNA at the synthase involves the hydrolysis of one high-energy bond: ATP → AMP + PP_i. The function of the ribosome in binding of aminoacyl-tRNA and translocation along the mRNA template involves the hydrolysis of two additional phosphate bonds: 2 GTP → 2 GDP + 2 P_i. Since the standard free energy of formation of each of these high-energy bonds is about +7.3 kcal and that of the peptide bond approximately +5.0 kcal, the net $\Delta G^{\circ\prime}$ for the formation of a peptide bond in translation may be calculated easily:

$$\begin{aligned} \Delta G^{\circ\prime} &= (3 \times -7.3) + (+5.0) \\ &= -16.9 \text{ kcal} \end{aligned}$$

This $\Delta G^{\circ\prime}$ would represent a strong thermodynamic pull of the overall reaction of peptide bond formation in the direction of synthesis. *In vitro,* the rapid removal of products of the hydrolysis from the reaction

$$\text{ATP} \longrightarrow \text{AMP} + \text{PP}_i \longrightarrow 2\,\text{P}_i$$

$$\text{GTP} \longrightarrow \text{GDP} + \text{P}_i \longrightarrow \text{GTP}$$

and the resulting displacement from equilibrium provides what is believed to be an even stronger pull of the reaction toward product [recall the relation $\Delta G = \Delta G^{\circ\prime} + RT\gamma n$ (products)/(reactants)].

The thermodynamic favorability and consequent driving of reactions of peptide bond formation resulting from the coupling of translation to the flux of ATP and GTP explains the need for the huge discrepancy between energy invested and conserved in the synthesis of each peptide bond. At the level of the function of the ribosome specifically, hydrolysis of GTP is believed to be responsible for driving the process of selection of aminoacyl-tRNA (cognate over noncognate) and the directionality of translocation.

Overall, the energy expended in synthesis of each peptide bond must also overcome the unfavorable entropy associated with producing a polypeptide of defined rather than random sequence (recall the relation $S = R\gamma n\omega$).

7. The genetic code is nonoverlapping with translation involving the sequential reading of adjacent triplet codons in a particular "frame" specified by the point of translation initiation. Insertion of deletion of a single nucleotide in a gene corresponding to a particular protein product results in a shift in its translation reading frame from the point of insertion or deletion onward. The result, in terms of the primary structure of the polypeptide, of such a shift in frame is

```
AUG  CGU  (U)CC  AGC  ACU  UUA  ACC  UUU  AUC, etc.
Met  Arg    Ser  Ser  Thr  Leu  Thr  Phe  Ile ,  etc.

AUG  CGU  CC A  GC A  CU U  UA A  CC U UU A UC, etc.
Met  Arg  Pro   Ala   Leu   Stop
```

In this instance, the deletion of a U in the third codon results in the ribosome reading the lower set of codons and synthesizing a product with a completely different amino acid sequence past the point of mutation. From this example it is readily apparent that for a "frameshift" mutation to affect a protein's function, the mutation must be somewhere between the point of translation initiation and a portion of the gene encoding region(s) of the protein important for that function. Point mutations, on the other hand, must, if they are to exert an effect on function, fall within a portion of the gene corresponding to such an essential region of the polypeptide product. Since the site of effect of a frameshift mutation is in most instances some distance away from the site of mutational change, secondary compensatory frameshift mutations restoring the original reading frame and protein function may occur at some distance from the site of the first, while point mutations, which usually exert their effect at the site of mutational change, generally require changes in the originally mutated codon in order to achieve recovery of function of a mutant gene product.

8. The number of codons resulting from a random assortment of two bases, A and U, is 2^3, or 8. Each of these codons, listed in the following table, will occur in a random copolymer of A and U with a frequency proportional to the probabilities of occurrence of its substituent bases.

	Codon Frequency	Amino Acid
AAA	$\frac{1}{2} \times \frac{1}{2} \times \frac{1}{2} = \frac{1}{8}$	Lys
AAU	$\frac{1}{2} \times \frac{1}{2} \times \frac{1}{2} = \frac{1}{8}$	Asn
AUA	$\frac{1}{2} \times \frac{1}{2} \times \frac{1}{2} = \frac{1}{8}$	Ile
UAA	$\frac{1}{2} \times \frac{1}{2} \times \frac{1}{2} = \frac{1}{8}$	—
UUA	$\frac{1}{2} \times \frac{1}{2} \times \frac{1}{2} = \frac{1}{8}$	Leu
UAU	$\frac{1}{2} \times \frac{1}{2} \times \frac{1}{2} = \frac{1}{8}$	Tyr
AUU	$\frac{1}{2} \times \frac{1}{2} \times \frac{1}{2} = \frac{1}{8}$	Ile
UUU	$\frac{1}{2} \times \frac{1}{2} \times \frac{1}{2} = \frac{1}{8}$	Phe

The triplet UAA is a termination signal for which there is no corresponding tRNA; therefore, of the amino acids incorporated, each codon would specify one-seventh of the total. The average polypeptide produced upon translation of the copolymer would therefore contain several amino acid residues in the proportion Lys (1/7), Asn (1/7), Ile (2/7), Leu (1/7), Tyr (1/7), and Phe (1/7).

9. On the basis of the information provided in table 29.1, the amino acid sequence corresponding to the template given may be deduced as follows:

AUG	GUC	GAA	AUU	CGG	GAC	ACC	CAU	UUG	AAG	AAA	CAG	AUA	GCU	UUC	UAG	UAA
Met	Val	Gln	Ile	Arg	Asp	Thr	His	Leu	Lys	Lys	Glu	Ile	Ala	Phe	Ter	Ter

10. The wild-type codons can in each case be deduced by simple inspection of table 29.1. For example, in part (c)

Leu
↙ ↓ ↘
Ser Val Met

The relationship Leu → Met allows only two possible identities for the wild-type codon: UUG and CUG, while the relationship Leu → Ser is consistent with two Leu codons also: UUA and UUG. The overlap between these conditions unambiguously identifies the original Leu codon as UUG. Similar treatment of the other relationships specified provides for the following codon identification:

(a) Gln (CAG) → Arg (CGG) → Trp (UGG)

(b) Glu (GAA) → Lys (AAA) → Ile (AUA)

(c)

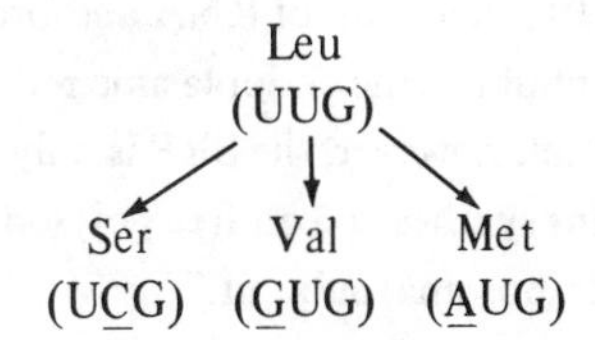

(d)

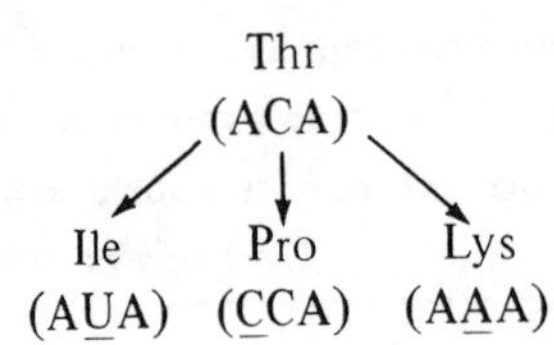

11. Many proteins are modified after they are translated because of where they are targeted in the cell, and also to make them functional. For example, the protein hormone insulin is synthesized as a preprotein, which is targeted to the Golgi. The targeting sequence is removed to form a proprotein that must be cleaved to remove the C-peptide to form the active insulin hormone. Insulin is made as a preproprotein. Cytochrome c_1 is another protein that is made as a preprotein. This protein is targeted to the mitochondria, where it is found in the intermembrane space. It has two signal sequences that are removed to form the final protein. Proteins targeted to the lysosome have mannose-6-phosphate, which allows these proteins to bind to a specific receptor and be targeted to the lysosome. If the mannose residues are not phosphorylated, the lysosomal proteins are secreted outside the cell (I-cell disease). Collagen is another protein that undergoes extensive processing. The precursor collagen has large peptides removed from the N- and C-terminals of 150 and 250 amino acids, respectively. Hydroxylation of specific proline residues takes place in the endoplasmic reticulum while extensive glycosylation takes place. The processed collagen forms a triplex helix procollagen, which is packaged into a secretory vesicle. When the procollagen is secreted outside the cell, the ends of the protein are removed to form collagen. There are many examples of posttranslational processing of proteins, but these examples give some idea of the role protein processing plays in expression.
12. PEST sequences are regions rich in proline, serine, and acid residues that are common to short-lived eukaryotic proteins. They may either be recognized directly by the protein degradation machinery, or promote unfolding of the proteins in which they are contained, indirectly leading to degradation. Regulatory proteins tend to be rather short-lived, whereas structural proteins have relatively extended lifetimes. Consequently, it would be more likely that the former would contain PEST sequences than the latter.
13. The catalytic RNPs (RNA-protein complexes) have been shown to catalyze splicing of RNA. The chemistry of these reactions involves the breaking of one phophoester bond and the formation of another (phophoester exchange reactions). The formation of a peptide bond by the peptidyl transferase activity involves the breaking of the ester linkage between the tRNA and the amino acid with the formation of the peptide bond. Again we see the breaking of an equivalent bond to form another. Perhaps it is not surprising that this reaction may be catalyzed by the ribosome (an RNA-protein complex).
14. The protein toxins such as ricin and α-sarcin are enzymes that can inactivate ribosomes by an N-glycolytic cleavage of an adenine residue or by a specific nuclease cleavage, respectively. They can work at such low concentrations because they are catalytic. Another unique property is their ability to cross the cell membrane and attack the large ribosome. The site of attack on the 60S ribosomal subunit is probably the region where various translation factors interact with the ribosome.
15. The signal recognition particle (SRP) is made up of RNA and protein (RNP) and appears to require GTP for its activity in targeting nascent polypeptides to the endoplasmic reticulum membrane. For these reasons it is tempting to call it a ribosomal subunit; however, the SRP is only transiently bound to the ribosome and is released after docking. Also, proteins synthesized on free polysomes do not require the SRP. Thus, it is probably not valid to call the SRP the "third ribosomal subunit."
16. Normal globin assembles rapidly into tetrameres that bind heme, while globin containing the valine analog aminochlorobutyrate is unable to form tetrameres or bind heme, and thus remains essentially in a monomeric form. The instability of the protein containing the analog is thought to be due to its inability to be assembled into the far stabler tetramer. A possible mechanism that would explain this instability is suggested by the role of ubiquitin in the ATP-dependent pathway of protein degradation in eukaryotes. Ubiquitin is known to be

covalently linked to proteins destined for rapid degradation. Although the structural features of the unstable proteins that are recognized in this tagging process are not well defined, it has been demonstrated that ubiquitination occurs more rapidly with denatured or unfolded proteins. On the basis of this observation, it would seem likely that unassembled globin would be a preferred target for reaction with ubiquitin, resulting in its rapid degradation.

30 Regulation of Gene Expression in Prokaryotes

Problems

1. What set of data originally led Jacob and Monod to suggest the existence of a repressor in *lac* operon regulation?

2. In a cell that is z^-, what would be the relative thiogalactoside transacetylase concentration, compared with wild type, under the following conditions?
   (a) After no treatment.

   (b) After addition of lactose.

   (c) After addition of IPTG.

3. Consider a negatively controlled operon with two structural genes (*A* and *B,* for enzymes A and B), an operator gene (*O*), and a regulatory gene (*R*). The first line of data below gives the enzyme levels in the wild-type strain after growth in the absence or presence of the inducer. Complete the table for the other cultures.

	Uninduced		Induced	
Strains	**Enz A**	**Enz B**	**Enz A**	**Enz B**
Haploid strains:				
(1) $R^+O^+B^+$	1	1	100	100
(2) $R^+O^cA^+B^+$				
(3) $R^-O^+A^+B^+$				
Diploid strains:				
(4) $R^+O^+A^+B^+/R^+O^+A^+B^+$				
(5) $R^+O^cA^+B^+/R^+O^+A^+B^+$				
(6) $R^+O^+A^-B^+/R^+O^+A^+B^+$				
(7) $R^-O^+A^+B^+/R^+O^+A^+B^+$				

4. In a diploid situation would you expect a crp^+ to be dominant to a crp^-? Referring to *lac* expression, describe the phenotypes of crp^-, crp^+, and crp^+/crp^-.

5. Although *E. coli* promoters generally conform to a rather well-defined consensus sequence, no perfect match to this consensus has ever been observed in a naturally occurring promoter. Suggest an explanation.

6. What is the advantage of using IPTG as an inducer of the *lac* operon?

7. The lac repressor has an "on" rate constant for the binding of the *lac* operator (when cloned into λ) of about $5 \times 10^{10} M^{-1}s^{-1}$. This value is much greater than the calculated diffusion-controlled process, which is about 10^8 $M^{-1}s^{-1}$ for a molecule the size of the lac repressor. Explain why this repressor binding works better than expected.

8. Explain how histidine biosynthesis is controlled in *E. coli.*

9. A mutation in the *trp* leader region is found to result in a reduction in the level of *trp* operon expression when the mutant is grown in rich medium. However, when the mutant is grown in a medium lacking glycine, a stimulation in the level of *trp* enzymes is observed. Explain these observations. What would you anticipate would be the effect of growing the mutant in a medium lacking both glycine and tryptophan?

10. In *E. coli* there are no pools of free rRNAs or ribosomal proteins floating around in the cell even when the bacteria are grown at different growth rates. Explain how *E. coli* coordinates the biosynthesis of the ribosome.

11. Explain why rRNA synthesis is slow to resume when amino acids are added back to a *spoT* mutant in *E. coli.*

12. Describe the principal differences between patterns of control of gene expression used by bacterial host and bacteriophage systems.

13. In the infection of the host cell by λ, cI repressor favors lysogeny, while cro repressor favors the lytic cycle, even though both repressors bind to the same sites. Explain the basis of their different effects.

14. How is the synthesis of the CAP protein regulated? What is unusual about this regulation?

15. Gene regulatory proteins in bacteria were predicted (before their precise structure was known) to interact in the major groove of DNA by a two-site model of binding. Describe the data that showed this model to be correct.

Solutions

1. The existence of a repressor in *lac* operon regulation was first suggested by the results of studies of merodiploids. In merodiploids of the type i^+z^-/Fi^-z^+, Jacob and Monod were able to demonstrate that the i^+ (inducible) allele is dominant to the i^- (constitutive) allele when on the same chromosome (*cis*) or on a different chromosome (*trans*) with respect to the z^+ allele. The fact that the i^+ allele was dominant in *trans* indicated to Jacob and Monod that *i* gene mutations belong to an independent cistron that governs the expression of *z*, *y*, and *a* genes through the production of a diffusible cytoplasmic component, the lac repressor.
2. The level of *lac* operon expression in a cell, as indicated by the intracellular concentration of thiogalactoside transacetylase and regardless of the *lac Z* allele, is determined by the activity of lac repressor. After no special treatment, both wild-type and *lac* Z^- strains might be expected to exhibit a low level of *lac* operon expression, owing to the unhindered function of repressor in the absence of inducer. After addition of lactose, however, wild-type and *lac* Z^- strains would be expected to synthesize thiogalactoside transacetylase to differing extents. Lactose, although the natural substrate of the operon, is not the inducer of the operon *in vivo*. Instead, allolactose, an intermediate in lactose metabolism formed by the action of the limited amount of β-galactosidase present in uninduced cells, is thought to be the natural inducer. Of the two strains, only the wild-type (*lac* Z^+) produced a functional β-galactosidase capable of converting lactose to the inducer. Therefore, only this of the two strains would be impaired for activity of the lac repressor and exhibit an elevated level of *lac* operon expression. After addition of IPTG, both wild-type and *lac* Z^- strains would be expected to exhibit high induced levels of thiogalactoside transacetylase. IPTG, a structural analog of allolactose, though not a substrate of the operon, may bind and inactivate the lac repressor without prior metabolism by β-galactosidase.
3. Given the operon structure and mode of regulation described, the activities of operon products for the various mutant classes may be estimated at the following values:

	Uninduced		Induced	
Strains	**Enz A**	**Enz B**	**Enz A**	**Enz B**
Haploid strains:				
(1) $R^+O^+A^+B^+$	1	1	100	100
(2) $R^+O^cA^+B^+$	1–100	1–100	100	100
(3) $R^-O^+A^+B^+$	100	100	100	100
Diploid strains:				
(4) $R^+O^+A^+B^+/R^+O^+A^+B^+$	2	2	200	200
(5) $R^+O^cA^+B^+/R^+O^+A^+B^+$	2–101	2–101	200	200
(6) $R^+O^+A^-B^+/R^+O^+A^+B^+$	1	2	100	200
(7) $R^-O^+A^+B^+/R^+O^+A^+B^+$	2	2	200	200

Mutations of the O^c type, resulting in a reduced affinity of the operator region for repressor, would be expected to be *cis* dominant and characterized in both haploid and diploid strains by a constitutive expression of the *cis* operon. The uninduced level of expression would depend on the degree of residual affinity of the operator for repressor. Mutations of the R^- type, resulting in a deficiency of repressor activity, would be expected to be recessive and characterized in a haploid strain only by a constitutive expression of the operon. The uninduced level of expression in an R^- mutant would be expected to be comparable to that in the induced condition.

4. In the absence of glucose, cAMP complexed with the catabolite activator protein (CAP) may bind a region of *lac* DNA 5′ to the start of the *lac* transcript. This binding promotes formation of an open complex between RNA polymerase and the *lac* promoter region as a prelude to transcription of the operon. Mutations in the gene encoding CAP, designated *crp*$^-$, destroy the ability of CAP to participate in cAMP-induced activation of *lac* operon expression. The nature of the deficiency in such mutants, elimination of the activity of a *trans* acting product, makes it likely that the wild-type allele *crp*$^+$ would be dominant to the *crp*$^-$ in a diploid condition.

   In order for the *lac* operon to be expressed at an appreciable level, both the presence of a specific inducer, and the activity of the cAMP-CAP activating complex are required. In a *crp*$^-$ haploid, deficiency of the latter leads to permanent catabolite repression of the operon even in the absence of glucose. In both *crp*$^+$ haploid and *crp*$^+$/*crp*$^-$ diploid strains, the functional CAP contributed by the wild-type allele provides for normal catabolite repressibility and inducibility by lactose.

5. It is generally accepted that the extent of homology of a promoter to the consensus –35 and –10 sequences is an important determinant of promoter strength. It might, therefore, be reasonably expected that genes for which transcripts are needed in great amount would possess promoters with sequences very close to the consensus. Although this is in some cases true, it is not generally the case. The explanation for the less than perfect match of most promoters to the consensus sequence is to be found in the need to regulate transcription. Transcriptional regulation is achieved in many instances by the selective improvement of the affinity of specific promoters for RNA polymerase. Such selective improvement is well illustrated in the case of regulation of the *lac* operon. The *lac* promoter is by all accounts a rather weak one; its match to the consensus sequence is as follows:

	–35	–10
Promoter consensus sequence	TTGACA	TATAAT
Wild-type *lac* promoter	TTTACA	TATGTT
lac UV5	TTTACA	TATAAT

   The strength of the wild-type *lac* promoter, and hence the level of expression of the *lac* operon, may, however, be significantly increased as a result of binding of the catabolite activator protein (CAP) upstream of the –35 region. This is because of the additional affinity of the promoter for RNA polymerase provided by the latter's interaction with CAP. A mutant containing two base replacements in the –10 region of the *lac* promoter (UV5), which dramatically improves the homology of the promoter to the consensus sequence, increases its inherent strength to such a degree that it no longer requires CAP binding for high-level transcription. As a consequence, expression of the *lac* operon in this mutant is no longer subject to regulation by catabolite repression, a situation that is clearly not advantageous to the cell when both glucose and lactose are available as carbon sources.

6. Isopropyl-β-D-thiogalactoside (IPTG) binds very tightly to the lac repressor and is not broken down by β-galactosidase. It is completely stable in *E. coli* and in extracts. Therefore, very little IPTG can induce high levels of *lac* expression.

7. The repressor binds to its operator much better than predicted by a simple bimolecular reaction limited by diffusion in three dimensions. Thus, the lac repressor must find its operator by some other mechanism. One possibility is the binding of the repressor nonspecifically to the DNA and then searching in one dimension (binding to the DNA and then sliding along until the promoter is reached). Also, a long section of DNA would

not be randomly distributed but would form a loose ball of DNA that would define a domain much smaller than the solution in the test tube. When the repressor was released from the DNA it could more quickly find another strand of DNA to bind (effectively giving a much higher concentration of DNA). The effect of both mechanisms would be to allow as few as 10 repressor molecules per cell to prevent transcription of the *lac* operon.

8. Histidine biosynthesis is controlled at two levels in *E. coli:* feedback inhibition at the enzyme level where histidine inhibits the first enzyme in the pathway, and attenuation regulation at the gene level. The *his* operon has no repressor and is regulated at the gene level exclusively with an attenuator. If histidine levels are high in the cell, the histidine biosynthetic pathway is inhibited at the first enzyme step. The leader peptide, which has histidines in it, is synthesized, and the ribosomes do not stall at the *his* codons, allowing the hairpin loop followed by a sequence of U residues to form and terminate transcription. Both of these mechanisms allow a high level of regulation of histidine biosynthesis in *E. coli.*

9. Given that the mutation is located within the *trp* leader region, most likely its effect would be mediated through an influence on attenuation of transcription. The *trp* attenuator system is comprised of a leader sequence that codes for a peptide containing consecutive tryptophan codons that can be organized into two alternative stem-loop secondary structures (see figures 30.19 and 30.21). One of these alternative secondary structures includes a stem-and-loop that signals termination of transcription, while the other precludes formation of this stem-loop, thus providing for readthrough by polymerase into the main body of the *trp* operon. This choice between these two potential secondary structures is conditioned by the concentration of trp-$tRNA^{Trp}$ in the cell. In the presence of low concentrations of trp-$tRNA^{Trp}$ (*trp* starvation), ribosomes translating the leader peptide become stalled at one or other of the consecutive tryptophan codons. This prevents formation of the terminating secondary structure, leading to readthrough. In contrast, when the concentration of trp-$tRNA^{Trp}$ is high (tryptophan excess), ribosomes are not stalled at this position, with the result that the terminating secondary structure is formed, and readthrough is eliminated.

The observation that *trp* operon expression is stimulated in the mutant by glycine may be explained in the context of this attenuation mechanism, in the way illustrated in the figure.

This explanation may be confirmed by testing the effects of starving the mutant for arginine or alanine. Low levels of arg-$tRNA^{Arg}$ stimulate readthrough in the wild type but should not influence *trp* operon expression in the mutant, whereas starvation for alanine should enhance readthrough only in the mutant.

Growing the mutant in a medium lacking both glycine and tryptophan would result in an increase in the level of *trp* operon expression compared to a medium lacking glycine only. This would be due to the relief from repression by the trp repressor system, which is independent of attenuation and would remain responsive to tryptophan levels in the mutant as in the wild type.

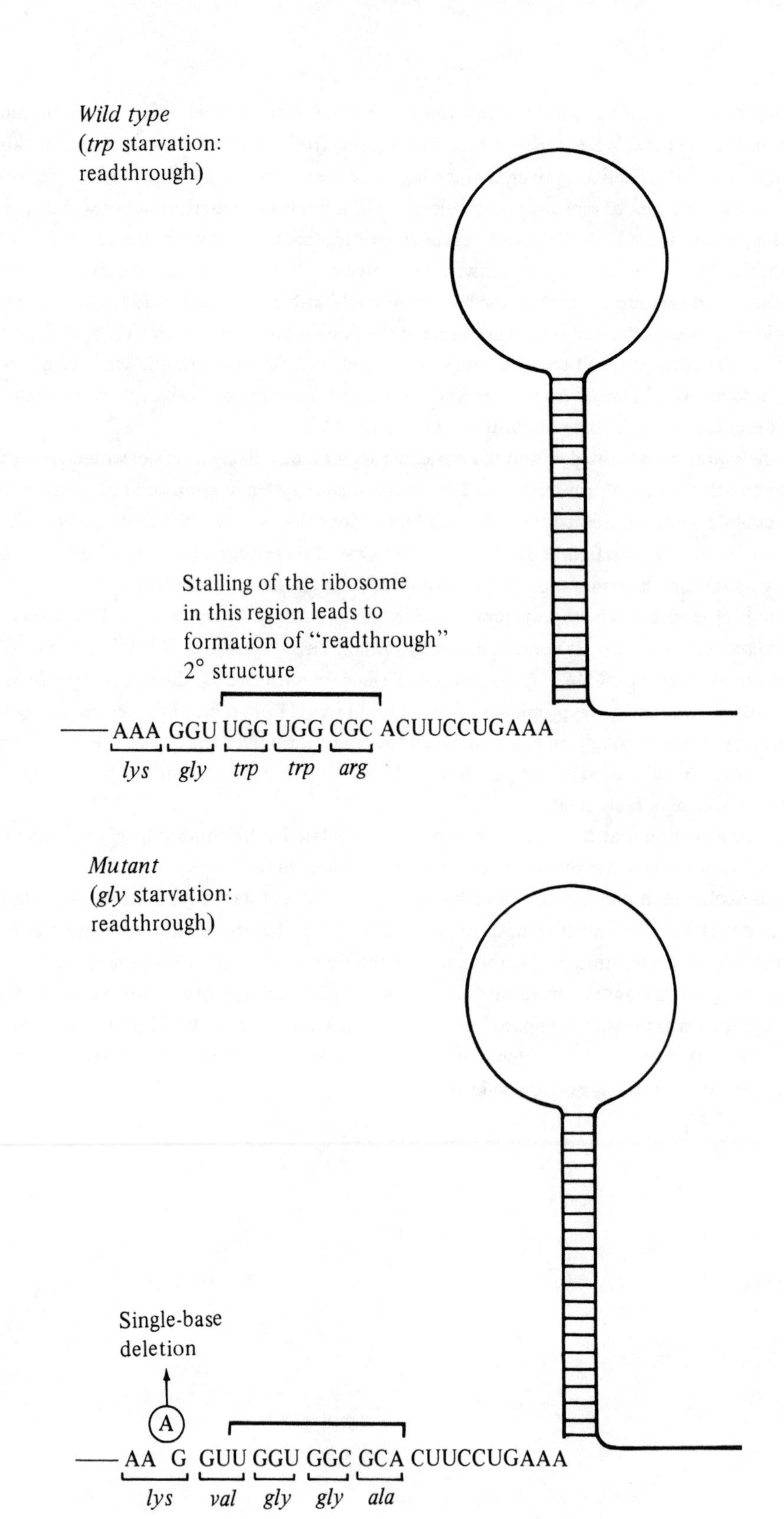
Wild type
(trp starvation:
readthrough)
Stalling of the ribosome
in this region leads to
formation of "readthrough"
2° structure
AAA GGU UGG UGG CGC ACUUCCUGAAA
lys gly trp trp arg
Mutant
(gly starvation:
readthrough)
Single-base
deletion
A
AA G GUU GGU GGC GCA CUUCCUGAAA
lys val gly gly ala

10. The synthesis of ribosomal proteins is regulated in *E. coli* by translational regulation, i.e., free ribosomal proteins inhibit the translation of their own mRNA. As long as rRNA is being made, these proteins bind to the rRNA and the translation of the ribosomal proteins continues. The genes for the ribosomal proteins are clustered in a number of operons that produce polycistronic mRNAs. One of the simplest operons is P_{L11} which codes for proteins L1 and L11. L1 is the regulatory protein and can bind to the 23S rRNA or to the 5′ end of its own polycistronic mRNA. If the levels of L1 increase, it binds to its own mRNA and inhibits translation of both L1 and L11 proteins. This mechanism keeps the levels of L1 and L11 in register with the amount of rRNA. The other ribosomal proteins are regulated in a similar manner.

11. When *E. coli* is starved for amino acids, the synthesis of rRNA is inhibited (stringent response). The inhibition of rRNA synthesis correlates with the levels of guanosine tetraphosphate (ppGpp) formed. The *spoT* gene codes for an enzyme that breaks down ppGpp and allows ribosomes to be made again. Normally, when amino acids are added back to starved cells the levels of ppGpp fall rapidly. But in the *spoT* mutant, the levels of ppGpp fall slowly and it takes a long time for rRNA synthesis to continue and cell growth to increase.

12. Viral gene expression is primarily controlled by the activation or suppression of synthesis or viral gene products that commit the virus to specific pathways. Viral genes are usually required only at certain times during the lytic cycle. Their expression is controlled in a programmed and essentially irreversible way. This is perhaps best illustrated in the case of the regulatory mechanism that determines whether bacteriophage enters the lysogenic or lytic cycle (see the solution to problem 13).

In contrast, bacterial regulatory proteins are present in the cell most of the time. Their activity is regulated usually as a function of the presence and concentration of small effector molecules such as cAMP, L-arabinose, or lactose, and in this way is responsive to the nutritional environment. This maintains bacterial genes in a potentially active state, so that they can be quickly expressed or turned off in response to changing conditions. The processes of bacterial control of gene expression, whether mediated by transcriptional or translational regulation, are therefore of necessity inherently reversible.

13. The choice between entering the lytic or lysogenic cycle for bacteriophage λ is determined by the balance between the amounts of cI and cro proteins that are made in the very early stages of infection. Both proteins bind to the same sites in the operator region that regulate the activity of the P_L and P_R promoters. The cI protein, at low concentrations, represses the activity of the P_R and P_L promoters, and activates the P_{RM} promoter. At high concentrations, cI represses the activity of all three promoters. In contrast, the cro protein, at low concentrations, represses the P_{RM} promoter, thus inhibiting the synthesis of cI protein. At high concentrations, cro represses the activities of both P_{RM} and P_R. It can be seen, therefore, that cro and cI can affect each other's synthesis as well as their own when they interact with the same sites in the operator region, and that both proteins show strong concentration effects in their mode of action. How this is achieved can be best understood in terms of the detailed interactions between cro and cI and the same sites in the operator DNA and the effect these interactions have on the expression of the viral gene functions.

The cI and cro proteins both can bind at three sites, o_{R1}, o_{R2}, and o_{R3}, in the right operator region of bacteriophage λ. The affinity of the three sites for cI or cro is not identical, nor is the effect of binding of either protein at the three sites the same. cI binds to sites o_{R1} and o_{R2} with approximately equal affinities, and to o_{R3} with a much lower affinity. Binding of cI at o_{R1} and o_{R2} has two major effects: one is to inhibit the activity of the P_R promoter, which controls the expression of cro, cII, Q, and 6S RNA gene products required for lytic infection. The other effect is to stimulate the activity of the P_{RM} promoter, which controls the expression of the gene encoding cI. Thus cI functions as its own activator. At relatively low cI concentrations, the o_{R3} site is not occupied. At high concentrations of cI, the o_{R3} site becomes filled, resulting in the repression of the activity of the P_{RM} promoter. This ensures that cI concentrations in the cell are kept at levels that permit repressor maintenance (and thus, maintenance of the lysogenic state), and yet are low enough to permit rapid prophage induction under changing physiological conditions.

In contrast, the *cro* gene product binds to o_{R3} with the highest affinity, and binds to o_{R1} and o_{R2} with less affinity. Binding of *cro* at the o_{R3} site can thus occur at low concentrations of *cro*, resulting in the same effect as when *cI* is bound at that site: The activity of the P_{RM} promoter is repressed, thus inhibiting the synthesis of *cI*. At high concentrations, *cro* can fill the o_{R1} and o_{R2} sites, which also provides for autoregulation. The main effect of binding of *cro* at the o_{R3} site, which occurs at low *cro* concentrations, is to inhibit the synthesis of *cI*: this commits the phage to the lytic pathway, by allowing for the expression of viral gene functions required for phage growth, and whose activity is blocked by the presence of *cI*. During prophage induction, inactivation of *cI* by the recA proteolytic function leads to the synthesis of *cro*, which blocks further *cI* synthesis and leads to an irreversible cascade of events culminating in the production of phage particles and lysis of host cells.

It has been stated that the balance between cI and cro expression is regulated by the levels of both proteins present in the early stages of infection. More accurately, this balance is also determined by the levels of cII protein early during infection (cII activates the P_{RE} and P_{int} promoters). The balance of this whole set of interactions is determined by concentration effects and by the different binding affinities of the two proteins for the operator sites. This allows for their antagonistic effects, although they interact with the same sites in this region of the viral genome. Such as mechanism provides for finely tuned regulation of the course of infection (lytic or lysogenic), while allowing for rapid reversal of the lysogenic state in conditions favoring prophage induction.

14. The synthesis of the CAP protein is regulated by the synthesis of a divergent RNA from a promoter located close to the *crp* gene (the gene coding for the CAP protein). This RNA is complementary to 10 of the first 11 nucleotides of the *crp* mRNA and allows a ρ-independent terminator to form terminating transcription. When the levels of CAP-cAMP are high, the antisense RNA is made and CAP synthesis is turned off.

15. The two-site model for binding of regulatory proteins was suggested by the DNA binding site having a twofold axis of symmetry, which matches the symmetry of the regulatory protein. This symmetry matching is a recurring theme in DNA-protein interactions. The final proof of the two-site model came with the cocrystallization of the regulatory protein and its DNA-binding site. The protein binds on one side of the DNA in two adjacent major grooves. The hydrogen-bonding groups exposed in the major groove presents many possibilities for interactions with the amino acid side chains of the protein (see figure 30.15 in the text for the repressor binding). A common element of these proteins that bind to specific sites in DNA is the *helix-turn-helix* motif.

31 Regulation of Gene Expression in Eukaryotes

Problems

1. On the basis of what you know about genes in prokaryotic and eukaryotic cells, define a gene. Make sure that your definition is brief and concise. Does your definition have any limitations or problems?

2. Why is attenuation control in eukaryotes unlikely?

3. A mutation in *GAL4*^{c} of yeast leads to constitutive galactose fermentation in haploid cells. Describe the nature of this mutation at the molecular level. Would you expect expression of *GAL* genes to be responsive to glucose repression in this mutant?

4. Explain how starvation of yeast for any one of ten amino acids derepresses the synthesis of more than 30 enzymes in nine different amino acid biosynthetic pathways (general amino acid control).

5. How would you expect a deletion of *HMLE* to affect the expression of mating-type genes in yeast? Compare this effect with the deletion of the α_2 gene from MAT_{α}. Consider both homothallic and heterothallic backgrounds.

6. Expression of some genes in eukaryotic cells can be induced by treatment with 5-azacytidine. Explain how this happens.

7. Briggs and King were able to grow a differentiating embryo from an egg of *Rana pipiens* whose chromosomes were replaced with a single diploid nucleus from another embryo. What does this tell us about the state of the nucleus in the developing embryo?

8. Hemophilia is an X-chromosome-linked disorder. Bearing in mind the phenomenon of X-chromosome inactivation, suggest an explanation for the observation that females who are heterozygous for the defective gene are essentially asymptomatic.

9. High salt weakens the interaction of histones with DNA but has little affect on the binding of many regulatory proteins. Explain this observation in terms of how these molecules interact with DNA.

10. List some characteristics that distinguish active from inactive chromatin.

11. In the *Xenopus* oocyte a large number of ribosomes are made in a short time to handle the rapid demand for cell growth during cleavage stages. How is this large amount of rRNA made in such a short time?

12. When mammalian cells in culture are treated with the antifolate methotrexate (see chapter 20), cells can be selected that are resistant to high levels of this toxic compound. The enzyme dihydrofolate reductase becomes elevated about 1,000-fold in these resistant cells. What mechanism can account for such a large amount of this enzyme being made and how does this protect the cell from the toxic effects of methotrexate? How could you test for the molecular mechanism of this drug resistance?

13. Given that specific subsets of homeotic genes are required for the development of specific segments of the *Drosophila* embryo, suggest a possible mechanism whereby the necessary spatially restricted pattern of homeotic gene expression might be achieved. Incorporate the observed effect of homeotic mutations upon segment morphology into your model.

14. Explain how a DNA sequence (enhancer sequence) located 5,000 base pairs from a gene transcription start site can stimulate transcription even if its orientation is reversed.

15. Speculate on some possible models to explain tissue-specific alternative splicing such as that seen with variants of tropomyosin.

16. Discuss the types of structural motifs found in eukaryotic transcription factors.

Solutions

1. We use the term "gene" often in biology and biochemistry and take for granted that we understand what a gene is. Yet, if you asked 50 scientists their definition of a gene you would probably get 50 different answers. Our concept of a gene has evolved dramatically from the original concepts proposed by Mendel, especially with the modern approaches used in molecular biology. One definition is: *the DNA (or RNA) sequences necessary to produce a peptide (or RNA).* Some viruses have RNA as their genetic material and some genes do not code for a protein but make a functional RNA such as tRNA or rRNA. This definition may be too general. Would it include enhancer sequences necessary for transcription located thousands of base pairs upstream or downstream from the gene? Does this definition include introns and all the regulatory sites located in the promoter region? How would polyproteins be viewed (a single peptide that is processed into a number of peptide products)? How would you explain genes produced from the same DNA region that, because of two different reading frames, generate two different polypeptides? How about different types of DNA structure such as DNA bending, Z-DNA, and higher-order chromatin structure in eukaryotes? Obviously, a single final definition of a gene remains uncertain, but the definition given covers most of our current concepts of a gene.
2. Attenuation control (found in bacteria) is dependent on the formation of a transcriptional-translational complex (as the mRNA is being formed ribosomes bind and protein synthesis starts). In eukaryotes, transcription occurs in the nucleus, the mRNA is processed (capping, poly A addition, and intron removal), and then it moves into the cytosol where ribosomes bind and translation starts. The compartmentalization of these processes physically separates transcription from translation, and the extensive processing of the mRNA would make it unlikely that a type of attenuation control will be found in eukaryotes. (Of course, attenuation control could occur in the mitochondria or chloroplast.)
3. In yeast, the immediate regulator of the *GAL* genes is the product of the GAL4 locus, a protein that activates *GAL* gene transcription. The product of a second locus, GAL80, which complexes directly with the GAL4 product, regulates the activity of this protein, so that in the absence of inducer, GAL4 activation of transcription is prevented. The addition of galactose induces dissociation of the GAL4-GAL80 complex, thereby allowing activation of *GAL* gene expression.

A mutation in the *GAL4* gene leading to constitutive expression of the *GAL* genes cannot involve deficiency in the product's function as an activator of transcription. Such a deficiency would result in a phenotype of permanent repression of *GAL* gene expression. Instead, it is likely that the mutation responsible interferes in some way with the complexing of GAL4 protein with its negative regulator, the product of the *GAL80* gene.

In yeast, glucose inhibition of galactose metabolism involves two distinct levels of control; repression of transcription of the *GAL* genes, and inactivation of GAL enzymes already synthesized. Although little is known about the mechanism underlying catabolite inactivation, the glucose effect on *GAL* transcription is thought to involve preventing activation of transcription by the GAL4 product. Assuming that the defect leading to constitutive *GAL* expression is as described above, glucose would still be expected to have this effect on GAL4 activation in the mutant. Therefore *GAL* gene expression would remain responsive to glucose.

4. During amino acid starvation in yeast, 30 enzymes in nine different biosynthetic pathways are expressed. These coregulated genes have multiple copies of a *cis*-acting sequence (TGACTC) located far upstream of the genes. The *GCN4* gene makes a protein that acts as a positive regulator by binding to the TGACTC sequences. Thus, all these genes are activated by a single protein. Interestingly, the *GCN4* gene expression is regulated at the translational level by *trans*-acting factors that allow the *GCN4* mRNA to be translated.

5. In wild-type haploid yeast only one of the HML_α and HMR_a sets of genes are expressed at any one time. This exclusivity of expression is conditioned by a transposition mechanism that removes either set of genes from an inactive "storage" chromosomal context (HML_α and HMR_a loci) to an active one (the *MAT* locus). The inactivity of HML_α and HMR_a genes at their "storage" loci is due to the close proximity of "silencer" elements (*HMLE* for HML_α and *HMRE* for HMR_a) that repress transcription of surrounding genes. Since the simultaneous expression of HML_α and HMR_a genes, which occurs in the diploid, prevents mating, it is essential that these "storage" loci be tightly repressed, in order to ensure that haploid cells display appropriate mating behavior. A deletion of *HMLE* would remove the element repressing transcription of HML_α at the "storage" location, and result in the constitutive expression of HML_α genes from this locus. The consequences of such constitutivity would depend on the mating type of the mutant. If HML_α were at the *MAT* locus (MAT_α), the arrangement characteristic of the **α** mating type, the cell exhibits the behavior expected of the **α** mating type. However, if HMR_a were at the *MAT* locus (MAT_a), the arrangement giving rise to the *a* mating type, the cell would resemble the diploid, expressing both *a* and **α** genes simultaneously, and thus be sterile.

Mutants that began as **α**-type would become sterile at a rate conditioned by the frequency of transposition of the HMR_a to the *MAT* locus. In homothallic strains this frequency is very high, approximately once per cell division, while in heterothallic strains it is considerably lower.

Both HMR_a and *HML* exert their effect on mating behavior by controlling the expression of sets of genes specific to each mating type. *A*-specific genes are expressed only by *a* cells, while **α**-specific genes are expressed only by **α** cells. The way in which products of *MAT* (i.e., *HMR* at the *MAT* locus) are believed to regulate expression of *a*- and **α**-specific sets of genes is shown schematically as follows:

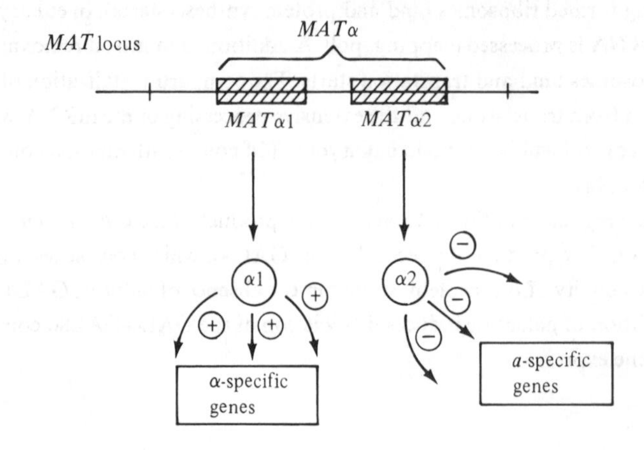

As can be seen from the figure, deletion of the gene encoding α2 from MAT_{α} would result in a failure to inhibit expression of the *a*-specific genes. Haploid mutants that contained such a deletion would therefore resemble the diploid, and be sterile. If the α2 gene contained by the HML_{α} locus were similarly mutated, mutants of this type would be unable to switch to the α-type. However, transposition of HMR_{a} to the *MAT* locus would allow for expression of the *a* mating type. Again the frequency of transposition would condition the rate of switching between sterile and *a*-type, the frequency being far greater in homothallic strains than in heterothallic strains.

6. When cells are treated with 5-azacytidine, this nucleoside analog of cytidine is incorporated into the DNA, and after a number of rounds of replication generates hypomethylated regions in the DNA. Since the analog has a nitrogen in the 5 position, this base cannot be methylated. Some genes are inhibited by DNA methylation in their promoter regions; when these regions become hypomethylated these genes can be transcribed and expressed.
7. The results of the nuclear transplantation experiments of Briggs and King indicated that nuclei, up to a certain stage in embryonic development, remain totipotent, in the sense that they can support the normal development of enucleated differentiating embryos. They found that blastula nuclei, although already "committed" to differentiated pathways, were capable of being "reprogrammed" by a proper environment to allow for some degree of dedifferentiation. Gastrula nuclei, in contrast, were found to support normal development only at a much reduced efficiency. From this they were able to conclude that these later-stage nuclei had undergone some measure of irreversible differentiation, with the result that they had lost, at least to some degree, their totipotency.
8. In contrast to males, who contain only one X chromosome per cell, females contain two X chromosomes per cell, and thus are essentially heterozygous for all X-linked loci. In the female, one of these X chromosomes is inactivated in each cell; however, because this inactivation is essentially random (either X is inactivated with the same frequency) and occurs at a late stage of development, females remain effectively heterozygous for X-linked loci (although not at the level of individual cells). As a consequence of this "mosaic" makeup (on the average, half the cells of the female contain an inactivated normal chromosome, and half an inactivated defective chromosome), the female does not exhibit the severe symptoms of hemophilia observed in the male, who is necessarily homozygous for the defective gene.
9. Histones bind to DNA by electrostatic interaction of basic amino acids (arginine and lysine) with the negative charge on the phosphate. These electrostatic (negative-positive charge) interactions are weakened by high salt, which will allow the histone proteins to disassociate from the DNA. Many regulatory proteins may initially interact in a nonspecific fashion with DNA until they locate their high-affinity binding sites. Once these proteins bind at their specific recognition sites, the major interaction is through hydrogen bonding and hydrophobic interactions. These hydrophobic interactions are stabilized by high salt.
10. Transcriptionally active chromatin is usually more swollen than inactive chromatin. In addition, the following characteristics have been used to differentiate between active and inactive chromatin in eukaryotic cells.
    (a) Active chromatin is more susceptible to DNase I degradation than inactive chromatin. This can be shown for genes that are active only in certain tissues, and inactive in others. The regions that are more susceptible to enzyme digestion usually encompass a large domain around a transcriptionally active gene or cluster of genes. A particular gene expressed in a given tissue can be shown to be sensitive to DNase I digestion in that tissue, but not in a tissue where it is not expressed. For example, the globin genes are sensitive to DNase I digestion in chromatin isolated from erythrocytes, but not in chromatin isolated from oviduct.
    (b) Active chromatin is often found to be associated with modified histones. Histones have been found to be subject to methylation and acetylation, and to phosphorylation in the case of histone H1. Histone H2A can also be found associated with ubiquitin. Ubiquitinated H2A is found preferentially in active chromatin. Such modifications have not been directly correlated with other characteristics of active chromatin, such as DNase I sensitivity.

(c) High mobility group (HMG) proteins are found preferentially associated with active chromatin. It has been shown that HMG proteins can impart DNase I sensitivity to chromatin in reconstitution experiments. When chromatin from erythrocytes was stripped of its HMG proteins, globin genes lost their preferential sensitivity to DNase I digestion. Reconstitution with HMG proteins (which could be isolated from any tissue, including those not expressing the globin genes) caused the globin genes to regain their susceptibility to DNase I digestion. When chromatin from brain (which does not express the globin genes) was reconstituted with HMG proteins, no sensitivity of the globin genes to DNase I digestion was observed. It thus appears as if HMG proteins will bind and confer sensitivity to DNase I digestion to chromatin packaged in a state different from that of inactive chromatin, perhaps due to the binding of specific regulatory factors. HMG proteins do bind to transcriptionally inactive chromatin in such reconstitution experiments but do not confer DNase I sensitivity upon it.

(d) DNA methylation correlates with inactive chromatin. Most cytosine residues in CG dinucleotides are found to be methylated in vertebrate DNA. In some cases, it has been shown that tissue-specific or developmentally regulated genes have methylated CG residues at times during development when, or in tissues where, they are not expressed, and become demethylated (or hypomethylated) when they are expressed.

11. One strategy that is employed to generate sufficient amounts of a product required at a specific stage of development is gene amplification. An example of this strategy is to be found in the sea urchin, which achieves a high rate of histone synthesis during embryogenesis by producing a large number of copies of the histone genes. Another example of gene amplification is provided by *Xenopus laevis.* The amounts of ribosomal RNA in frog eggs can be greatly increased by amplification of the rRNA genes. The amplified DNA is extrachromosomal, located near the nucleolus, transcribed during oogenesis, and subsequently discarded. The 5S rRNA genes of *Xenopus laevis* fall into two families. There are approximately 20,000 copies of a 5S RNA gene expressed during oogenesis, and about 400 copies of a 5S RNA gene expressed in somatic cells and growing oocytes. The expression of 20,000 copies of the oocyte gene provides for very high levels of 5S RNA during oocyte growth.

    An alternative strategy that is used to provide large amounts of required products early in development is the accumulation of mRNAs or proteins prior to need. This strategy is best exemplified by *X. laevis,* which stores large amounts of histone proteins and histone mRNAs in the egg to be used in embryogenesis, when the demand for histones is high.

12. When cells resistant to high levels of methotrexate are assayed for dihydrofolate reductase, the levels of this enzyme are elevated about 1,000-fold. Methotrexate binds very tightly to dihydrofolate reductase and normally would be very toxic to rapidly growing cells. When large amounts of enzyme are made, the methotrexate is bound, leaving some enzyme activity to make tetrahydrofolate and allow cell growth. Large amounts of dihydrofolate reductase are made in these resistant cell lines by amplifying the dihydrofolate reductase gene, generating a large gene dose that produces a great amount of dihydrofolate reductase mRNA. The large amount of this specific mRNA then produces a lot of the dihydrofolate reductase enzyme. The amount of dihydrofolate reductase genes could be determined by probing with a specific DNA sequence made using recombinant DNA techniques or synthetic DNA made on a "gene machine." DNA could be isolated from normal and resistant cells and hybridized to the labeled probe. The DNA from the resistant cells would bind more probe than the DNA isolated from normal cells when an equal amount of DNA was used (the "dot blot" procedure would be one of the easiest methods of hybridization analysis).

13. Although the precise regulatory relationships between homeotic genes remain unclear, it is apparent that a hierarchy of interactions exists among gap, pair-rule, and segment polarity genes. Each gene class is influenced by the action of earlier-acting genes that control larger units of pattern, and by some members of the same class. Thus gap genes that are expressed early influence the pattern of expression of pair-rule, (e.g., fushi terazu, *ftz*), segment polarity (e.g., engrailed, *en*) and homeotic genes, (e.g., ultrabithorax, *ubx*), which are expressed later, in a hierarchical fashion. While segment polarity genes do not affect the expression of gap or pair-rule genes that

occupy a higher position in the hierarchy, some gap genes, and presumably genes at other levels in the hierarchy, have been found to be mutually negative regulators of each other.

The phenomenon of mutual negative regulation offers perhaps the most plausible explanation for the observed effect of homeotic mutations on segment morphology. To date, a number of such interactions between homeotic genes have been described. The first of these involved genes of antennapedia (*Antp*) and *ubx* complexes. Mutation of the *Ubx* gene, which is expressed posterior to *Antp*, was found to result in the ectopic expression of *Antp* in those segments where *Ubx* was normally expressed, suggesting that *Ubx* represses *Antp* directly or indirectly in these segments. Subsequent studies have indicated that similar mutually negative interactions occur generally among homeotic genes.

A schematic outline of the hierarchy of interactions leading to the integrated partitioning and specification of the *Drosophila* embryo is

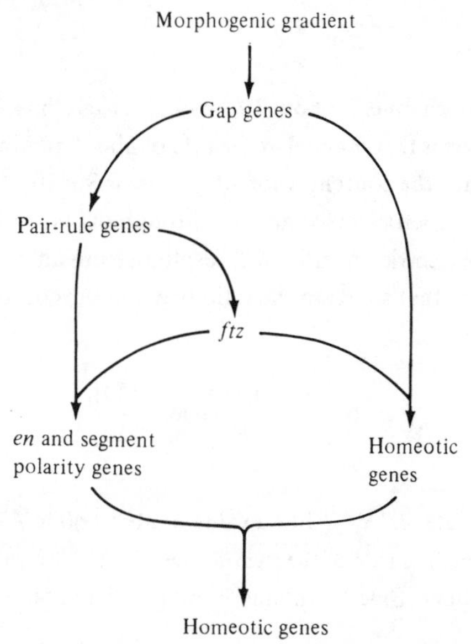

14. *Cis*-acting sequences acting over great distances have been found in eukaryotes and more recently in prokaryotes. These sequences have been called "enhancers" because they enhance the expression of their associated gene. These enhancer sequences can bind to tissue-specific *trans*-acting proteins and the DNA can loop to allow the enhancer-protein complex to interact with the promoter region of the gene and RNA polymerase II. These enhancer sequences can act thousands of base pairs from the promoter and are functional in either orientation ($5' \rightarrow 3'$ or $3' \rightarrow 5'$).

15. Tropomyosin, an important protein in striated muscle, is found in different forms in various tissues. Seven splice sites can be found in the unspliced precursor RNA. Alternative splicing can produce variants of this protein in different tissues from the same organism. *Trans*-acting factors (proteins or RNA) found in specific tissues could influence the selection of splice sites and generate the tissue-specific variants of this muscle protein.

16. Most transcription factors have two-domain structures, one for DNA binding and another for protein binding. The DNA-binding domains involve various structural motifs that interact with the DNA in the major groove. The type of structure previously discussed in prokaryotes is the helix-turn-helix motif, which is also found in some eukaryotic transcription factors. A more common motif in eukaryotes is the zinc finger (zinc forms a tetrahedral complex with histidines and cysteines). Another type of structure found is the leucine zipper (a highly conserved stretch of amino acids with net basic charge followed by a region of four leucine residues at intervals of seven amino acids). All of these motifs found in these transcription factors evolve interactions in the major groove of the DNA with either an α-helix segment (more common) or a two-stranded antiparallel β sheet.

32 Mechanisms of Membrane Transport

Problems

1. (a) Using Fick's law, show that the diffusion coefficient D has the dimensions of area per unit time.

   (b) The diameter of a pore channel is about 10^{-9} m and its length is 4×10^{-9} m. In planar membrane bilayers, glucose traverses this channel at the rate of about 50 molecules per channel per second at room temperature when the concentration of glucose is 3×10^{-6} M on one side of the membrane. Calculate the diffusion coefficient for glucose through the pore channels under these conditions.

2. The Nernst equation relates the electric potential $\Delta\Psi$ resulting from an unequal distribution of a charged solute across a membrane permeable to that solute to the ratio between the concentration of solute on one side and on the other:

$$m\ \Delta\psi = \frac{-2.3RT}{F} \log \frac{[So]_1}{[So]_2}$$

where m is the charge on the solute, $2.3RT/F$ has a value of about 60 mV at 37°C, and $[So]_1$ and $[So]_2$ refer to the concentrations of solute on either side of the membrane. Consider a planar phospholipid bilayer separating two compartments of equal volume. Side 1 contains 50-mM KCl and 50-mM NaCl, while side 2 contains 100-mM KCl.

   (a) If the membrane is made permeable only to K^+, e.g., by addition of valinomycin, what will be the magnitude of $\Delta\Psi$?

   (b) If the membrane is made permeable to H^+ and K^+, in which direction will H^+ initially flow?

   (c) If the membrane could be made selectively permeable to both K^+ and Cl^-, what would be the value of $\Delta\Psi$ and the ion concentrations on both sides of the membrane at equilibrium? (Hint: Initially, K^+ would diffuse down its concentration gradient accompanied by an equivalent amount of Cl^-. Equilibrium would be established when the potentials due to K^+ and Cl^- were equal to each other and to the overall membrane potential.)

3. Membrane vesicles of *E. coli* that possess the lactose permease are preloaded with KCl and are suspended in an equal concentration of NaCl. It is observed that these vesicles actively, although transiently, accumulate lactose if valinomycin is added to the vesicle suspension. No such active uptake is observed if KCl replaces NaCl in the suspending medium. Explain these results in light of what you know about the mechanism of lactose transport and the properties of valinomycin.

4. Intracellular vacuoles in the yeast *Saccharomyces cerevisiae* are membrane-bounded organelles that are known to concentrate within them a variety of basic amino acids, including arginine (net charge =+1). Vesicles prepared from these vacuoles lack an electron-transport chain, and arginine uptake into them is dependent on extravesicular ATP. A membrane potential $\Delta\Psi$ has no effect on ATP-dependent arginine uptake in the absence of a proton gradient, while proton ionophores and dicyclohexylcarbodiimide (a known inhibitor of the F_1/F_0 ATPase) greatly inhibit accumulation of arginine by this system. Upon addition of ATP in the absence of arginine, the intravesicular pH of these vesicles drops. Describe a mechanism for the energization of arginine transport in this system, taking into account all these observations.

5. In *E. coli*, lactose is taken up by means of proton symport, maltose by means of a binding (ABC-type) protein-dependent system, melibiose by means of Na^+ symport, and glucose by means of the phosphotransferase system (PTS). Although this bacterium normally does not transport sucrose, suppose that you have isolated a strain that does. How would you determine whether one of the four mechanisms just listed is responsible for sucrose transport in this mutant strain?

6. In some instances, the efflux of a radioactively labeled transport substrate out of preloaded cells or vesicles is transiently stimulated by addition of the same nonradioactive transport substrate to the outside. This phenomenon is known as trans-stimulation and occurs with transport systems that are reversible (i.e., can operate in either direction). Can you think of an explanation for trans-stimulation in view of what is known about the molecular mechanisms of transmembrane transport?

7. Outline a molecular mechanism by which, and the conditions under which, an H^+ symport system (such as the *E. coli* lactose permease system) might operate to actively accumulate a metabolite such as lactose.

8. Predict the effects of the following on the initial rate of glucose transport into vesicles derived from animal cells that accumulate this sugar by means of Na^+ symport. Assume that initially $\Delta\Psi = 0. \Delta pH = 0$ (pH = 7), and the outside medium contains 0.2-M Na^+, while the vesicle interior contains an equivalent amount of K^+.

(a) Valinomycin.

(b) Gramicidin A.

(c) Nigericin.

(d) Preparing the membrane vesicles at pH 5 (in 0.2-M KCl), resuspending them at pH 7 (in 0.2-M NaCl), and adding 2,4-dinitrophenol.

9. You are growing some mammalian cells in culture and measure the uptake of D-glucose and L-glucose (see data in chart). What type of transport is observed with these sugars? (Hint: plot V versus [sugar] and $1/V$ versus 1/[sugar].) Explain the significance of these data.

[Sugar] (mM)	V(mM cm s^{-1}) × 10^7 D-glucose	L-glucose
0.100	166	4.8
0.167	252	8.0
0.333	408	16
1.000	717	50

10. If the cells in problem 9 are treated with $HgCl_2$ (mercury reacts with -SH groups in proteins), the rate of transport of D-glucose is the same as L-glucose. What is indicated about D-glucose transport?

11. The translocation of K^+ was studied using an artificial membrane system (this membrane system had a phase transition T_m of 41°C) with valinomycin and gramicidin. The results showed that translocation with gramicidin

was high over a broad temperature range, while valinomycin translocated K^+ at a high rate only above 41°C. What models of translocation are suggested by these data for each of these polypeptide antibiotics?

12. While there are many different and diverse types of transport systems in cells, these systems seem to be evolutionarily related and can be accommodated within a unified model. Outline the overall mechanism that is similar in these transport systems.

Solutions

1. (a) Rearranging equation (5) from the text, we have

$$D = \frac{V \cdot 1}{[So] - [So']}$$

V for diffusion across a membrane has the dimensions of number of molecules per unit area per unit time [e.g., mole/($cm^2 \cdot s$)]. Therefore,

$$\text{Dimensions of } D = \frac{\left(\frac{\text{mole}}{cm^2 \cdot s}\right) \cdot (cm)}{\left(\frac{\text{mole}}{cm^3}\right)} = \frac{cm^4}{cm^2 \cdot s} = \frac{cm^2}{s}$$

(b) From the data given, the area of the cross section of a porin pore is 0.78×10^{-14} cm^2. Fifty molecules of glucose per second equals 0.83×10^{-22} moles per second. Therefore,

$$D = \left(-\frac{0.83 \times 10^{-22} \text{ mole}}{0.78 \times 10^{-14} \ cm^2}\right) \cdot \left(\frac{4 \times 10^{-7} \ cm}{3 \times 10^{-9} \text{ mole}/cm^3}\right)$$
$$= 1.4 \times 10^{-6} \ \frac{cm^2}{s}$$

To be more precise, the diameter of a glucose molecule is about half the diameter of a porin pore (0.5×10^{-9} m). Thus the effective pore diameter is only about 0.5×10^{-9} m if we subtract the diameter of glucose. The effective cross-sectional area for diffusion is actually 0.19×10^{-14} cm^2. Taking this into account,

$$D = 5.7 \times 10^{-6} \ \frac{cm^2}{s}$$

This value is close to the diffusion coefficient for glucose in dilute aqueous solution, which is about 7×10^{-6} cm^2/s at room temperature and suggests that glucose is more or less freely diffusible in the pore channel.

2. (a)

$$\Delta\psi = -2.3\frac{RT}{F}\log\frac{[K^+]_1}{[K^+]_2} = -60\log\frac{50}{100}$$
$$= -60(-0.3) = +18\text{ mV}$$

(b) From side 1 to side 2 because the potential due to K^+ diffusion makes side 1 positive relative to side 2, and H^+ will flow down this electrical gradient.

(c) Let X equal the millimolar concentrations of K^+ and Cl^- that are lost from side 2 and gained by side 1. At equilibrium,

$$-2.3\frac{RT}{F}\log\frac{[K^+]_1}{[K^+]_2} = +2.3\frac{RT}{F}\log\frac{[Cl^-]_1}{[Cl^-]_2}$$

Therefore,

$$\frac{[K^+]_2}{[K^+]_1} = \frac{[Cl^-]_1}{[Cl^-]_2} \quad \text{or}$$
$$\frac{(100 - X)}{(50 + X)} = \frac{(100 + X)}{(100 - X)}$$

Solving for X gives 14.3 mM. Therefore, at equilibrium,

Side 1: 64.3 mM K^+, 50 mM Na^+, 144.3 mM Cl^-
Side 2: 85.7 mM K^+, 85.7 mM Cl^-

To calculate $\Delta\Psi$, one can use either the K^+ or Cl^- gradient:

$$\Delta\Psi = -60\log\frac{64.3}{85.7} = +60\log\frac{114.3}{85.7}$$
$$= 60(0.125) = 7.5\text{ mV}$$

(This potential is such that side 1 is positive and side 2 is negative.) For a further discussion of diffusion potentials and Donnan equilibria, of which this is an example, see chapter 35.

3. Valinomycin will cause an electrogenic flow of K^+ out of the vesicles, down its concentration gradient. Thus a $\Delta\Psi$ will be induced across the membrane (see problem 2), interior negative. This $\Delta\Psi$ then drives uptake of lactose via the well-known H^+-symport system. As H^+ is accumulated in the vesicles along with lactose, $\Delta\Psi$ is collapsed, and lactose can no longer be actively accumulated. This explains the transience of the phenomenon.

4. The observations suggest that these vesicles contain F_1/F_0 ATPase with the F_1 portion oriented on the outside. Since a $\Delta\Psi$ alone has no effect on arginine transport, the process must be electroneutral. ATP hydrolysis by the ATPase would pump protons into the vesicles, and this stimulates arginine uptake. Proton ionophores prevent proton accumulation, and thus, arginine transport is dependent on a ΔpH, interior acidic. The simplest mechanism is therefore an H^+-arginine exchanger (antiporter).

5. This problem could be approached in a number of ways, but the simplest of these will be described for illustrative purposes.

(a) If sucrose transport were energized via H^+ symport, membrane vesicles should transport sucrose in the presence of an electron donor, such as D-lactate, which would set up a ΔP. Proton ionophores should inhibit sucrose uptake under these conditions.

(b) If Na^+ symport were involved, active uptake should be absolutely dependent on extravesicular (or extracellular) Na^+, stimulated by a $\Delta\Psi$ (interior negative) and insensitive to ΔpH (at constant $\Delta\Psi$).

(c) If there were a sucrose binding protein necessary for transport, cells subjected to cold osmotic shock, spheroplasts, and plasma membrane vesicles should all be defective in sucrose transport regardless of the energy source tested.

(d) If a PTS for sucrose were present, crude extracts of cells should phosphorylate sucrose by a reaction dependent on PEP, but not on ATP.

6. Consider a typical "carrier" that has its single substrate binding site exposed to either the inside or the outside of the cell but never to both simultaneously. Efflux of substrate from inside to outside would involve a conformational change in the protein, switching the binding site accordingly. Initially, in the absence of added substrate to the outside, the concentration of substrate outside the cell or vesicle would be negligible. Since the return of the carrier to the "inside" conformation would be necessary for it to participate in another efflux step, addition of nonradioactive substrate to the outside should stimulate efflux by increasing the rate of switching of the reversible carriers back to the "inside" conformation. In other words, the transport-associated conformational change in either direction is accelerated by substrate binding.

7. The lactose permease transports both lactose and H^+ concomitantly. The binding of H^+ and lactose to the carrier on the outside of the membrane presumably triggers a conformational change that deposits both molecules on the inside. Accumulation of lactose should continue as long as there is a ΔP (interior negative) or until the inside concentration of lactose becomes high enough that the efflux rate balances the influx rate.

8. (a) Stimulation of glucose transport because a $\Delta\Psi$ (interior negative) would result from the electrogenic flow of K^+ out of the vesicles.
   (b) Inhibition, because gramicidin would collapse the Na^+ gradient present initially.
   (c) No effect (neither $\Delta\Psi$ nor the Na^+ gradient should be affected).
   (d) Stimulation (H^+ would flow down the pH gradient out of the vesicles creating a $\Delta\Psi$, interior negative).

9.

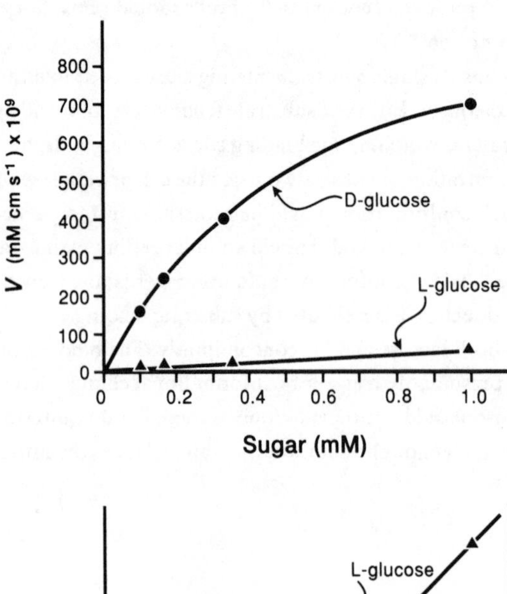

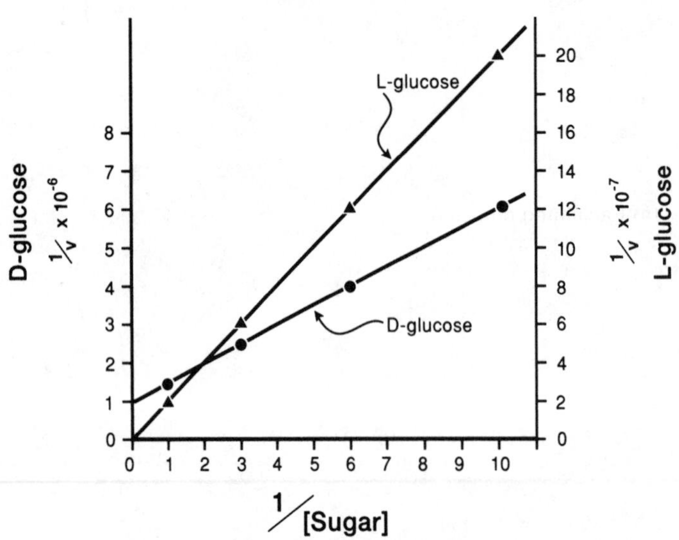

It is clear from the data that the D-glucose is taken up by the cells by facilitated transport (permease), while L-glucose is moving into the cell by passive transport. The D-glucose has a maximum velocity of uptake of about 1×10^9 mM cm s^{-1} (Y intercept in plot; analogous to V_{max}) and is saturable. L-glucose is not saturable and is moving through the membrane by passive diffusion. Actually, the data on L-glucose could be used as a control for passive diffusion and could be subtracted from the D-glucose data to obtain the true values for the transport system.

10. The effect of Hg^{2+} on the transport of D-glucose is to denature the protein transporter leaving only the passive transport of glucose. This evidence also supports the experiment in problem 9 that there is carrier-mediated transport of glucose in these cells and that the transporter is protein. Since the rate of transport of D-glucose in the mercury-treated cells is the same as L-glucose, this slow rate of transport is not carrier-mediated but due to simple diffusion through the cell membrane.

11. The transport of K^+ by valinomycin only occurred at a high rate above 41°C, which is the "melting temperature" of this membrane system. At the higher temperature the membrane is more fluid and allows the carrier to diffuse through the bilayer (mobile carrier). Since the antibiotic gramicidin transports K^+ well at all temperatures studied and is insensitive to the physical state of the phospholipid bilayer, this ionophore must form a static pore through the bilayer.

12. Although there are many seemingly different permeases, they all seem to have a hydrophilic channel (pore) that is gated. The translocation involves a conformational change in the protein that in facilitated diffusion serves to equilibrate the concentrations of the transport substrate on both sides of the membrane. In active transport, the binding affinity of the transport substrate is higher outside than inside the cell. These common features of many permeases is consistent with the evolutionary relatedness of all transport systems found in cells and can allow them to be described by a unified model.

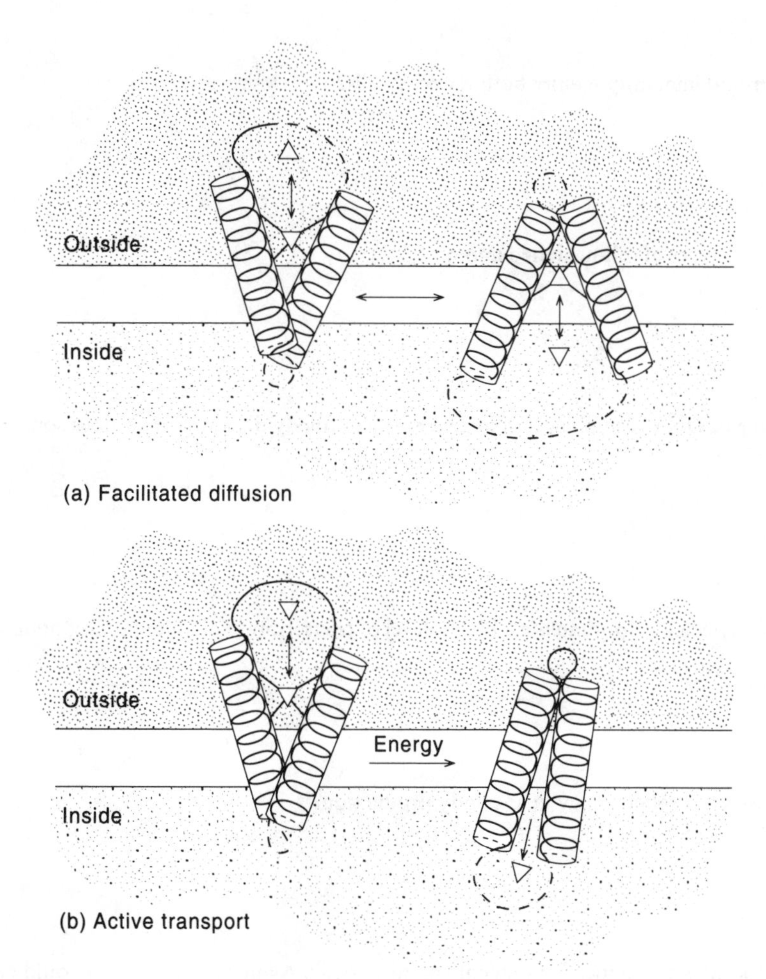

(a) Facilitated diffusion

(b) Active transport

33 Immunobiology

Problems

1. Where do T cells and B cells *mature* in the body? Where do the mature T and B cells *reside* in the body?

2. Why does humoral immunity require both B and T cells?

3. What are the two types of cell-mediated immunity?

4. Explain the role of DNA recombination in generating antibody diversity. How does somatic mutation contribute to diversity?

5. How can DNA splicing help stimulate a higher level of transcription of an immunoglobulin gene?

6. What role does RNA splicing play in antibody production?

7. You have two strains of genetically nonidentical mice (strain A and strain B). How could an adult mouse from strain A be made to tolerate a transplant from strain B?

8. How could all the immune recognition proteins have evolved from a single common ancestor protein?

9. If an antibody and its antigen are mixed in a test tube at some optimum concentration, a precipitate is formed. If either the antibody or the antigen is in great excess, no precipitate is formed. Explain these results from what you know about antibody structure.

10. Recently, antibodies with catalytic activity have been produced. How would you develop a catalytic antibody for a specific reaction? (Hint: How does an enzyme work?)

11. If you inject a pure protein into a rabbit and raise antibodies to a high titer, the antibodies produced will be a mixed population with various binding constants. Why would a single protein produce many different antibodies?

12. If you added a thiol compound such as 2-mercaptoethanol at a high concentration (1 mM) to an antibody and antigen mixture, no precipitation complex would form even if the concentration of antibody and antigen were optimal. Explain this observation. (Hint: How do thiol compounds work?)

Solutions

1. There are two different types of white cells (lymphocytes) that take part in the immune response: B cells and T cells. The B lymphocytes mature in the bone marrow and then migrate to various lymphoid organs where they reside until they are stimulated by antigens. When exposed to antigens these white cells proliferate and produce antibodies, which are secreted. The T cells mature in the thymus and then migrate to different areas of the body, including lymphoid organs. T lymphocytes also proliferate when exposed to antigens and produce specific effector molecules, which remain bound to the cell surface. The T cells do not secrete antibodies.
2. It has been known for a long time that if the thymus is removed from animals they show a reduced reaction to foreign antigens. Thus, the T cells must be important in mediating the B-cell response. The helper T cells present the antigen to the B cells in a concentrated form. This focusing of the antigen causes the immunoglobulin (Ig) membrane receptors to cluster together on the B-cell membrane, forming what is called a CAP structure. The

formation of the CAP structure triggers the proliferation of the B cells. When the thymus is removed, these animals are depleted in T cells and the focusing of the antigen does not occur (very few helper T cells are left).

3. It was known before the role of T and B cells was determined that certain immune responses could be transferred with white blood cells but not with serum (cell-mediated immunity). In the delayed hypersensitivity response the T_D cells react with antigens and secrete lymphokines that attract and activate macrophages and other leukocytes. This then leads to the slowly developing inflammatory response. Another cellular response involves the killer T cell (T_K), which reacts with antigen bound to cells and causes their lysis. Both of these responses are important in our immune system.
4. The recombination of antibody genes from three unlinked gene families produces much of the diversity seen in antibodies. All antibody polypeptide chains are derived from split genes. Additional variation is generated in regions of immunoglobulins that are responsible for antigen recognition by a mechanism of hypermutation that causes a high mutation rate. The number of different antibodies produced is almost without limit.
5. DNA splicing or recombination can activate the newly rearranged gene by activating the promoter with an enhancer element. This DNA splicing brings an enhancer close to the promoter just when antibody production becomes an important part of the cell's function. These Ig enhancers act in a tissue-specific fashion and work when placed on either side of the gene.
6. The mRNAs for the heavy-chain regions can be spliced differently so that membrane-bound or secreted forms of the antibody can be produced. The intermembrane part of the heavy chain is encoded on a separate exon. Before antigenic stimulation, most of the antibody produced is membrane bound, while after stimulation antibody is secreted. These two mRNAs are produced by alternative pathways of RNA splicing to give rise to the two forms of the antibody, membrane-bound or secreted.
7. The mouse from strain A could be irradiated with whole-body radiation to destroy the immune system. Then the bone marrow and thymus from strain B could be transplanted into the irradiated mouse, giving it a competent immune system that is isogeneic with strain B tissue. If you were working with newborn mice, you could inject cells from strain B mice into strain A mice. This would make the strain A mice tolerant to transplants from strain B mice when they are adults.
8. There is striking similarity among all immune recognition proteins; they all contain one or more immunoglobinlike domains with about 100 amino acids that are folded into characteristic structures made of two antiparallel sheets that are stabilized by conserved disulfide structures. This suggests that a common ancestral protein made up of many exons was duplicated and mutated to generate the wide variety of immunoglobins seen today.
9. The antibody is bivalent with two binding sites for antigen. The antigen may have only a single antigenic determinant (univalent antigen), which would not result in the formation of a large complex, or it might be multivalent with multiple antigenic determinants (more common), which would lead to the formation of large complexes that will precipitate. If excess antigen is present, then the antibody will be titrated so that one antigen will be binding in each antibody binding site (bivalent) and no complex will form. Also, when excess antibody is present, the multivalent sites on the antigen will be saturated and again no complexes will form and precipitate.

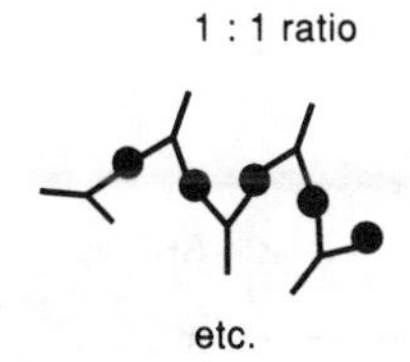

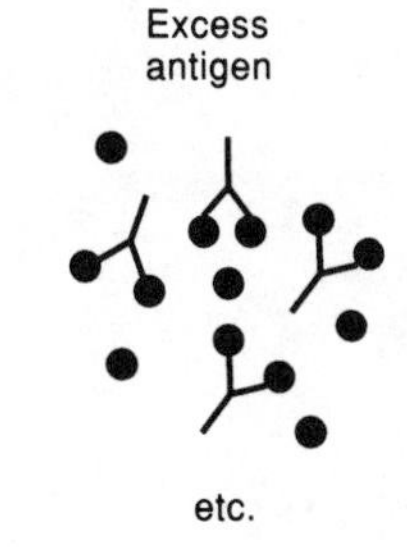

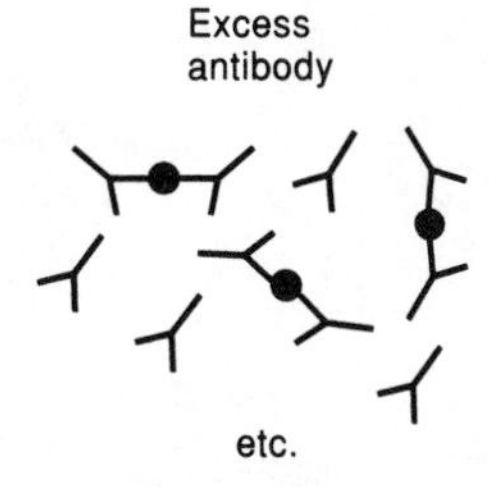

10. Antibodies with catalytic activity have been produced by using a hapten for the transition-state intermediate. An enzyme works by binding substrates and stabilizing the transition-state intermediate. If a stable transition-state analog is bound to a proteinlike serum albumin, antibodies can be produced that will have catalytic activity. Recently, recombinant DNA techniques have been used to produce binding proteins that have catalytic activity. With these techniques, enzymes can be designed for reactions that have no known natural enzyme activity in cells.
11. When a protein is injected into a rabbit it can produce polyclonal antibodies. The antigen (pure protein injected) may have many determinants on its surface. Each of these determinants will produce a plasma cell clone that will produce antibodies against that structural domain. These different determinants and antibodies produced will have different affinities of binding. It is possible to produce monoclonal antibodies against a single determinant on the antigen by using hybridoma techniques.

12. The high concentration of a thiol compound such as 2-mercaptoethanol would reduce the disulfide bonds holding the heavy chains together and holding the light chains to the heavy chains, allowing the disassociation of the chains. The reduced antibody would not have the necessary structure to bind to the antigen.

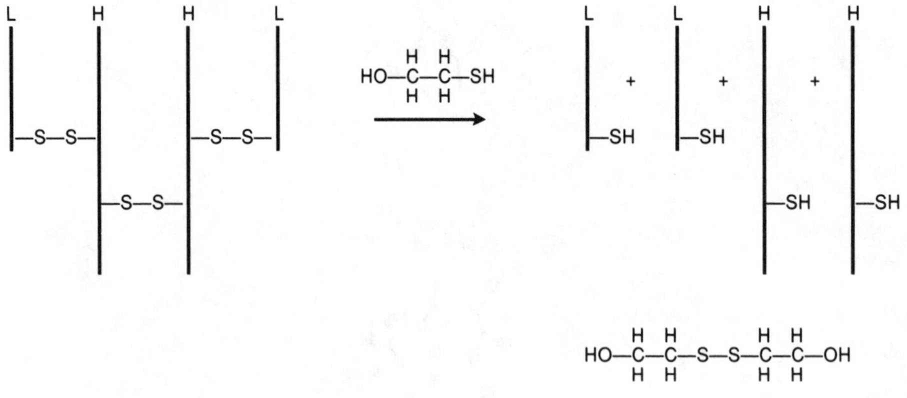

L H H L

—S—S— —S—S—

—S—S—

$$\text{HO—}\underset{\text{H}}{\overset{\text{H}}{\text{C}}}\text{—}\underset{\text{H}}{\overset{\text{H}}{\text{C}}}\text{—SH} \longrightarrow$$

L + L + H + H

—SH —SH —SH —SH

$$\text{HO—}\underset{\text{H}}{\overset{\text{H}}{\text{C}}}\text{—}\underset{\text{H}}{\overset{\text{H}}{\text{C}}}\text{—S—S—}\underset{\text{H}}{\overset{\text{H}}{\text{C}}}\text{—}\underset{\text{H}}{\overset{\text{H}}{\text{C}}}\text{—OH}$$

34 Carcinogenesis and Oncogenes

Problems

1. What are some of the differences between a cancer cell and a normal cell?

2. Consider the following statements: "The vast majority of cancers are preventable" and "Cancer is a disease of aging, i.e., predominantly older people get cancer." Do you agree with either statement? Explain your answer.

3. The methionine analog L-ethionine (which has an ethyl instead of a methyl group attached to the sulfur) causes liver cancer in rats. This analog is a nonmutagenic carcinogen that inhibits cellular methylation. How might this analog cause cancer? (Hint: What role does DNA methylation play in gene expression?)

4. You have just gotten your first job in Houston, Texas, and are presented with the following observations:
   (a) West of Houston there is a very high incidence of a variety of skin cancers.
   (b) South of Houston the incidence of skin cancer is low.
   (c) The region west of Houston was settled by immigrants from Germany who are mostly farmers, while the region south of Houston was settled, in large part, by Hispanics, who also do mostly outdoor work.
   The Texas Health Department asks you to explain why there is such a high incidence of skin cancer west of Houston. (Give a biochemical explanation.)

5. Renato Dulbecco suggested that DNA viruses that cause cancer do so by a mechanism similar to lysogeny of *E. coli* by λ virus, that is, by integration of the viral genome into the host cell DNA. How can RNA viruses cause cancer?

6. Speculate on where and how retroviral oncogenes originated.

7. Why is an oncogene like *myc* found associated with many different kinds of cancer, while the oncogene *sis* in found only in cancers with the PDGF cell-surface receptor?

8. What are some of the ways a cellular protooncogene can be converted to an active oncogene?

9. How does over expression of the *src* gene cause cancer?

10. The AIDS virus (HIV) can cause many types of cancers (Kaposi's sarcoma, B-cell lymphoma, cancers of the rectum and tongue, etc.), yet HIV does not have a viral oncogene. Explain how HIV can cause cancer.

Solutions

1. Cancer cells are easy to recognize when compared with normal cells. The cancer cell divides faster, is less well organized, and does not have the same type of structure as normal cells in the surrounding tissue. Cancer cells lack much of the growth regulation seen in normal cells. Cancer cells do not recognize territorial boundaries and eventually invade and disrupt vital organs, leading to death. When cancer cells are grown in tissue culture, they do not display the normal contact inhibition of growth seen in untransformed or normal cells. Normal cells will form a confluent monolayer of cells and then stop growing, but cancer cells may reach cell densities twenty times higher than untransformed cells. They will form layers of cells that sometimes look like small tumors and are visible without magnification. Normal cells are much harder to grow in culture and have more complex growth requirements. Many transformed cells isolated from tissue culture can give rise to tumors when injected into isogeneic animals. Many cancer cells or transformed cells will produce fetal antigens that sometimes can be used as markers for specific types of cancer, for example, colon cancer.
2. The types and frequency of different cancers varies in different populations because of diet and habits such as smoking. Migrant populations usually adopt the customs and diet of the area where they live and show the same frequency and types of cancers. There is little genetic influence. Instead, various carcinogens in food and smoke, and physical agents such as ultraviolet light and ionizing radiation (x-rays, gamma rays, exposure to radon, etc.) are responsible for most cancers. Recently, viruses have been implicated in many types of cancer (hepatitis B and C leads to liver cancer, Epstein-Barr virus can cause Burkitt's lymphoma, HTLV-1 and HTLV-2 are human retroviruses that can cause leukemia and other types of cancer). Many of these environmental agents cannot be totally eliminated, but improvements in diet, eliminating smoking, reducing exposure to sunlight and radon, and other changes could lower the frequency of certain types of cancer. Since many of the causes of cancer are mutagenic chemicals or physical agents that eventually lead to cancer, it is not surprising that the incidence of

cancer rises dramatically with age; thus cancer is basically a disease of aging. When the DNA is damaged by some mutagenic agent, twenty years might pass before the consequence becomes apparent. For example, tanning in the sun during one's youth may lead to skin cancer when one reaches 50 years of age. In conclusion, both the statements are true.

3. Ethionine is taken up by cells and metabolized to the *S*-adenosylmethionine (SAM) analog *S*-adenosylethionine (SAE). SAE is a substrate analog for various methyltransferase and inhibits cellular methylation. It is thought that the inhibition of methylation causes a change in the methylation pattern in DNA, resulting in hypomethylation on the 5′ flanking sequences of some genes. Many genes involved in development require hypomethylation for expression. For example, some of the genes important in development can become oncogenes when expressed at the "wrong" time in development. It is very common in cancers to see the expression of fetal genes. While ethionine is a nonmutagenic carcinogen, it does change methylation patterns, which can affect gene expression and lead to cancer.
4. Perhaps the Texas Health Department is trying to blame the high incidence of skin cancer on the air pollution generated by the chemical industries located on the Houston Ship Canal. However, as a biochemist, you realize that the high incidence of skin cancer is caused by the immigration of light-skinned people to an area of high ultraviolet light intensity. These people face the additional hazard of working outdoors. The high incidence of cancer is not seen in the region south of Houston because Hispanic people have darker skin, which screens out much of the ultraviolet light. If many pyrimidine dimers are formed upon exposure to ultraviolet light, there may not be enough time to repair them before the next round of DNA replication. This lack of repair then leads to mutations, which many years later can lead to skin cancer. Australia was settled by light-skinned people from England and Scotland, and because of the country's closeness to the equator these immigrants have one of the highest skin cancer levels in the world. In the United States, the incidence of skin cancer doubles about every 300 miles as one drives south. Even in the deep south, however, black people rarely get skin cancer. As the ozone layer thins, the occurrence of skin cancer should increase as more ultraviolet light passes through the atmosphere.
5. RNA viruses, like retroviruses, use reverse transcriptase to make a DNA copy that integrates into the host cell genome and forms a provirus. If the viruses insert close to a cellular oncogene, it may be turned on by viral enhancers and generate a transformed cell, which may cause cancer. Some animal retroviruses have picked up cellular oncogenes that have become part of the virus (*src* gene) and can readily transform cells in culture.
6. Retroviral oncogenes arose from cellular protooncogenes, which are normal genes expressed during development. These genes could have been picked up by the virus because of the distance to the viral insertion site in the genome or because of transsplicing that generated an mRNA which was converted into DNA by reverse transcriptase. Many of these viruses were grown in tissue culture for years; thus, anything that would increase the growth of the cultured cells (transformation) would select for viruses carrying the oncogene. Many of these oncogenes may have been selected by laboratory conditions and may not reflect what is happening in the "wild" virus infecting the host animal.
7. The oncogene *myc* may be a protooncogene that functions commonly in cell proliferation and may explain why it is found associated with a wide variety of tumors. *Sis* is found specifically with cells that possess the PDGF (platelet-derived growth factor) membrane receptor. Since these cells produce their own growth factor, they are continuously stimulated to proliferate in an uncontrolled manner. These observations explain why some oncogenes are expressed in many types of cancers while other oncogenes are found only in specific types of tumors.
8. Cellular protooncogenes can be converted to oncogenes by a number of mechanisms. The gene product can be made in excessive amounts (*src, jun, sis*), or normal amounts can be made but the protein altered so that it behaves differently (*H-ras* or *K-ras*). The *myc* oncogene is similar in amount and structure to the cellular gene but is produced at the wrong time in the cell cycle. The expression of the oncogenes is accompanied by abnormal expression of the gene products, leading to a transformed cell.

9. The *src* gene can cause cancer because it produces a tyrosine phosphokinase activity that adds a phosphate group to the tyrosine in various proteins. Since there is a high level of expression of this phosphokinase, the levels of protein phosphorylation are elevated, activating many proteins involved in growth and development.
10. The HIV virus can destroy the immune system by killing the T4 cells, which are central to the immune response. Some of the rare cancers that develop in AIDS patients may develop because the person lacks immune surveillance. Some cancer cells can be detected by the immune system as foreign and targeted by killer T cells. If the patient's immune system cannot detect these cancer cells, the cells will grow and spread, leading to death.

35 Neurotransmission

Problems

1. (a) Calculate the membrane potential $\Delta\Psi$ across a resting nerve-cell membrane in the presence of tetrodotoxin. (Assume that the resting permeability to Na^+ is due to a small, steady-state level of "open" Na^+ channels.)

   (b) What would the value of $\Delta\Psi$ be if the resting axonal membrane were permeable only to Cl^-? (Assume an axoplasmic Cl^- concentration of 50 mM.)

2. Explain why a nerve-cell membrane exhibits an "all or none" response (i.e., action potential) independent of the magnitude of an electric or chemical stimulus (above a threshold value).

3. List the criteria for demonstrating that a particular compound acts as a neurotransmitter in a given system, assuming that the mechanism of synaptic transmission in most instances is analogous to that found in cholinergic systems.

4. In reconstituted transport systems, it is often important to demonstrate that the rate of transport is similar to that observed *in vitro,* that is, that the transport protein is fully functional in the reconstituted state. For systems in which *in vivo* fluxes are very rapid (e.g., cation flux through the acetylcholine receptor), it is often difficult to measure these rates directly in reconstituted vesicles. Thallous ion (Tl^+) is known to pass readily through the "open" state of the acetylcholine receptor. It also very efficiently quenches the fluorescence emission of the fluorophore 8-aminonaphthalene-1,3,6-trisulfonte (ANTS), which is relatively impermeable to phospholipid bilayers.

   (a) Using this information, outline a series of experiments to measure Tl^+ fluxes in reconstituted proteoliposomes containing purified acetylcholine receptors.

(b) Actual Tl^+ fluxes into proteoliposomes containing an average of two acetylcholine receptor channels per vesicle in the presence of agonists have been measured to be 200 moles/(liter · s). If the average inner diameter of such vesicles is 400 Å, what is the number of Tl^+ ions transported per second by each activated acetylcholine receptor channel? How does this value compare with the rate of Na^+ flux measured *in vivo?*

5. Describe in your own words, and critique, the evidence that voltage-sensitive K^+ channels may be regulated by a "ball-and-chain" mechanism.

6. Patients with Parkinson's disease are treated with DOPA and inhibitors of dopa decarboxylase. This treatment will raise levels of dopamine in certain areas of the brain. However, dopamine given as a drug has no effect! How does this treatment of Parkinson's disease work? (Hint: Think about the blood-brain barrier and membrane permeability.)

7. Draw a chemical reaction mechanism for the reaction of the nerve poison parathion with acetylcholinesterase (see fig. 35.15*b*). How do these types of inhibitors prevent repolarization of the postsynaptic membrane?

8. What are the data which support a gated-pore-type model for ion permeability conferred by the acetylcholine receptor, as opposed to a carrier-mediated mechanism?

9. Earlier in this text, and in this chapter as well, we have cited evidence for protein structure-function relationships that has come by applying the powerful tools of molecular genetics. From what you have learned so far, what do you think are both the advantages and limitations of this kind of approach? (Use proteins described in this chapter as examples.)

Solutions

1. (a) Using the Goldman equation (equation (35.1) in the text) and the values in table 35.1, and setting the permeability of Na^+ equal to zero, we obtain

$$\Delta\psi = 60 \log \frac{20 + 0.45\,(50)}{400 + 0.45\,(560)} = 60 \log(0.065) = -71 \text{ mV}$$

(b) Using the Nernst equation (equation (35.2) in the text) and the values in table 35.1, we find that

$$\Delta\psi = 60 \log \frac{50}{560} = 60 \log (0.089) = -63 \text{ mV}$$

2. The conformational state of an Na^+ channel is dependent on $\Delta\Psi$ across the membrane. Current pulses or chemical stimuli that depolarize the membrane by about 20 mV lead to an abrupt opening of all Na^+ channels. Depolarizations below this value are not sufficient to open the channels, whereas no more channels can be opened by greater current pulses than are activated at the threshold depolarization value. This results in an "all or none" action potential response, since the magnitude of the potential spike is related to the number of open Na^+ channels, which abruptly changes at values of $\Delta\Psi$ around –40 mV.

3. These criteria include but are not limited to

(1) The ability of exogenously added putative transmitter to excite the postsynaptic cell in the same manner as occurs when the presynaptic cell is stimulated electrically.

(2) The presence of high concentrations of the presumed transmitter in the presynaptic nerve terminal.

(3) The release of the compound from the nerve terminal upon stimulation of the presynaptic cell.

(4) The presence of specific receptors for the compound in the postsynaptic membrane.

4. (a) Since ANTS is membrane-impermeable, it can be trapped in reconstituted proteoliposomes during their preparation. If Tl^+ is added to the outside of such vesicles loaded with the fluorophore ANTS, a decrease in fluorescence emission of ANTS would accompany Tl^+ flow down its concentration gradient into the vesicles. The rate of fluorescence emission decay in the presence of agonists such as acetylcholine should then be proportional to the rate of Tl^+ entry into vesicles through acetylcholine receptor channels (when corrected for the same value obtained from vesicles devoid of receptors). Since those rates should have half-times on the order of milliseconds (if they resemble physiological rates), the experiments would have to be conducted in an instrument designed to measure the kinetics of rapid processes, such as a stopped-flow device.

(b) The average volume of one vesicle is about 3.3×10^{-20} liter from the information given. This yields 6.6×10^{-18} mole Tl^+ transported per vesicle per second. If each vesicle has an average of two functional channels, 2×10^6 ions per second per channel are transported. Fluxes measured for Na^+ *in vivo* are on the order of 10^7 ions per channel per second (see the text). Thus the reconstituted system approaches the rates of ion flux measured in undisrupted cells.*

5. The ball-and-chain model or "stopper-on-a-string" mechanism can explain most of the experimental data on the K^+ channel. Previous experiments had shown that the N terminal is involved in inactivation, so numerous mutations in this region were tested in the *Xenopus* oocyte expression system. The first 20 amino acid residues form a "ball" with a hydrophobic core and positively charged amino acids on the surface. These positively charged residues probably help bind the ball to the inside surface of the pore by interacting with the negatively charged amino acids near the opening. The next 20 amino acids act as a "chain" that can swing the ball in or out

*Source: This experimental approach is described in W. C.-S. WV, H.-P. H. Moore, and M. A. Rafter, *Proc. Natl. Acad. Sci. (USA)* 78:775, 1981

of the pore opening. Even the free ball can inactivate a mutant that lacks the ball if high concentrations are added. This model (figure 35.12) will allow scientists to design experiments to determine the details of this mechanism. Similar approaches to other ion channels will determine whether this is a common mechanism for all these pores.

6. DOPA is given to Parkinson's disease patients to raise levels of dopamine in the brain. If dopamine was given directly it would not pass the blood-brain barrier. DOPA, the precursor, is a zwitterion that can pass the blood-brain barrier. When DOPA is given, one of the dopa decarboxylase inhibitors is given to prevent the DOPA from being converted to dopamine in the blood and tissues in the body. This allows serum levels of DOPA to remain high and pass the blood-brain barrier. The dopa decarboxylase inhibitors also do not pass the blood-brain barrier, allowing decarboxylation of DOPA to dopamine in the brain. Thus, by using the decarboxylation inhibitors, much lower doses of DOPA can be given, reducing the side effects of this treatment.

7.

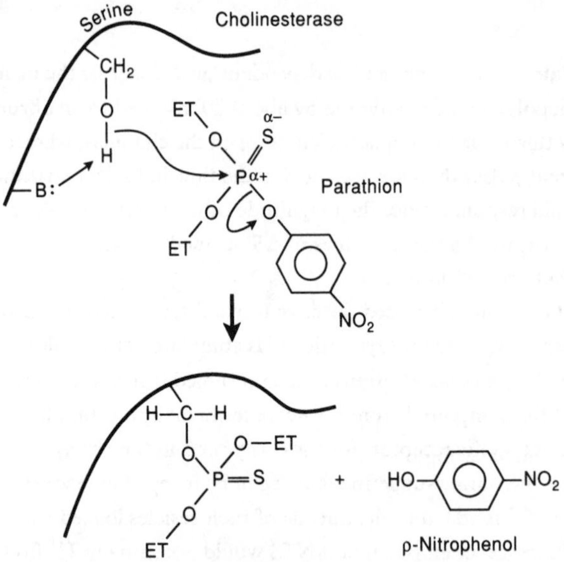

Serine
Cholinesterase
Parathion
p-Nitrophenol

When acetylcholinesterase is inhibited by compounds like parathion, the levels of acetylcholine in the synaptic cleft remain high and do not allow the depolarizing signal to be switched off. The acetylcholine has to be hydrolyzed to allow the membrane to become repolarized to allow the nerve to continue working.

8. The acetylcholine receptor appears to be acting by a gated-pore-type mechanism rather than a carrier-mediated one. Some of the main data pointing to this gated-pore mechanism is the high Na^+ flow seen in the open channel. The carrier-mediated model could not account for such a high flux of Na^+. Also, the detailed structure of the polar channel found in the receptor fits the pore model as seen in Na^+ and K^+ channels.

9. Molecular genetics (sequencing the cDNA and site-directed mutagenesis), together with molecular modeling, have been powerful tools in determining the structure-function relationships found in the ion-channel proteins and have suggested mechanisms of how these ion-pores are working. The major limitation in the molecular genetics approach is misinterpretation of the site-directed mutagenesis experiments. When an amino acid residue in the protein sequence is changed, the results observed in the function of that protein may not always be simply interpreted. For example, a single residue change may actually have an impact on the structure of the protein some distance away. Perhaps replacing hydrophilic residues within the pore with fewer polar amino acids may have a more global impact on the structure, and the change may not represent an impact on the local microenvironment of the channel. However, site-directed mutagenesis experiments, when carried out with proper controls, are becoming some of the most powerful techniques in looking at structure-function relationships.

36 Vision

Problems

1. What chemical characteristics would you look for in a compound that absorbs visible light?

2. The spectrophotometers in biochemistry laboratories are very sensitive because they use a photomultiplier tube as a detector and convert photons of light into an electrical signal. The signal subsequently is multiplied to produce a strong signal. How do our eyes multiply the signal derived from light to increase sensitivity?

3. Draw traces showing how the optical absorbance of a fresh suspension of rod outer segment disks might change as a function of time when the suspension is excited with a short flash of light at 37° C. Show the absorbance at (a) 545 nm and (b) 480 nm. Select the time scale for each trace judiciously, so that the traces illustrate the kinetics of the major absorbance changes that occur at the two wavelengths. (You may need to use two traces with different time scales at each wavelength to show both the initial absorbance change and its decay.)

4. If rhodopsin is illuminated at 500 nm at 77° K, the absorbance of the sample at 500 nm decreases. If the sample is then illuminated at 500 nm (still at 77° K), the absorbance at 500 nm increases again. Explain.

5. Why is the absorption spectrum of metarhodopsin II so different from that of metarhodopsin I?

6. Draw traces showing how the membrane potential of a rod changes with time when the cell is excited with (a) one photon, and (b) two photons. Assume that each photon converts one molecule of rhodopsin to bathorhodopsin and on to metarhodopsin II. (The probability of conversion actually is only about 0.67.)

7. Provide evidence and a model for the involvement of a cGMP as a mediator in visual transduction.

8. A frog retina is excited with a flash of polarized light that passes through the rods end-on. The flash causes optical absorbance changes at 500 nm, which are measured with polarized light that also passes through the rods end-on.
   (a) Draw traces showing the kinetics of the absorbance changes measured with light polarized parallel to the excitation polarization and with light polarized perpendicular to the excitation polarization. The traces should cover the time period up to 100 μs after the excitation.

   (b) Similar experiments are done with a retina that has been treated with glutaraldehyde, which causes cross-linking of the rhodopsins in the membrane. Draw traces showing the kinetics of the absorbance changes expected in this case. Assume that the cross-linking immobilizes the rhodopsin molecules but does not otherwise affect their photochemical transformations.

9. In solution, 11-*cis*-retinal maximally absorbs light near 380 nm, but in rhodopsin the peak is at 500 nm. Explain the spectral shift in the retinal bound to the protein.

10. How do our eyes detect color? Why do we not see color in dim light?

Solutions

1. Compounds that absorb visible light and are used in light-detection systems have many alternating double bonds (conjugated double bonds), such as retinal, or even as in chlorophyll found in plants. This allows visible light to be absorbed and the compound to become excited to a higher energy state where it can undergo oxidation (chlorophyll) or isomerization (retinal).
2. The absorption of light by rhodopsin can be amplified by a factor of almost one million by using an enzymatic cascade using the G protein transducin. Photoactivated rhodopsin interacts with transducin, allowing the binding of GTP to one of the subunits in transducin. The GTP-transducin subunit complex then activates a phosphodiesterase that hydrolyzes cyclic GMP. Cyclic GMP keeps the Na^+ channel open. One rhodopsin can

activate many phosphodiesterases, which can quickly lower cyclic GMP levels. This again is an example of the importance of cascade mechanisms in biochemistry.

3.

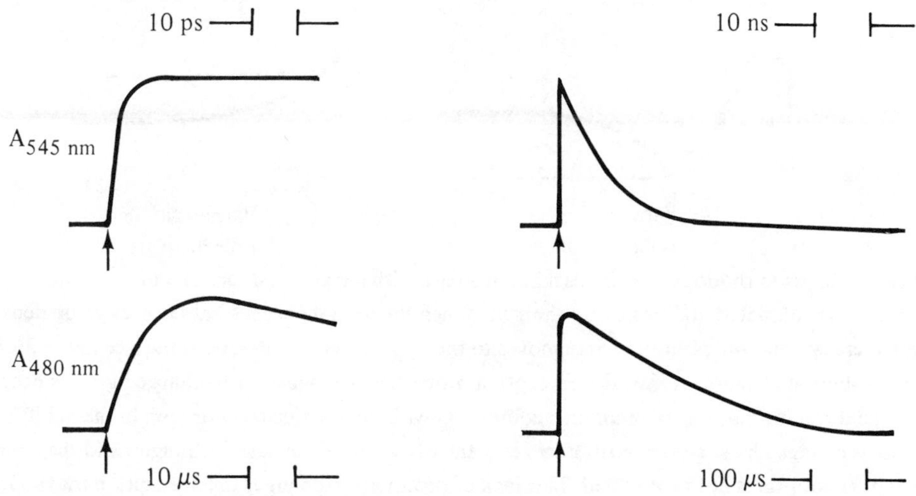

The absorbance changes at 545 nm are due mainly to the formation and decay of bathorhodopsin; those at 480 nm, mainly to metarhodopsin I. (Vertical arrows indicate flashes.)

4. Rhodopsin has an absorption maximum at 500 nm. Illumination at this wavelength converts rhodopsin to bathorhodopsin, which absorbs maximally at longer wavelengths (543 nm). At 77° K, bathorhodopsin does not decay to lumirhodopsin. Illumination at 500 nm thus causes the absorbance at 500 nm to decrease and the absorbance at 540 to 550 nm to increase. Light at 550 nm is absorbed mainly by bathorhodopsin and can convert the bathorhodopsin back to rhodopsin. (The same excited state can be generated by exciting either rhodopsin or bathorhodopsin. No matter how it is generated, the excited state decays to bathorhodopsin about 70% of the time and to rhodopsin about 30% of the time. Continuous illumination at any given wavelength will set up a photostationary state in which the mixture of rhodopsin and bathorhodopsin depends on the wavelength: Longer wavelengths are more likely to be absorbed by the bathorhodopsin in the mixture and so favor the net formation of rhodopsin.)

5. The Schiff's base linkage between the retinal group and opsin is protonated in metarhodopsin I but not in metarhodopsin II.

6.

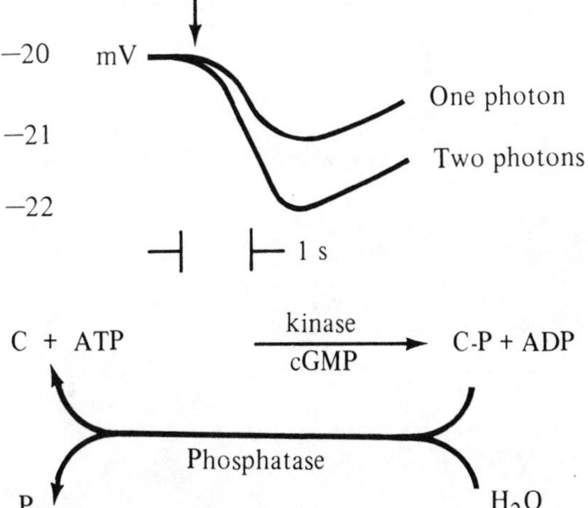

7.

$$C + ATP \xrightarrow[\text{cGMP}]{\text{kinase}} C\text{-}P + ADP$$

Phosphatase: $C\text{-}P + H_2O \rightarrow C + P_i$

C = inactive Na channel in outer segment plasma membrane; C – P = channel activated by phosphorylation. The phosphorylation is catalyzed by a kinase that requires cGMP. Inactivation of the channel is catalyzed by a phosphatase that is independent of cGMP or is stimulated by cGMP. (This scheme is speculative; you may think of a better one. See the text for evidence.)

8.

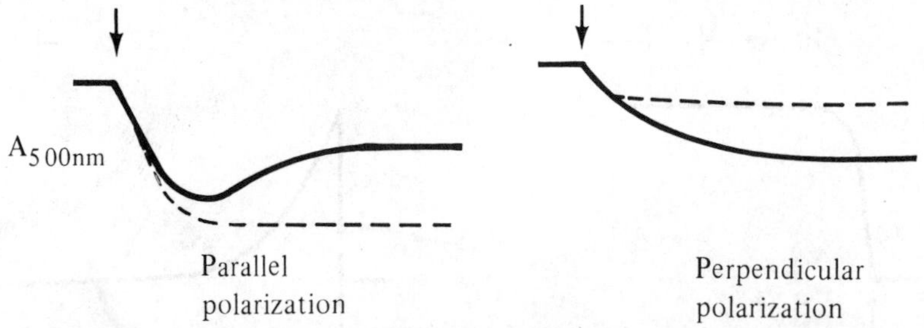

A_{500nm}

Parallel polarization

Perpendicular polarization

9. When retinal binds to rhodopsin, it forms a Schiff's base with a lysine residue. The nitrogen atom in the Schiff's base linkage is protonated and positively charged. When the retinal absorbs light, the electron density of the nitrogen increases and the positive charge moves to the opposite end of the molecule (see figure 36.7 in the text). This redistribution of charge makes the absorption properties very sensitive to charged groups nearby. There is a glutamic acid residue that forms an anionic counterion with the protonated nitrogen in the Schiff's base. If this glutamate is changed by site-directed mutagenesis, the pK value of the base is changed and the Schiff's base in the mutant rhodopsin is not protonated. This lack of protonation changes the absorption maximum from 500 nm to 380 nm, the absorption maximum seen with free retinal in solution.

10. There are three types of cone cells that absorb blue, green, and yellow light, thereby allowing us to see color. These cone cells are not connected to bipolar cells like the rod cells. Rod cells are connected to bipolar cells that are connected to each ganglion cell, allowing summing of the signals from the rods. Since the cone cells are not connected, they are not sensitive to dim light like the rod cells. Thus, we do not see color in dim light.